23 Springer Series in Solid-State Sciences

Edited by Manuel Cardona and Peter Fulde

Springer Series in Solid-State Sciences

Editors: M. Cardona P. Fulde H.-J. Queisser

Physics in One Dimension

Proceedings of an International Conference
Fribourg, Switzerland, August 25–29, 1980

Editors:
J. Bernasconi and T. Schneider

With 176 Figures

Springer-Verlag Berlin Heidelberg New York 1981

Dr. *Jakob Bernasconi*

Brown Boveri Research Center
CH-5405 Baden-Dättwil, Switzerland

Dr. *Toni Schneider*

IBM Research Center
CH-8803 Rüschlikon, Switzerland

Series Editors:

Professor Dr. Manuel Cardona
Professor Dr. Peter Fulde
Professor Dr. Hans-Joachim Queisser

Max-Planck-Institut für Festkörperforschung, Heisenbergstrasse 1
D-7000 Stuttgart 80, Fed. Rep. of Germany

ISBN 3-540-10586-7 Springer-Verlag Berlin Heidelberg New York
ISBN 0-387-10586-7 Springer-Verlag New York Heidelberg Berlin

Offset printing: Beltz Offsetdruck, 6944 Hemsbach/Bergstr. Bookbinding: J. Schäffer oHG, 6718 Grünstadt.
2153/3130-543210

Preface

In 1966, E.H. Lieb and D.C. Mattis published a book on "Mathematical Physics in One Dimension" [Academic Press, New York and London] which is much more than just a collection of reprints and which in fact marked the beginnings of the rapidly growing interest in one-dimensional problems and materials in the 1970's. In their Foreword, Lieb and Mattis made the observation that "... there now exists a vast literature on this subject, albeit one which is not indexed under the topic "one dimension" in standard indexing journals and which is therefore hard to research...". Today, the situation is even worse, and we hope that these Proceedings will be a valuable guide to some of the main current areas of one-dimensional physics.

From a theoretical point of view, one-dimensional problems have always been very attractive. Many non-trivial models are soluble in one dimension, while they are only approximately understood in three dimensions. Therefore, the corresponding exact solutions serve as a useful test of approximate mathematical methods, and certain features of the one-dimensional solution remain relevant in higher dimensions. On the other hand, many important phenomena are strongly enhanced, and many concepts show up especially clearly in one-dimensional or quasi-one-dimensional systems. Among them are the effects of fluctuations, of randomness, and of nonlinearity; a number of interesting consequences are specific to one dimension.

In a very important respect, however, the situation has changed considerably over the past decade. Today, studies of one-dimensional systems are characterized by a close coupling between theory and experiment because of the rapidly increasing experimental work on real materials with quasi-one-dimensional structures and properties. These materials often correspond remarkably closely to certain idealized one-dimensional models, and present great challenges not only to physicists, but also to chemists and metallurgists. Thus, the present situation is adequately reflected in the absence of the word "mathematical" in the title of these Proceedings.

After two years of planning by a number of physicists who recognized common interest among their generally disjointed subfields, the first general meeting on "Physics in One Dimension" was held in Fribourg, Switzerland, 25-29 August, 1980. In contrast to the more specialized conferences on one-dimensional conductors (Siofok 1976, New York 1977, Dubrovnik 1978, Helsingør 1980), the focus was quite generally one ONE DIMENSIONALITY and its consequences for various physical properties of different classes of materials. The main topics covered by the program were Nonlinear Phenomena, Disorder and Localization, Lattice Dynamics, Magnetic Chain Systems, Conducting Polymers, and Quasi-One-Dimensional Conductors.

The success of the conference exceeded all expectations. The 170 participants came from 21 different countries (Europe, ca.120, USA, ca.40); two thirds were affiliated with universities, and one third with industrial and governmental research institutes. A high level of attendance was maintained throughout the whole week, and the many discussions evidenced the similarity of concepts which have arisen in vastly different physical situations.

The present Proceedings include the 26 invited lectures and a small fraction (15 papers) of the nearly 100 poster contributions. This rather restrictive selection of contributed papers was necessary to keep the price of the Proceedings to a reasonable level, and we should like to emphasize that all of the poster papers contributed significantly to the success of the conference. The individual articles are grouped according to the main topics of the conference, and the invited papers are marked as such in the Table of Contents.

We are convinced that these Proceedings will prove to be a valuable guide to the important and active subject of one-dimensional physics, and we hope that they will shed some light onto unifying features and interrelations between the different subfields.

Finally, on behalf of the Organizing Committee

D. Baeriswyl, H. Beck, J. Bernasconi, U.T. Höchli, E. Mooser, T. Schneider, H. Thomas, and H.R. Zeller,

we have the pleasure of expressing our sincere gratitude to a long list of individuals and organizations.

Without the cooperation and the financial support of our sponsors

European Physical Society
Swiss Physical Society
University of Fribourg
Swiss National Science Foundation
United States National Science Foundation
Ecole Polytechnique Fédérale de Lausanne
Troisième Cycle de la Physique en Suisse Romande
Brown Boveri Research Center, Baden
IBM Zurich Research Laboratory, Rüschlikon
Laboratories RCA Ltd., Zurich
European Research Office of the United States Army
United States Office of Naval Research

the realization of this conference would have been impossible. We are particularly indebted to the Rector of the University of Fribourg, B. Schnyder, and his staff for their warm hospitality and for their skillful technical assistance. The Ecole Polytechnique Fédérale de Lausanne (through E. Mooser) kindly agreed to administer the European grants, and the University of California Los Angeles (through R. Orbach) acted as intermediary for travel support from NSF.

The International Advisory Committee

A.R. Bishop, S.A. Brazovskii, J. Hori, S.W. Lovesey, E. Mooser, R. Orbach, T.M. Rice, M. Steiner, H. Thomas, and H.R. Zeller

was consulted at various states, and we wish to thank its members for their cooperation and their valuable suggestions concerning the program.

We are especially grateful to our conference secretary, Miss Karin Frei, for her invaluable help both before and during the conference. Thanks are also due to U. Bitterli, Dilys Brüllmann and Dianne Kunz of the IBM Zurich Research Laboratory, Publications Department.

Special thanks also go to J.A. Krumhansl for his competent Closing Address in Fribourg, a summary of which will be reproduced in EUROPHYSICS NEWS (the Bulletin of the European Physical Society).

The rapid publication of these Proceedings, is due to the excellent cooperation shown by all the authors and by our publisher, Springer-Verlag.

Baden-Dättwil
Rüschlikon
October 1980

J. Bernasconi
T. Schneider

Contents

---

* denotes invited papers

Part IV. *Polymers*

Part V. *Quasi-One Dimensional Conductors*

Part I

Introductory Lecture

How to Reduce Practically Any Problem to One Dimension

Daniel C. Mattis

Physics Department, University of Utah
Salt Lake City, UT 84112, USA

1. Introduction

A large number of interesting problems of theoretical physics are exactly soluble in one spatial dimension (1D). These include studies in statistical mechanics of interacting or of random systems (solved by diagonalization of a transfer matrix), many-body problems (solved by a Bethe ansatz), nonlinear dynamics (leading to solitons), magnetism (reducible by the Jordan-Wigner transformation to a many-fermion problem), etc. Some of these studies apply to physically interesting systems such as polymers, magnetic molecules, and energy transport in long chain molecules, yet others might appear to be of academic interest only. Surprisingly, even some far-fetched work in 1D may ultimately find important physical applications in three spatial dimensions. In this paper I shall endeavor to demonstrate what sorts of problems, arising in arbitrary dimensions, can be successfully reduced to 1D where they can be analyzed and ultimately solved.

The insight comes from a method which was first applied to the tri-diagonalization of hermitean matrices by Lanczos [1] in the 1950's, and later by Haydock, and Haydock, Heine and Kelly [2] to the study of electronic density-of-states in imperfect solids. Recent applications have included a series of articles on Anderson localization, mobility edges, etc. in 2D and 3D disordered solids by Stein and Krey and others [3], and a simple derivation of Wigner's famous result, the semicircular density-of-states of a totally random Hamiltonian, by Mattis and Raghavan [4]. Applications to configuration interactions in atoms and molecules have been suggested [5], and refinements have been proposed such as tri-diagonal matrices in which the matrix elements are themselves block matrices [6]. In the present paper we shall review a little bit of this rapidly expanding body of work, then present a new application: to the many-body problem.

In the tri-diagonalization scheme an Hamiltonian is studied in two stages: First, an initial state ϕ_0 is selected for its symmetry properties, or perhaps merely for simplicity. The action of H on ϕ_0 then generates a new state ϕ_1, the action of H on ϕ_1 creates--in addition to ϕ_0 and ϕ_1--a new state ϕ_2, etc. The result is an orthonormal set of states ϕ_n and a tri-diagonal hermitean matrix M_{mn} with $m,n = 0,1,2,...$ and $M_{mn} = 0$ unless $m = n$ or $n \pm 1$. Such a matrix is isomorphic to a finite or semi-infinite linear chain with nearest-neighbor coupling, the archetype normal-mode problem in 1D. Second, one obtains the eigenstates, the "normal modes" of this tridiagonal matrix. These constitute the "generalized partial waves" of the original Hamiltonian for the sector determined by the symmetry eigenvalues of the chosen ϕ_0.

Repetition of this procedure, starting with a ϕ_0' belonging to a different set of eigenvalues (i.e., to a different symmetry class) from ϕ_0 will yield a new chain M_{mn}' with a new set of generalized partial waves. Continuation of this process, with ϕ_0',..., reveals the underlying symmetries of the most complicated or even random systems. Always the important step lies in the selection of the initial ϕ_0 and the construction of M_{mn} therefrom. For the second step, the diagonalization of the M_{mn}, is a routine exercise in the mathematical physics of 1D, for which by now there exists substantial documentation [7].

I shall illustate this--and thereby justify the title of this paper--by two vastly different examples. In the first, we analyze Wigner's Hamiltonian with random matrix elements for which it might seem that no symmetry can be found. On the contrary, we shall see that because of the equal probabilities all initial ϕ_0 's are, if properly selected, equivalent. In a second example we treat a many-body problem for which tri-diagonalization offers a new avenue of analysis,valid equally in weak and in strong coupling.

2. Totally Random Hamiltonian [4]

If bonds connecting electron states at atoms located at R_i and R_j are random variables, as in the case in random alloys or in liquids or amorphous solids, where R_i and R_j are themselves random variables, then the band structure must be extracted from an Hamiltonian

$$H = \sum K_{ij} c_i^{\star} c_j \tag{1}$$

(omitting spin indices), and the density of states (dos) at the ith site is

$$\rho_i(E) = \sum_q |\phi_q(R_i)|^2 \, \delta(E-\varepsilon_q), \tag{2}$$

where q labels the eigenstates of (1), $\phi_q(R_i)$ are their amplitudes at the ith site and ε_q their energy. For an homogeneously random medium one requires the averaged quantity

$$\rho(E) \equiv \frac{1}{N} \sum_i \rho_i(E). \tag{3}$$

Direct diagonalization of the random matrix K_{ij} is out of the question. Suppose there are P independent matrix elements over which to average, with $P = O(N)$ or $O(N^2)$. If each K_{ij} can have two distinct values, then the evaluation of eq. (3) requires summing over 2^P configurations, an impossibility in cases of interest where $P >> 100$.

The procedure of tri-diagonalization was originally introduced into solid state theory by Haydock et al. [2] in cases, such as this, where the simplification of translational invariance is lacking. This does not solve the problem directly. Rather, it decomposes the Hamiltonian into its irreducible partial wave representations, all expressible as one-dimensional, nearest-neighbor problems. One selects a site, say the ith, for which to compute eq. (2). In the Hilbert space, the starting vector is ϕ_0, defined to have all entries zero except the ith which is 1. Then

$$\tilde{K} \cdot \phi_0 = m_0\phi_0 + m_{01}\phi_1 , \tag{4}$$

where ϕ_1 is a normalized vector orthogonal to ϕ_0. Proceeding in the same way with ϕ_1, we generate ϕ_2, etc. As the new vectors are related to the old by a unitary transformation, the new triagonal matrix $\tilde{M}$ (of which m_{ij} are the matrix elements, $m_{ii} \equiv m_i$ and $m_{i,i\pm1}$ being the nonvanishing elements) is hermitean if the original K (matrix elements K_{ij}) was. Proofs are given in ref. [1], and are omitted here.

After a certain number of steps, say Q steps, the linear chain terminates ($m_{Q,Q+1} = 0$). The value of Q may depend on the initial R_i, but will generally be smaller than N.

Now consider the "long-range matrix", each element of which is either $+1/N^{1/2}$ or $-1/N^{1/2}$ with equal probability, for $i \geqslant j$. For $i < j$ we have $K_{ij} = K_{ji}$. The first few matrix elements are obtained by simple hand calculation: starting with $\phi_0 = (1,0,0,0,\ldots)$ we have $m_0 = K_{00} = \pm1/N^{1/2}$. The matrix element $m_{01} = m_{10}$ is independent of the distribution of K_{ij}'s or choice of i, and has the value $(1 - 1/N)^{1/2}$. Continuing out to finite n, we find m_n and $m_{n,n+1}$ to be averaged over increasingly complicated random walks. However, in the limit $N \to \infty$ their averages and most probable values coincide (central limit theorem) and reduce to

$$m_n = 0 \quad \text{and} \quad m_{n,n+1} = m_{n+1,n} = 1 \qquad (N \to \infty) \; . \tag{5}$$

All sample random matrices of the type defined above and all choices of the ith site, to within a set of measure zero, lead to the same tri-diagonal form (5). Therefore a computation of one yields an average over all of them. Proceeding, we use the well-known eigenfunctions and eigenvalues of such a Toeplitz matrix,

$$\psi_k(n) = (2/N)^{1/2} \sin k(n+1), \qquad \varepsilon_k = 2 \cos k \; , \tag{6}$$

with the all-important initial amplitude $\psi_k(0) = (2/N)^{1/2} \sin k$. The dos, eqs. (2) and (3), is

$$\rho_i(E) = \sum_k (2/N) \sin^2 k \; \delta(E - 2\cos k) = \frac{2}{\pi} \int_0^\pi dk \; \sin^2 k \; \delta(E - 2\cos k)$$

$$= \frac{1}{\pi}\left(1 - \frac{1}{4} E^2\right)^{1/2} , \quad \text{for } |E| \leq 2 \; , \tag{7}$$

and $\rho_i(E) = 0$ otherwise. It is the famous semicircular dos first obtained by Wigner, as discussed in some detail in Mehta's book [8], where it is derived by use of some rather more arduous mathematics.

3. Magnetic Impurity in a Metal

To understand the Kondo effect, which is a sort of prototypical many-body manifestation, one studies the effects of a magnetic impurity (e.g. Fe) in a nonmagnetic host metal (e.g. Cu). The essential physics in this has been distilled by a number of authors, notably J. Friedel, P. W. Anderson, and P. Wolff, in simple terms: the host metal is characterized by the Fermi sea of conduction electrons $|F\rangle$ which interact only at one site, that of the impurity. In Wolff's model [9], which we examine here, only two-body "Coulomb" interactions, parameterized by a coupling constant U, are considered. As is well known [10], this model is exactly soluble in the weak-coupling limit because of two useful features: the electronic p-, d- and higher partial waves about the impurity site all decouple, so the interaction concerns only the s waves. The resulting one-dimensional problem is written as a quadratic form in Tomonaga operators, then diagonalized by scattering-theoretic methods [10]. However, the solution obtained either in this fashion or by the Hartree-Fock approximation [9], becomes singular once U exceeds a value U_c which is of the order of the bandwidth or electronic kinetic energy. One may conclude from this either that a local moment is formed [9,10] or that the perturbation expansion fails at strong coupling [11].

We seek here to resolve this by an independent evaluation of the ground state, using tri-diagonalization. To my knowledge, this is the first application of tri-diagonalization to the many-body problem, but no special difficulties are encountered. Other applications may require modifications such as those which are discussed at the conclusion of the paper.

The prototype free electron Hamiltonian H_0 of a metal with a half-filled conduction band is given as

$$H_0 = \sum \varepsilon_k c^*_{k\sigma} c_{k\sigma} + W_0 \tag{8}$$

where we shall assume that the one-particle kinetic energies ε_k are distributed according to a symmetric density of states $\rho(\varepsilon)$ with the Fermi level at $\varepsilon = 0$. W_0 is chosen for computational convenience to make H_0 vanish on the Fermi sea. Thus,

$$H_0 |F) = 0 \ . \tag{9}$$

The interaction Hamiltonian will be chosen in the symmetric form $H' = 1/4\ U\Omega$ where

$$\Omega = \Omega^\dagger = (2n_{0\uparrow}-1)(2n_{0\downarrow}-1) \ . \tag{10}$$

The impurity occupation-number operators are $n_{0\sigma}$, given by

$$n_{0\sigma} = \frac{1}{N} \sum_{k,k'} c^*_{k\sigma} c_{k'\sigma} \ . \tag{11}$$

For the half-filled symmetric band $(F|\ n_{0\sigma}\ |F) = 1/2$ and $\Omega^2 = 1$. By symmetry, certain expectation values of odd powers of the interaction vanish:

$$(F|\ \Omega^{2p+1}\ |F) = 0 \ , \ p = 0,1,\ldots \tag{12}$$

The ground state must be a singlet [12] and must reduce to $|F)$ when $U \to 0$. It therefore includes $|F)$, and armed with this knowledge we start with $\phi_1 = -\Omega\ |F)$, obtaining $m_0 = 0$, $m_{01} = -\ 1/4\ U$. Further calculations yield:

$$m_1 = (F|\ \Omega H_0 \Omega\ |F) \tag{13}$$

$$m_{12} = -\ [(F|\ \Omega(H_0 - m_1)^2 \Omega\ |F)]^{1/2} \tag{14}$$

$$\phi_2 = \frac{1}{m_{12}}\ (H_0 - m_1)\Omega\ |F) \tag{15}$$

$$m_2 = \frac{1}{m_{12}^{\ 2}}\ (F|\ \Omega(H_0^3 - 2H_0^2 m_1 + H_0 m_1^2)\Omega\ |F) \tag{16}$$

etc. Matrix elements $m_{23}, m_{33}, \ldots$ become fairly complicated and depend on U, although they are analytic in U.

Diagonalization of any submatrix of $\tilde{M}$ yields a variational upper bound on the ground state energy E_0. Even the simplest 2×2 constructed with ϕ_0 and ϕ_1 yields a useful approximation $E_0^{(2)}$ to E_0

$$E_0^{(2)} = \frac{1}{2} m_1 - [(\frac{1}{2} m_1)^2 + (\frac{1}{4} U)^2]^{1/2} \quad . \tag{17}$$

This, in the weak-coupling limit, becomes $-(1/4\ U)^2/m_1 + \ldots$ which has the correct U^2 dependence but differs by a numerical factor from the exact result obtainable by second-order perturbation theory. Inclusion of the higher states $\phi_2, \ldots$. easily cures this. In the opposite limit of U large, (17) has the expansion $-1/4\ U + 1/2\ m_1 - 2(1/2\ m_1)^2/U + \ldots$. The leading term is exact, and the rest correspond to an expansion in increasing powers of U^{-1}, with approximate coefficients. As a practical procedure for obtaining $E_0(U)$ at all U, we should compute the sequence $E_0^{(r)}$ with $r = 2,3,\ldots$ This should yield reasonable convergence at all U.

The failure of Hartree-Fock [9] or of the Tomonaga-operator theory [10] at large U, or indeed of any weak coupling expansion [11] in powers of U, arises, then, <u>not</u> from any qualitative or physical change in the nature of ground state as U is increased but rather from the limited radius of convergence owing to singularities in the complex U plane. In (17) we observe branch cuts on the imaginary axis starting at $\pm i(2m_1)$. Assuming that these singularitites do not move much as we proceed through the sequence

$r = 2,3,\ldots$ of better approximations, we can now identify U_c as this radius of convergence with $U_c \approx 2m_1$.

A complete analysis of the model should also involve the excitation spectrum, both the eigenvalues and the eigenstates, out of which we can construct the thermodynamics and the transport-theoretic scattering cross section. One starts with $\phi_0' = c^*_{k\sigma}|F)$ to calculate the energy with an extra particle, or with $c_{k\sigma}|F)$ if $\underline{k}$ labels a state occupied in the Fermi sea. From this new starting point one obtains a new $\phi_1', \phi_2', \ldots$ and the relevant M'_{mn}. Evidently there will be a shift of $\pm\varepsilon_k$ in E_0'. In addition, there will be a complex shift of $O(1/N)$, the imaginary part of which yields the scattering lifetime. At the time of writing, I have not yet completed this this analysis.

4. General Many-Body Problem

The well-known prototype of the interacting electron gas, the Hubbard model, is a lattice on which there is a Wolff-model interaction at every site. The interaction Hamiltonian takes the form,

$$H' = \frac{U}{N} \sum_{k,k',q\neq 0} c^*_{k+q\uparrow}\, c^*_{k'-q\downarrow}\, c_{k'\downarrow}\, c_{k\uparrow} \quad . \tag{18}$$

Again, by adjusting the zero of energy we can set $m_0 = 0$ for a choice $|F)$ of ϕ_0. The calculations of successive tri-diagonal matrix elements differ from the previous section in that the links $(m_{01}, m_{12}, \ldots)$ are all $O(N^{1/2})$ whereas the diagonal terms $(m_1, m_2, \ldots)$ are still $O(1)$. Careful analysis leads to sensible results: a ground state energy that is extensive $(O(N))$ and displays the proper limiting dependence on U at both weak and strong couplings. These calculations are particularly tedious because of the entanglement of terms $O(1)$ and $O(N)$. A formalism, such as that of Feynman diagrams, would greatly help the application of tri-diagonalization techniques to this class of models, by disentangling the various contributions in the thermodynamic limit $N \to \infty$. Such a formalism is not yet available.

J. Moser has written [13] that the study of "hidden symmetries in dynamical systems" such as those that give rise to solitons, etc. will have "far-reaching significance in the fields of physics, engineering and mathematics." In tri-diagonalization, the appropriate selection of the set of initial states ϕ_0, $\phi_0', \ldots$ and the construction of the tri-diagonal matrices therefrom is tantamount to the discovery of the symmetries in the problem, and may open an important area of investigation as it clarifies the nature of these "hidden symmetries". What makes all this possible, of course, is the ease with which nowadays we solve problems in one dimension.

Footnotes and References

1. C. Lanczos, J. Res. N.B.S. 45, 255 (1950). See also: G. Strang, "Linear Algebra and its Applications", Academic, New York, 1975, pp. 276 et seq.
2. R. Haydock, J. Phys. A7, 2120 (1974), ibid A10, 461 (1977); R. Haydock, V. Heine and M.J. Kelly, J. Phys. C5, 2845 (1972) and C8, 2591 (1975).
3. J. Stein and U. Krey, J. Magn. and Magn. Mat. 6, 196 (1977); Solid State Comm. 27, 797 (1978), ibid., 1405 (1978), Z. Phys. B34, 287 (1979); also R. Haydock, Phil. Mag. B37, 97 (1978).
4. D.C. Mattis and R. Raghavan, Phys. Lett. 75A, 313 (1980).
5. R.F. Hausman, Jr. and C.F. Bender, in "Methods of Electronic Structure Theory", H. F. Schaefer III, Ed., Plenum, New York 1977, p. 319.
6. K.S. Dy, Shi-Yu Wu and T. Spratlin, Phys. Rev. B20, 4237 (1979).
7. See various chapters in E.H. Lieb and D.C. Mattis, "Mathematical Physics in One Dimension", Academic, New York, 1965; or P.J. Dean, Rev. Mod. Phys. 44, 127 (1972).
8. M.L. Mehta, "Random Matrices", Academic, New York, 1967.
9. P.A. Wolff, Phys. Rev. 124, 1030 (1961).
10. D.C. Mattis, Ann. of Phys. (N.Y.) 89, 45 (1975).
12. D.C. Mattis, Phys. Rev. Lett. 19, 1478 (1967).
13. J. Moser, Am. Scientist 67, 1689 (1979).

Part II

Solitons

An Overview of Soliton Mathematics

David W. McLaughlin

University of Arizona and Los Alamos Scientific Laboratory
Los Alamos, NM 87545, USA

I. Introduction

I have been asked to provide some background about "soliton mathematics". I will describe an overview of the subject which indicates how some of the different mathematical developments fit together and compliment each other. In this lecture emphasis will be placed on the mathematics of solitons, rather than on their remarkable properties [1,2] or their physical applications [3,4,5,6].

Because of space and time constraints, general features will only be illustrated by examples (Kortweg de Vries or sine-Gordon equations). For many other soliton equations, the current state of the art is summarized in Table 2. All details and calculations have been left to the references. When possible, review articles have been cited rather than the original technical papers since my purpose is tutorial rather than historical.

II. Mathematical Formations

II.A Nonlinear Compatibility Equations for Linear Systems

The key to integrating soliton equations is the realization that they arise as compatibility conditions for linear systems. For example, the Korteweg de Vries (KdV) equation,

$$q_t = 6qq_x \quad q_{xxx} , \tag{II.1}$$

is the compatibility condition for the linear system [8]

$$\left.\begin{aligned} -\psi_{xx} + q\psi &= \lambda\psi \\ \psi_t &= [-4\,\partial_{xxx} + 3(q\partial_x + \partial_x q)]\psi \end{aligned}\right\} \tag{II.2a,b}$$

That is, the eigenfunction ψ exists if its cross derivatives are equal, $\partial_t\psi_{xx} = \partial_{xx}\psi_t$, and equality occurs if the potential q solves the KdV equation (II.1).

Linear systems such as (II.2) impose some hidden linearity upon their nonlinear compatibility conditions. During the last fifteen years such linear systems have been used to deduce rather complete information about the "nonlinear wave" q and its "nonlinear wave equation". In particular, such nonlinear wave equations possess complete sets of solutions.

Many nonlinear equations of physical interest arise as compatibility conditions for linear systems. We illustrate several uses of these linear systems with the KdV and sine-Gordon examples.

II.B Typical Uses of Linear Systems in Studying Solutions of their Compatibility Conditions

II.B.1 Infinite Families of Conservation Laws

Let ψ and ϕ denote two solutions of the KdV linear systems (II.2a,b). Then the quadratic product $\Psi \equiv \psi\phi$ solves the linear system

$$\left.\begin{aligned} [-\partial_{xxx} + 2(q\partial_x + \partial_x q)]\Psi &= 4\lambda\partial_x\Psi \\ \Psi_t &= -2q_x\Psi + 2(q+2\lambda)\Psi_x \end{aligned}\right\} . \qquad \text{(II.2'a,b)}$$

This linear system for Ψ also carries KdV as its compatibility condition. (In fact, (II.2) and (II.2') are equivalent, though distinct, realizations of the same abstract linear system.) A solution $\Psi(x,t;\lambda)$ of linear system (II.2') generates an infinite number of conservation laws for the KdV equation. First, one rewrites (II.2'b)

$$\partial_t[\Psi] = \partial_x[-2(q-2\lambda)\Psi] + 4q\Psi_x ,$$

and then uses (II.2'a) to write the last term $q\Psi_x$ as a perfect derivative. The result is an infinite family (indexed by λ) of KdV conservation laws,

$$\partial_t[\Psi] + \partial_x[6(q-2\lambda)\Psi - 2\Psi_{xx}] = 0 \ \forall \ \lambda \in \mathbb{R} . \qquad \text{(II.3)}$$

Expanding Ψ asymptotically near $\lambda \simeq \infty$ yields an infinite collection of KdV conservation laws which are polynomial in q and its derivatives [1,7]. This infinite family of conservation laws forces solutions of the KdV equation to behave very regularly. Indeed, this infinite family eventually leads to "action-angle" co-ordinates for the phase space where q resides.

II.B.2 Normal Modes of the Field q can be Explicitly Generated by the Linear Systems

Next, we use the linear system to generate waves which contain exactly N degrees of freedom (N solitons or, more generally, N-phase soliton wave trains). We work with the sine-Gordon equation. The linear system

$$\vec{\Psi}_{\substack{x\\t}} = \begin{bmatrix} 0 & i(\lambda \pm \frac{e^{iq}}{16\lambda}) & i(\lambda \pm \frac{e^{-iq}}{16\lambda}) \\ 2i(\lambda \pm \frac{e^{-iq}}{16\lambda}) & \frac{-iw}{2} & 0 \\ 2i(\lambda \pm \frac{e^{iq}}{16\lambda}) & 0 & \frac{iw}{2} \end{bmatrix} \vec{\Psi} \qquad \text{(II.4a,b)}$$

carries the sine-Gordon equation as its compatibility condition. (This system is the sine-Gordon analogue of (II.2'a,b) for KdV.) The existence of a solution $\vec{\Psi}$ of linear system (II.4) forces q to satisfy the sine-Gordon equation. Additional properties of $\vec{\Psi}$ (other than existence) force additional restrictions on q. For example, if $\vec{\Psi}$ is polynomial in λ, q will possess a finite number of degrees of freedom [8].

We briefly outline the construction of this finite degree of freedom wave. See the survey [9] for details. First, notice $(\Psi_1^2-\Psi_2\Psi_3)$ is a constant of motion for linear system (II.4a,b). We assume this constant is polynomial in λ of degree 4N, with prescribed zeros $\{\lambda_j\}$:

$$(\Psi_1^2-\Psi_2\Psi_3) = \prod_{1}^{2N} (\lambda^2-\lambda_j^2) \ . \qquad \text{(II.5)}$$

After making an Ansatz for $\vec{\Psi}(x,t;\lambda)$ which is polynomial in λ, *simple* calculations provide an explicit formula for the wave q,

$$e^{-iq(x,t)} = \frac{\prod_{1}^{N} \mu_j(x,t)}{\prod_{1}^{2N} \lambda_k} \ .$$

This wave is determined by the input parameters $\{\lambda_j\}$ and the N degrees of freedom $\{\mu_j(x,t)\}$. A transformation maps the $\{\mu\}$ variables to new "angle" variables $\{\theta\}$, in terms of which the solutions $\theta_j(x,t)$ depend *linearly* on x and t,

$$\theta_j(x,t) = \kappa_j(\vec{\lambda})x + \omega_j(\vec{\lambda})t + \overset{\circ}{\theta}_j \ .$$

Physically, the parameters κ_j and ω_j denote the wave numbers and temporal frequencies of the wave q. These physical constants are uniquely determined by the input parameters $\{\lambda_j\}$.

In general, these N degree of freedom waves are quasi-periodic wave trains, with N spatial and N temporal periods. By allowing the parameters $\{\lambda_j\}$ to pairwise coalesce, one obtains the N-soliton wave forms. (For N = 1 and N = 2 the situation is depicted in Figures 1 and 2.)

Physically, these N-soliton waves are probably the most important configurations. (See, for example, BISHOP's lecture). They provide the basic description of fields which have a low density of solitons. The more general wave train configurations are candidates for a description of fields with a high density of solitons.

Using the linear system we have constructed normal modes for nonlinear waves which possess a finite number N of degrees of freedom. Are these modes complete as $N \to \infty$?

II.B.3 Linear Systems Yield Complete Sets of Normal Modes

Rather than work with special classes of solutions, it is more natural to study initial value problems. Are the normal modes sufficient to solve the initial value problem for arbitrary initial data which is subject to natural boundary conditions?

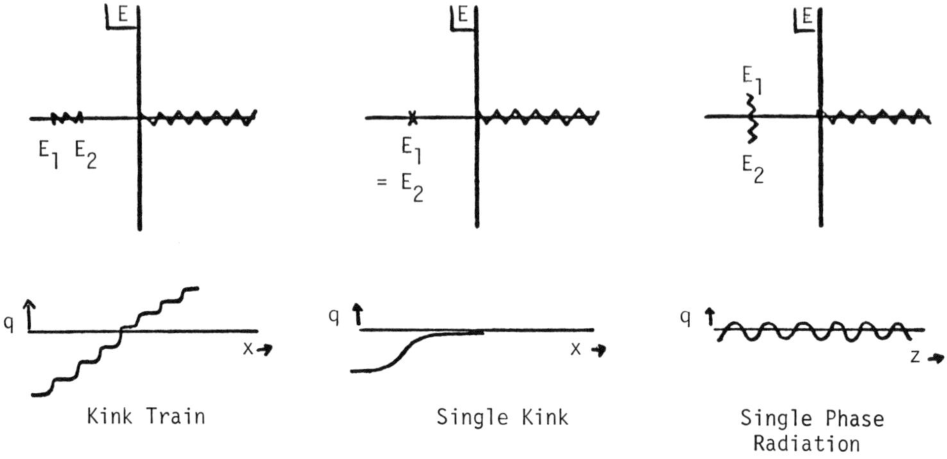

Fig.1 N = 1 Spectral and Spatial Configurations ($E \equiv \lambda^2$)

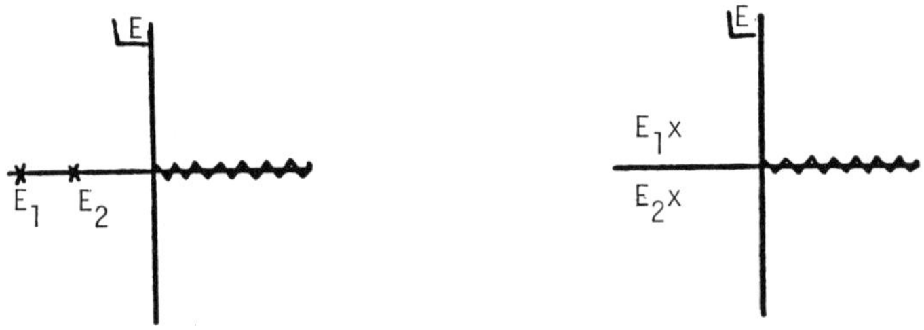

$$q = 4 \tan^{-1}\left(\frac{\sinh[vt/\sqrt{1-v^2}]}{v \cosh[x/\sqrt{1-v^2}]}\right) \qquad q = 4 \tan^{-1}\left(\frac{\tan \nu[(\cos \nu)t]}{\cosh[(\sin \nu)x]}\right)$$

Kink - Antikink Breather

Fig.2 N = 2 Soliton Configurations ($E \equiv \lambda^2$, $E_3 = E_1$, $E_4 = E_2$)

For example, consider the KdV equation (II.1) with initial data $q(x,t=0) = \mathring{q}(x)$ subject to one of three boundary conditions:

(i) *Periodic*: : $\mathring{q}(x+L) = \mathring{q}(x)$

(ii) *Vanishing at* ∞: $\mathring{q}(x) \to 0$ as $|x| \to \infty$

(iii) *Step* : $\mathring{q}(x) \to \begin{matrix} c & \text{as } x \to +\infty \\ 0 & \text{as } x \to -\infty \end{matrix}$.

In the periodic case, McKEAN and TRUBOWITZ [10] have given a *rigorous* mathematical proof that the normal modes are complete as $N \to \infty$. In case (ii), when the wave vanishes at infinity, a complete set of normal modes exists. In this case, one uses (II.2a)(of the linear system (II.2) for ψ) to define a transform from q as a function of x to $\hat{q}$ as a function λ; then the second member (II.2b) is used to show the KdV temporal evolution of $\hat{q}$ is trivial. (One should compare this procedure with the use of the Fourier transform from q(x) to $\hat{q}(k)$ to solve linear wave equations.)

The actual definition of the spectral representation $\hat{q}$ is given in terms of the spectral theory of (II.2a), the linear Schrödinger equation. As in one-dimensional Schrödinger quantum mechanics, one defines (for the potential q) bound state eigenvalues $\{\kappa_j\}$ and a reflection coefficient r(k), $k^2 = \lambda$. The spectral representation $\hat{q}$ is then given by

$$\hat{q} \equiv \{\kappa_j, m_j \text{ for } j = 1,2,\ldots,N \text{ and } r(k) \forall k \in \mathbb{R}\} , \qquad \text{(II.6)}$$

where m_j denotes a certain normalization for the bound state eigenfunctions. The potential q certainly determines the spectral representation $\hat{q}$ uniquely. The converse (that $\hat{q}$ uniquely determines q) follows from the integral equations of inverse scattering theory [1,11].

Once the equivalence of the spatial (q) and spectral ($\hat{q}$) has been established, one uses the second half (II.2b) of the linear system to calculate the trivial KdV time flow of $\hat{q}$:

$$\kappa_j(t) = \kappa_j(0), \; m_j(t) = e^{8\kappa_j^3 t} m_j(0), \; r(k,t) = e^{8ik^3 t} r(k,0) \; . \qquad \text{(II.7)}$$

Notice that κ_j and $|r(k)|$ are constants of the motion, while the phase of r(k) flows linearly in time.

Nonlinear normal modes *as defined through the spectral representation* $\hat{q}$, equation (II.6), are complete under vanishing boundary conditions [12,13]; similar completeness results hold for step boundary conditions [14].

Notice that spectral representation (II.6) splits into two parts, the discrete and continuous components. The discrete states label solitons in q. Indeed, with $r(k) \equiv 0$, the integral equations of inverse scattering theory reduce to linear algebraic systems whose solutions provide an alternative representation of the N-soliton states [1,11]. The continuous components r(k) label radiation within q [1,15]. The connection of these continuous radiation modes with the general N-phase quasi-periodic wave trains is not completely understood. (See, however [16].)

III. Algebraic and Geometric Structure in Soliton Systems

In Section II we have described soliton mathematics in terms of formal manipulations rather than in terms of mathematical structure. This situation has led theorists to investigate algebraic and geometric structures in the theory. A frequent goal of these investigations is to decide whether a given nonlinear equation is integrable. Although such studies have not solved this problem, they have produced many useful results of interest to participants of this conference.

I tend to organize such studies into two categories - (i) the structure of integration procedures through linear systems and (ii) the structure of solution manifolds for the nonlinear equations. (MANIN [17] gives an alternative classification in terms of the mathematical languages which are employed. In his survey, he discusses the merits and deficiencies of each language and, therefore, reasonable goals of each in unraveling the structure of soliton equations.)

III.A Algebraic and Geometric Structure in the Linear Systems

Consider a linear system of the form

$$\left.\begin{aligned} \psi_x &= A(x,t;\lambda)\psi \\ \psi_t &= B(x,t;\lambda)\psi \end{aligned}\right\} , \qquad \text{(III.1)}$$

where A and B are matrices. If these matrices belong to a Lie algebra, the fundamental matrix solution of this system belongs to the Lie group. By exploiting such algebraic symmetries, ZAKHAROV and co-authors [18,19,20] have continually managed to integrate equations of interest in basic physics - among them the principle chiral fields of model field theory.

The abstract Lie algebras to which the matrices A and B belong admit different realizations, which yield the eigenfunction representation, the squared eigenfunction representation, and so forth. In this manner the equivalences of different linear systems with the same compatibility conditions (see Section II) can be made transparent [21].

In the language of differential geometry, system (III.1) defines a linear connection on a principle bundle. The compatibility conditions demand that its curvature vanishes [22,23]. Such a language brings to mind gauge field theory. Indeed, a relation does exist between the linearization of the soliton equations and the linearization of the Euclidean Yang-Mills equations [24].

Of particular interest to participants of this conference is the concept of "gauge equivalent compatibility conditions". Using this concept, ZAKHAROV and TAKHATAJIAN [25] have established the complete equivalence of the cubic Schrödinger equation and a continuous chain of isotropic Heisenberg magnets,

$$iq_t + q_{xx} + 2|q|^2 q = 0 \Leftrightarrow \vec{S}_t = \vec{S} \times \vec{S}_{xx}, \ \vec{S}^2 = 1 .$$

With this sampling of the information which can be obtained from the mathematical structure of linear systems, we turn to the second category.

III.B Geometry of Solution Manifolds for the Nonlinear Equations

Of most interest to participants of this meeting is phase space geometry and its action-angle co-ordinatization. I will concentrate on these topics, and omit any discussion of algebraic geometry in the construction of N-phase waves [26,27]. There are two classes of phase spaces - (i) infinite dimensional spaces appropriate to natural boundary conditions and (ii) finite dimensional spaces appropriate to special classes of waves with finite degrees of freedom. I begin with an example of the former, KdV under vanishing boundary conditions at ∞.

In this case, the phase space is an infinite dimensional space of functions which vanish as $|x| \to \infty$. It possesses a Poisson bracket structure $\{,\}$, in terms of which the KdV equation takes the form of a Hamiltonian system

$$q_t = \{q,H\} ,$$

$$\{F,G\} \equiv \int [\frac{\delta F}{\delta q} \frac{\partial}{\partial x} \frac{\delta G}{\delta q}]dx , \quad H(q) \equiv \int [q^3 + \frac{q_x^2}{2}]dx .$$

The KdV time flow of $\hat{q}$, equation (II.7), shows that κ_j and $|r(k)|$ are constants of motion. Indeed, as functionals of q, they commute with the KdV Hamiltonian, $\{|r(k)|,H\} = 0$, and with each other, $\{|r(k)|,|r(k')|\} = 0$. Since the variables (κ_j and $|r(k)|$) constitute one-half the co-ordinates in the spectral representation (II.6), these commutation relations lead to an action-angle co-ordinate system for the phase space [28]. In particular, the action variables are defined by

$$J(k) = - \frac{k}{\pi} \ \ell n \ (1-|r(k)|^2) \ , \ J_j = \kappa_j^2 \ ,$$

and the angle variables θ are essentially the phase of the reflection coefficient (and the log of the normalization constants.) Since this complete set of co-ordinates satisfies the canonical Poisson bracket commutation relations, they provide a legitimate set of canonical variables which co-ordinatize the infinite dimensional phase space [29,30]. They are appropriate for a statistical mechanics of the field q, as BISHOP describes in his lecture.

Under periodic boundary conditions, the phase space is still infinite. Formally, one can define action-angle variables in analogy with the procedure for the finite dimensional, periodic, Toda lattice [16]. Although there is no mathematical rigor in this procedure, the variables do reduce to those of ZAKHAROV and FADDEEV in the infinite period limit. McKEAN and TRUBOWITZ [10] have studied the rigorous geometry of the periodic problem; however, they do not give action-angle variables. They study the manifold M of all smooth period 1 potentials with a fixed periodic spectrum of the Schrödinger equation (II.2a). Loosely speaking, M amounts to the manifold of all period 1 potentials with fixed action variables. Generically, this manifold is an "∞ dimensional torus". In a thorough study, they compute quantities such as its translational invariant volume element, its normal spaces (as spanned by squared eigenfunctions) and its tangent spaces (as spanned by derivatives of squared eigenfunctions). Abstractly, tangent and normal spaces are central objects in linearized perturbation theory; and indeed, in formal applied perturbation schemes, squared eigenfunctions are central [7,31,32,33]

We turn to nonlinear KdV waves with finite degrees of freedom. This class possesses two simple and elegant mathematical descriptions. If one makes a finite dimensional restriction of a general theory [34], such waves become finite collections of *one-dimensional harmonic oscillators* which are coupled through quadratic constraints. A natural canonical theory exists for such constrained systems [35].

A second approach to N phase KdV waves begins by constraining the initial data. A wave q(x,t) will be a KdV wavetrain provided its initial data $\mathring{q}(x)$ is constrained to solve the LAX-NOVIKOV equation [26]. This equation takes the compact algebraic form

$$[L,B] = 0 \ , \qquad \text{(III.2)}$$

where $L \equiv -\partial_{xx} + \overset{\circ}{q}$ and B denotes a linear combination of $B_0 \equiv \partial_x$, $B_1 \equiv -4\partial_{xxx} + 3(\overset{\circ}{q}\partial_x + \partial_x \overset{\circ}{q})$ (which generates the KdV time flow of ψ by (II.2b)), and higher order operators B_j (which generate the time flow of ψ for the "KdV Hierarchy"). (By the way, the *algebraic* equation (III.2) is the key reason that algebraic geometry enters the construction of N-phase KdV waves [27]. To generate self-similar waves, this algebraic condition is replaced by a second one, $[L,B] = L$ [36].) GEL'FAND-DIKII [37] treat (III.2) as a dynamical system (in x) in a finite dimensional phase space. They use very simple and explicit canonical transformations to identify action-angle variables for this system.

Since these mathematical descriptions are so explicit, we are tempted to use one of these formalisms for a statistical mechanics of N phase waves. The problem is that the mathematical descriptions impose constraints on the manifold of N-phase waves which are difficult to interpret physically. We have the unfortunate situation that constraints which are natural physically lead to complicated action-angle formalisms, while the simple mathematical formalisms involve physically unnatural constraints.

IV. Applications of Soliton Mathematics

While some theorists have concentrated upon mathematical structures in the soliton machinery, others have developed methods for using this machinery. In this section we briefly describe some of the successes of these methods and give some references.

The spectral transform can be used to deduce the properties of waves in (i) systems which are exactly integrable by soliton mathematics and (ii) systems which are nearly integrable by these methods. We discuss each class separately.

IV.A Completely Integrable Systems

A very early use of the spectral transform was the prediction directly from the initial data $\overset{\circ}{q}$ of the exact number and speeds of solitons which will eventually emerge from it [1,2]. This prediction is accomplished by calculating (or estimating) the number and locations of the bound state eigenvalues of $L = -\partial_{xx} + \overset{\circ}{q}$. These eigenvalues fix the number and speeds of solitons in q through its spectral representation $\hat{q}$.

A more recent application concerns shock waves on a Toda lattice. When a piston launches a shock on a conservative Toda lattice, the shock will be smoothed by a soliton wave train. The spectral transform under step boundary conditions can be used to calculate speed of the shock front in terms of the piston speed, and the separation of the solitons near the front [14]. Numerical experiments [38] have indicated the existence of a threshold in the piston velocity, below which the lattice far behind the shock thermalizes, and above which the region far behind the shock experiences very regular vibrations. A spectral transform under step boundary conditions establishes the existence and value of a strict threshold [39].

The spectral representation has identified solitons with more structure than the early solitons with their "no-scattering" property [1,2]. In addition to sine-Gordon breathers, which possess an internal degree of freedom, solitons have been found which fuse and others which decay into solitons of different types [40]. Such behavior has been found in a three-wave interaction of plasma physics. These three-wave equations have been rather completely analysed with the spectral transform [41].

Squared eigenfunctions can be used to study modulations of nonlinear wavetrains (dense packets of solitons). For example, the squared eigenfunction Ψ which was described in Section II permits us to identify the *nonlinear* group velocities of the KdV wavetrain by *inspection* [7,42]. These group velocities are the characteristic speeds of the modulation equations; hence, they describe the stability of slowly modulating wave trains [7,42,43].

With the spectral representation, one can study the large time behavior of the completely integrable wave [44,45,46,47]. For example, LAX and LEVERMORE [47] have used the scattering transform to establish the validity of the wave train modulation equations for KdV under a certain class of initial data (which includes initial data with a single hump).

IV.B Nearly Integrable Systems

When a system is close to one which can be exactly integrated by soliton mathematics, the spectral transform remains a useful tool. In such situations, this tool has been used in two ways - (i) numerically, to measure the nonlinear normal mode content of a given wave and (ii) analytically, to calculate the response of solitons to perturbations. I believe that theoretical physicists should make more use of both of these methods, *especially the numerical one.*

The idea is to use the spectral transform as a numerical measuring device to study a wave q. At various times t_j one calculates numerically the action variables, a numerical computation which is often simple to perform. In this manner, one measures *directly* the soliton and radiation content in the wave at time t_j. (Of course, the measurement is only meaningful when the system is close to integrable by that spectral transform which is being used. Otherwise the modes which are being measured would change rapidly with time.)

We have used this method on three practice problems - a mass defect in a Toda lattice [48], the annihilation of a breather by dissipation [49], and shocks on a Toda lattice [39]. Particularly in the first two cases, the numerical measurement of the spectral transform gave much sharper information than the spatial representation. In the mass defect experiment [48], a soliton strikes a light mass. The numerical transform showed that *three* soliton modes were fundamental participants in the reaction. In the breather annihilation experiment [49], the numerical measurement showed that *all* of the breather excitation was transferred into radiation modes *in a finite time*; the breather did not decay exponentially because of dissipation. In any case, when *direct* measurements of soliton modes are simple to perform, they are preferable to various indirect measurement schemes.

Soliton perturbation theory has established that sparse configurations of solitons behave as particles. It is rather easy to compute the response of solitons to perturbations [15]; and many scientists are using soliton perturbation calculations in their applications (see, for example, the volume [6]). Here I list only a few of the particle-like properties which emerge from such calculations: Solitons obey a form of Newton's law as they respond to external forces [31]. Two solitons of nearly equal speeds can "play leap-frog" in response to a perturbation [50]. A breather can phase-lock to an a.c. driver [31]. Systems can possess thresholds for a soliton-antisoliton pair to bind into a breather [15].

V. Completely Integrable Quantized Systems

Soliton mathematics can be used to study model quantum field theories. Although this topic may be viewed as another application of the spectral transform, I list it separately because it demands *substantial* extensions of the mathematical formalisms. Three areas have been investigated: (i) the use of classical soliton mathematics to study quantum fields semi-classically [51], (ii) the use of deformation theory of linear systems to characterize N-point functions of quantum fields (or classical statistical mechanics) near critical points in scaling limits [52],and (iii) the use of the associated linear systems to study the quantized fields directly [53]. Since I have not worked on quantized fields, I only refer the reader to the referenced surveys. I remark that category (ii) has lead both to very old (PAINLEVE, etc. [54] and to very new classical deformation theory,and that category (iii) is intimately connected with the "Bethe ansatz" method of classical statistical mechanics [53,55]. The present status of category (iii), as summarized by P. KULISH this summer at the Crete conference [56], is presented in Table 1.

Table 1

Stages in Quantum Inverse Scattering:

1. Construction of the quantum linear system.
2. Definition of the quantum scattering data.
3. R-matrix derivation of commutation relations between scattering data.
4. Representation of the operators (quantum states, bare vacuum, creation operators).
5. Trace formula derivation of quantum integrals of motion.
6. Excitation to different sectors: S-matrix.
7. Inverse problem (Green's functions).

A.	KdV	: No Results
B.	Nonlinear Schrödinger	: 1 through 7
C.	sine-Gordon	: See [73]
D.	Thirring Model (C-Numbers)	: 1 through 5
E.	Thirring Model (Fermi Fields)	: 1 through 4
F.	$\Box\phi = e^{2\phi} - e^{\phi}$	: 1 through 3
G.	Supersymmetric sine-Gordon	: 1 through 3
H.	A_n Model	: 1 through 3
I.	Toda Lattice	: 1 through 3
J.	XYZ and Spin 1/2 Ising	: 1 through 5
K.	Lattice Models with as a Space of Quantum States	:
L.	Discrete Nonlinear Schrödinger	: 1 through 4
M.	O(3) Nonlinear Sigma Model	: See [74]
N.	Gross-Neveu Chiral Model	: 1 through 6

Table 2

Linear System (Typical Eq^s)	N Solitons	N Wavetrains	Boundary Conditions: Infinite, Complete	Boundary Conditions: Infinite, Action Angle	Boundary Conditions: Periodic, Complete	Boundary Conditions: Periodic, Action Angle	References: Infinite	References: Periodic
1. Schrödinger (KdV)	√	√	Rigorous	√	Rigorous	Formal	[1],[13]	[10],[26]
2. Zakharov-Shabat** (Nonlinear Schrödinger, Modified KdV, sine-Gordon	Solitons don't exist. Holes do	√	√	√	*√	*√	[57],[58]	[26],[59] [9]
	√	√	√	√	Open	Very Formal		
3. 3rd Order System (Boussinesque 3-wave, $q_{tt}-q_{xx}=e^{2q}-e^{-q}$	√	√	Partial results	Partial results	Open	Open	[41],[61], [62]	[60]
4. Matrix Systems (N-Waves, Chiral Fields	√	(?)	Very few results				[18],[20], [63]	
5. Time Dependent Schrödinger $(u_t+uu_x+u_{xxx})_x=u_{yy}$	√	√	Partial results		Definitely not		[18]	[64]
6. Self Dual Yang-Mills	√ (N-instantors)		Open from this point of view				[24],[69]	
7. Discrete Systems a) Toda	√	√	√	√	√	√	[30],[11]	[16],[65],[66]
b) Nonlinear Schrödinger (with solitons)	√	*√	√	*√	Open	Open	[67]	[68]

Remarks about Table 2

1. This table is not intended to cover all equations which are integrable by soliton mathematics. It does include the most common ones, however. Some which have been omitted are Painleve eq. [70], Benney eq. [17], Long Wave-Short Wave eq. [71], Deep Water eq. [7].

2. $\overset{*}{\surd}$ means that I don't know of these formulas in the literature, but that they could certainly be developed by easy adaptions of known schemes.

3. ** The first row stands for the self-adjoint Zakharov — Shabat linear problem. The second row stands for the non-self-adjoint system. A similar distinction applies to the discrete nonlinear Schrödinger equation, but we have only listed the non-self-adjoint case.

4. "(?)" means that I am unsure of the present status.

VI. Conclusion

I trust this general overview of the mathematics of solitons, together with its references, will be useful to the participants of this conference. Obviously, such an overview is colored by my own preferences and prejudices, and I apologize to those authors whose work I have overlooked or oversimplified. I acknowledge many delightful working hours with my colleague and friend, H. FLASCHKA, whose views on soliton mathematics have substantially influenced my own. Also, I thank A. BISHOP for helping me select those aspects of the theory which will be of interest to participants of this conference, and F. TAPPERT for critically reading the manuscript.

References

1. R. M. Miura, SIAM Review 18, 412 (1976)
2. A. C. Scott, F.V.F. Chu, and D. W. McLaughlin, Proc. IEEE 61, 1443 (1980)
3. H. Flaschka and D. W. McLaughlin, eds., Rocky Mountain Journ. Math. (1978). This volume is devoted to solitons.
4. A.R. Bishop and T. Schneider,eds., *Solitons and Condensed Matter Physics*, Springer Series in Solid-State Sciences, Vol.8 (Springer, Berlin, Heidelberg New York 1978)
5. K. Lonngren and A. C. Scott, eds., *Solitons in Action*, (Academic Press, N.Y. 1977)
6. H. Wilhelmsson, ed., Physica Scripta 20 (1979). This volume is devoted to solitons.
7. D. W. McLaughlin, "Modulations of KdV Wavetrains", to appear in Proc. of Kiev Conf. on Solitons, V. E. Zakharov, ed. (1979)
8. C. S. Gardner, J. M. Green, M. D. Kruskal, and R. M. Miura, Comm. Pure. Appl. Math. 27, 97 (1974)
9. M. G. Forest and D. W. McLaughlin, to appear J. Math. Phys. (1981)
10. H. P. McKean and E. Trubowitz, Comm. Pure. Appl. Math. 29, 143 (1976)
11. M. Toda, Phys. Reports 18, 1 (1975)
12. L. D. Faddeev, Soviet Physics Dokl 3, 747 (1958)
13. P. Deift and E. Trubowitz, Comm. Pure. Appl. Math. 32, 121 (1979)

14. E. Ja. Hruslov, Math. USSR Sbornik 28, 229 (1976)
15. D. W. McLaughlin and A. C. Scott, Phys. Rev. A 18, 1652 (1978)
16. H. Flaschka and D. W. McLaughlin, Prog. Theor. Phys. 55, 438 (1976)
17. Yu. I. Manin, J. Sov. Math. 11, 1 (1979)
18. V. E. Zakharov and S. V. Manakov, "Soliton Theory" in *Soviet Scientific Reviews, Section A, Physics Reviews,* Vol. 1, ed. by I. M. Khalatnikov (1979)
19. V. E. Zakharov and A. B. Shabat, Funk. Anal. Pril. 13, 13 (1979)
20. V. E. Zakharov and A. V. Mikhaĭlov, Comm. Math. Phys. 74, 21 (1980)
21. J. Corones, lecture at Potsdam, N.Y. (1979)
22. M. Crampin, F. Pirani, and P. Robinson, Lett. Math. Phys. 2, 15 (1977)
23. F. Lund, in *Nonlinear Equations in Physics and Mathematics* edited by A. O. Barut (D. Reidel, Holland, 1978)
24. A. Belavin, in *Sov. Scient. Rev.* 1, (1979)
25. V. E. Zakharov and L. A. Takhtadzhyan, Teor. Mat. Fiz. 38, 26 (1979)
26. B. A. Dubrovin, V. B. Matveev, and S. P. Novikov, Uspekhi Mat. Nauk 31, 55 (1976)
27. I. M. Krichever, Usp. Mat. Nauk 32, 183 (1977)
28. V. E. Zakharov and L. D. Faddeev, Funk. Anal. Pril. 5, 18 (1971)
29. G.L. Lamb, Jr. and D.W. McLaughlin, "Aspects of Soliton Physics", in *Solitons*, ed. by R. Bullough and P. Caudrey, Topics in Current Physics, Vol. 17 (Springer, Berlin, Heidelberg, New York 1980) p. 65
30. H. Flaschka and D. W. McLaughlin,"Some Comments on Bäcklund Transformations, Canonical Transformations and the Inverse Scattering Method" in *Bäcklund Transformations, the Inverse Scattering Method, Solitons, and Their Applications,* ed. by R.M. Miura, Lecture Notes in Mathematics, Vol. 515 (Springer, Berlin, Heidelberg, New York 1976)
31. D. J. Kaup and A. C. Newell, Proc. R. Soc. Lond. A 361, 413 (1978)
32. J. P. Keener and D. W. McLaughlin, J. Math. Phys. 18, 2008 (1977)
33. J. P. Keener and D. W. McLaughlin, Phys. Rev. A 16, 777 (1977)
34. P. Deift, F. Lund, and E. Trubowitz, "Nonlinear Wave Equations and Constrained Harmonic Motion", Courant Inst. Preprint (1980)
35. C. Neumann, Journ. für Reine und Angew Math. 56, 46 (1859)
36. H. Flaschka, J. Math. Phys. 21, 1016 (1980)
37. I. M. Gel'fand and L. A. Dikii, Funk. Anal. Priz. 13, 8 (1979)
38. G. K. Straub, B. L. Holian, and R. G. Petschek, Phys. Rev. B 19, 409 (1979)
39. B. Holian, H. Flaschka, and D. W. McLaughlin, "Shock Waves in the Toda Lattice: Analysis", preprint Los Alamos Scientific Lab. (1980)
40. V. E. Zakharov and A. V. Mikhaĭlov, JETP Lett. 27, 42 (1978)
41. D. J. Kaup, A. Reiman, and A. Bers, Rev. Mod. Phys. 51, 275 (1979)
42. H. Flaschka, M. G. Forest, and D. W. McLaughlin, "Multiphase Averaging and the Inverse Spectral Solution of KdV", to appear Comm. Pure. Appl. Math. (1980)
43. M. G. Forest and D. W. McLaughlin, "Averaging sine-Gordon Wave Trains", in preparation
44. S. V. Manakov, Sov. Phys. JETP 38, 248 (1974)
45. V. E. Zakharov and S. V. Manakov, Sov. Phys. JETP 44, 106 (1976)
46. M. J. Ablowitz and H. Segur, J. Math. Phys. 17, 710 (1976)
47. P. D. Lax and C. D. Levermore, Proc. Nat. Acad. Science (U.S.A.), (1979)
48. W. Ferguson, H. Flaschka, and D. W. McLaughlin, "Nonlinear Normal Modes for the Toda Chain", Univ. of Arizona preprint (1980)
49. D. W. McLaughlin and E. A. Overman, "Breather Annihilation by Simple Dissipation", Univ. of Arizona preprint (1980)
50. A. C. Newell, J. Math. Phys. 18, 922 (1977)
51. L. D. Faddeev and V. E. Korepin, Phys. Reports 42, 1 (1978)
52. M. Jimbo and T. Miwa, "Aspects of Holonomic Quantum Fields-Isomonodromic Deformation and Ising Model", preprint Kyoto Univ. (1980)

53. L. A. Takhtadzhyan and L. D. Faddeev, Usp. Mat. Nauk 34, 13 (1979)
54. H. Flaschka and A. Newell, preprint Clarkson College (1979)
55. B. Sutherland, Rocky Mount. J. Math. 8, 413 (1978)
56. P. Kulish, Lecture at Crete Conference on Solitons (1980). Table 2 was compiled from notes taken by D. Holm.
57. H. Flaschka and A. C. Newell, in *Dynamical Systems, Theory and Applications*, ed. J. Moser (Springer-Verlag, Lecture Notes in Physics 38, N.Y., 1975)
58. L. A. Takhtadzhyan and L. D. Faddeev, Theor. Math. Phys. 21, 1046 (1974).
59. V. B. Matveev, "Abelian Functions and Solitons" Inst. Theor. Phys., Univ. Wroclaw, preprint No. 373 (1976)
60. H. P. McKean, in preparation
61. A. V. Mikhailov, JETP Lett. 30, 414 (1979)
62. A. G. Izergin and V. E. Korepin, "The Inverse Scattering Method Approach to the Quantum Shabat-Mikhailov Model", preprint Leningrad (1980)
63. I. V. Cherednik, Teor. Mat. Fiz. 41, 236 (1979)
64. I. M. Krichever and S. P. Novikov, Funk. Anal. Pril. 12, 41 (1978)
65. E. Date and S. Tanaka, Prog. Theor. Phys. Suppl. 59, 107 (1976)
66. P. van Moerbecke, Invent. Math. 30, 217 (1975)
67. M. J. Ablowitz and J. F. Ladik, J. Math. Phys. 17, 1011 (1976)
68. H. Flaschka, W. Ferguson, and D. W. McLaughlin, in Proc. Kyoto Conf. on Nonlinear Waves, ed. by M. Toda (1978)
69. I. M. Krichever, Funk. Anal. Pril. 13, 81 (1979)
70 M. J. Ablowitz, A. Ramani, and H. Segur, J. Math. Phys. 21, 1006 (1980)
71. A. C. Newell, Rocky Mount. Journ. Math. 8, 25 (1978)
72. M. J. Ablowitz and J. Satsuma, preprint Clarkson College (1980)
73. L. A. Takhtadzhyan and Sklyanin, Theor. Math. Phys. 40 (1979)
74. L. D. Faddeev and L. A. Takhtadzhyan, Leningrad preprint (1980)
75. H. Thacker and D. Wilkinson, Phys. Rev. D 19, 3660 (1979); 20 (1980)
76. J. Honerkamp, P. Weber, and A. Wiesler, Nuck. Phys. B 152, 260 (1979)

Statistical Mechanics of Solitons

Alan R. Bishop

Theoretical Division, Los Alamos Scientific Laboratory
University of California
Los Alamos, NM 87545, USA

1. Introduction

I was asked to talk about the statistical mechanics (SM) of solitons (in 1-d). This is an almost open-ended brief—indeed the "soliton" concept is one of the major unifying threads of low-d physics. Fortunately, at this Conference there are a number of interrelated soliton contributions. Here, I have decided to i) indicate natural connections with other invited papers; ii) survey what is known *rigorously* about soliton SM, what is conjectured, and what *important* open questions remain; iii) emphasize SM of solitons rather than of soliton Hamiltonians. By this I mean that I will mostly stress what is known rigorously about "solitons" as (nonlinear) elementary modes. I contrast this with purely numerical results [e.g., from the transfer integral operator (TIO) or molecular dynamics (MD)] without interpretation or with *approximate* mode interpretations—these are hopefully useful for experimental guidance and will be surveyed by SCHNEIDER. I will omit any discussion of *dynamics* here. Little is known rigorously; the many interesting questions concerning analytic, numerical or phenomenological approaches in Hamiltonian or Langevin frameworks will be surveyed by BÜTTIKER and SCHNEIDER.

2. Representative Models

It would be inappropriate to restrict ourselves to solitons in the strict mathematical sense (see McLAUGHLIN); the soliton paradigm is far more pervasive [1,2]. Nevertheless, strict solitons systems are valuable test models and also of deep interest to the mathematical physicist. We include several below.

A) *Nonlinear Klein-Gordon (KG) class.* Sustained attention has been given [1-3] to this class of nonlinear Hamiltonians supporting kink-soliton excitations. The general Hamiltonian is defined on a dimensionless one-component field $\{\phi_n\}$ (lattice sites labelled by $\{n\}$) and has the form

$$\mathcal{H} = \sum A\ell \left[\frac{1}{2}\dot{\phi}_n^2 + \frac{1}{2}c_0^2\ell^{-2}(\phi_{n+1} - \phi_n)^2 + \omega_0^2 V(\phi_n)\right] \quad . \tag{1}$$

A sets the energy scale and ℓ is the lattice spacing. The continuum (or "displacive") limit ($c_0/\omega_0 >> \ell$) is especially attractive theoretically,

$$\mathcal{H} \to A \int dx \left[\frac{1}{2}\dot{\phi}^2(x,t) + \frac{1}{2}c_0^2\,\phi_x^2(x,t) + \omega_0^2 V(\phi)\right] \quad . \tag{2}$$

In either (1) or (2) the local potential $V(\phi)$ is the sole source of nonlinearity. Its only restriction is that it have at least two *degenerate* minima (at say $\phi = \phi_{1,2}$), with (for simplicity) local symmetry about the minima. This is sufficient to admit a topologically stable kink solution to the equation of motion from (2)

$$\ddot{\phi} - c_0^2\phi_{xx} + \omega_0^2\, dV/d\phi = 0 \quad . \tag{3}$$

The single-kink (or antikink) excitation translating with velocity v carries the field from $\phi_{1,2}$ to $\phi_{2,1}$ over a distance $\propto d(1 - v^2/c_0^2)^{\frac{1}{2}}(d \equiv c_0/\omega_0)$, and is found easily [1-3]. The single-kink energy $E_K^{(v)}$ exhibits the same "relativistic" dependence [implicit in (3)]: $E_K^{(v)} = E_K^{(0)}(1 - v^2/c_0^2)^{-\frac{1}{2}} = [E_K^{(0)2} + p^2c_0^2]^{\frac{1}{2}}$; $E_K^{(0)} = M_K c_0^2$; $p = M_K v(1 - v^2/c_0^2)^{-\frac{1}{2}}$; $M_K = \sqrt{2}\, Ad \int_{\phi_1}^{\phi_2} d\phi|V(\phi)|^{\frac{1}{2}}$. Such kink solutions are in striking contrast to familiar small-amplitude ($<<|\phi_1 - \phi_2|$), approximate, harmonic ("linear-phonon") solutions $\phi(x,t) - \phi_{1,2} \propto \exp[i(kx - \omega_k t)]$ with continuum (Debye) dispersion

$$\omega_k^2 = \omega_0^2 + c_0^2k^2 = \omega_0^2(1 + d^2k^2) \quad . \tag{4}$$

The nonlinear KG class contains several examples which have long been practice models for "soliton" SM. For example, the "sine-Gordon" (SG) [$V(\phi) = 1 - \cos\phi$] is effectively considered in [4]; this and other periodic local potentials are studied in [5] [such as the Eshelby form $V(\phi) \propto (|\phi| - 1)^2$, periodically continued]. Again the "ϕ-four" form [6] $V(\phi) = (\phi^2 - 1)^2/8$ is the most familiar of Landau expansions but many other such unbounded potentials are studied, as are potentials supporting several kink types, e.g., the "double SG" [7]. The important recent developments (Sects.3,4) are the *proof* of an exact kink statistical mechanics at low temperature (T), and of its validity for the *whole* class (2). (Differences between periodic and unbounded potentials *are* important at high T or for dynamics.) The *discrete* lattice [c.f., (1)] also supports harmonic modes (with a trivially modified dispersion [1,2]) and kinks, but these interact increasingly strongly as d/ℓ decreases. Conveniently, however, many low-v kink properties (e.g., energy, width) are only weakly affected unless $d \sim \ell$[8].

One other excitation must be mentioned. *Anharmonic* perturbation theory about $\phi_{1,2}$ suggests a tendency towards the formation of spatially and temporally coherent excitations: particle-like envelopes with an oscillatory internal motion. We refer to these generically as "breathers". (Indeed linear SG wave trains are modulationally unstable to breather formation.) They are observed in MD simulations of ϕ-four or SG, and in deterministic simulations of several members of class (2) (corresponding to strong anharmonic effects as well as weak ones). Breathers are fundamentally different from kinks and lie at the heart of the interesting open areas in soliton SM. Breathers, ϕ_B, are only known (McLAUGHLIN) *analytically* in class (2) for SG, in which integrable (Sect. 5) case, they in principle exhaust the "anharmonic phonon" spectrum

$$\phi_B(x,t;\, v_B,\omega_B,x_0,\phi_0) = 4\tan^{-1}\psi_B \;, \quad \text{where} \quad [\gamma \equiv (1 - v_B^2/c_0^2)^{-\frac{1}{2}}]$$

$$\psi_B = \frac{\left(\dfrac{\omega_0^2}{\omega_B^2} - 1\right)^{\frac{1}{2}} \sin\left[\gamma\omega_B\left(t - \dfrac{v_B}{c_0^2}(x - x_0)\right) + \phi_0\right]}{\cosh\left[\gamma d^{-1}(x - x_0 - v_B t)(1 - \omega_B^2/\omega_0^2)^{\frac{1}{2}}\right]} \;. \qquad (5)$$

B) *Multicomponent generalizations*. A number of generalizations of class A to multicomponent fields have been studied. These are immediately relevant for problems with multicomponent-order parameters, leading to *coupled* nonlinear equations. Kink and other soliton solutions are known (sometimes analytically) for many of these, but little has been accomplished in incorporating them in a SM theory at the level we will report for class A. Two exceptions should be noted. First, for the Heisenberg ferromagnet with Ising anisotropy, see C. Second, a natural generalization of class A to a complex-order parameter with *2-fold* symmetry breaking (see [9,10]) exhibits both ϕ-four and SG-like kinks and a mean field *bifurcation*. SM of this model has revealed a possible role for an additional non-topological excitation indicated in numerical simulations [10].

C) *The Heisenberg ferromagnet and relations*. Spin Hamiltonians (classical and quantum) have gained renewed attention from soliton devotees in recent years for several reasons. Models of easy-plane ferro- or antiferromagnets with easy-plane magnetic fields (STEINER) have mostly been limited to SG approximations (possibly with periodic coupling, c.f., (1), ([11,12]; (Sect.3.2)). More generally we will refer here to the classical Heisenberg-Ising-XY Hamiltonian

$$\mathcal{H} = - J \sum_n \vec{S}_n \cdot \vec{S}_{n+1} - A \sum_n S_n^z S_{n+1}^z \quad , \tag{6}$$

where $\{\vec{S}_n\}$ are classical spin vectors $\vec{S}_n = S(\sin\theta_n \cos\phi_n, \sin\theta_n \sin\phi_n, \cos\theta_n)$ $(|\vec{S}_n| = S)$. The Landau equations of motion are $d\vec{S}/dt = \{\mathcal{H},\vec{S}\}$. The exchange anisotropy in (6) might be replaced or supplemented with local anisotropy, e.g., $A \sum_n (S_n^z)^2$. Major interest centers on the complete integrability of (6) in the *continuum* limit [12] and of the corresponding spin-½ quantum chain [c.f., (5)]. Here we simply give the flavor of soliton types by recording a few single-soliton forms we need. The continuum ferromagnet with uniaxial local anisotropy {Hamiltonian density $\mathcal{H}(x) = - J[1 - \frac{1}{2}(d\vec{S}/dx)^2] - A(S^z)^2$} has a static π-domain wall solution [13,14] (distance in lattice spacings)

$$S^z = S \tanh\left(\frac{x - x_0}{d}\right) \quad ; \quad S^\pm = S e^{\pm i\phi_0} \operatorname{sech}\left(\frac{x - x_0}{d}\right)$$

$$S^\pm = S^x \pm iS^y \quad ; \quad d = (J/2A)^{\frac{1}{2}} \quad . \tag{7}$$

These should be contrasted with familiar spin waves having continuum dispersion [c.f., (4)]

$$\omega_k = 2A + Jk^2 \quad . \tag{8}$$

Notice that there are no simple *dynamic* generalizations of these walls (unlike A). In fact the natural dynamic modes are *envelope* solitons in which (θ,ϕ) are intrinsically coupled. (The same is true for exchange easy-axis anisotropy.) General expressions are rather cumbersome [12,13] and will not be given, but the pulse amplitude $\to \pi$ ($\sim$bound wall-antiwall) as its translation velocity $\to 0$. In the isotropic Heisenberg limit, the pulse (velocity v) is

$$\begin{aligned} \cos\theta(x,t) &= 1 - 2b^2 \operatorname{sech}^2[b\tilde{\Omega}^{\frac{1}{2}}(x - x_0 - vt)] \\ \phi(x,t) &= \phi_0 + \Omega t + (v/2JS)(x - x_0 - vt) \\ &\quad + \tan^{-1}\{b(1 - b^2)^{-\frac{1}{2}} \tanh[b\tilde{\Omega}^{\frac{1}{2}}(x - x_0 - vt)]\} \end{aligned} \tag{9}$$

with $b^2 = 1 - v^2/(4JS\Omega)$; $\Omega = JS\tilde{\Omega}$. The pulse has energy $E = 16JS^3|M^z|^{-1} \sin^2(P/4S)$, with linear momentum $P = 4S \sin^{-1} b$ and z-component of angular momentum $M^z = -4bS\tilde{\Omega}^{-\frac{1}{2}}$. Importantly, there is a gauge equivalence [15] between the above continuum spin equations and the continuum nonlinear (cubic) Schrödinger equation (see McLAUGHLIN). Both systems are completely integrable (Sect.5), and the envelope soliton solutions to cubic Schrödinger require four action and angle variables [$\sim x_n$, v, ϕ_n, Ω in (9)] for specification, as for SG breathers (5). Amongst integrable Hamiltonians we also mention the Toda lattice (see McLAUGHLIN). Although a contrived lattice dynamics, it supports the third typical soliton variety, a *pulse*. In addition it is certainly the most studied *discrete* integrable Hamiltonian (c.f., Sect.5).

3. Exact Analytic Results: Transfer Integral Operator (TIO)

3.1 Statics: Nonlinear KG Class

The TIO, especially as applied to 1-d Hamiltonians with nearest-neighbor couplings (e.g., Sect.2), has a dense history [1,2,6,11,16] with extensive data (both published and unpublished). Recent attention to soliton phenomenology (Sect.4) has generally encouraged more analytic investigation, particularly at low T. We have described the TIO for nonlinear KG Hamiltonians in detail elsewhere [1,2]. A few central results will suffice here. We consider the discretized form (1) and write the *classical* partition function $Z = Z_{\dot{\phi}} Z_{\phi}$

$$Z_{\dot{\phi}} = (2\pi A\ell/\beta h^2)^{N/2} \tag{10}$$

$$Z_{\phi} = \sum_n \exp(-\beta A\omega_0^2 L\varepsilon_n) \quad . \tag{11}$$

Here $L = N\ell$ (N = number of lattice sites), h is Planck's constant. The quantities $\{\varepsilon_n\}$ in the "configurational" (i.e., potential-energy) partition function expression (11) are the eigenvalues of the TIO [1,2]. The TIO eigenfunctions $(\{\Phi_n(\phi)\})$ constitute a complete orthonormal set on $(-\infty, +\infty)$. In the thermodynamic limit only the lowest eigenvalue ε_0 is important to Z_ϕ or, for instance [2,16], the (potential) free-energy density $F_\phi \equiv -L^{-1}\beta^{-1}\ell n Z_\phi$

$$F_\phi \xrightarrow{L\to\infty} A\omega_0^2\varepsilon_0 \quad . \tag{12}$$

Equilibrium correlation functions, however, depend on both TIO eigenvalues *and* functions [2,16]: $C_g(x) \equiv \langle g[\phi(x)]g^*[\phi(0)]\rangle \xrightarrow{L\to\infty} \sum_n |\langle n|g(\phi)|0\rangle|^2 \exp(-x/\lambda_n)$. Here g is an arbitrary function of ϕ and the "n-th correlation length" $\lambda_n = [\beta A\omega_0^2(\varepsilon_n - \varepsilon_0)]^{-1}$. In some ("kink-sensitive") cases only the first excited level with nonzero matrix element is important [1,2,17]. In writing (11) and therefore (12), etc., we have assumed *periodic* boundary conditions $\phi_{N+1} = \phi_1$. These are readily varied within the TIO formalism [2,18], which is sometimes physically necessary, e.g., rigid or free boundaries with a fixed or floating "winding number" $[\equiv(\phi_N - \phi_1)/2\pi$ in SG] (for discommensuration arrays, see BAK). These considerations will usually not affect *intensive* quantities such as F_ϕ. Such boundary conditions will, of course, influence the kink-antikink density (Sect.4).

If we wish to interpret the formally exact TIO results in terms of "elementary" nonlinear modes (class A), the displacive limit $d \ll \ell$ affords most analytic control. To $O(\ell/d)$ the TIO can [2,16] be replaced by a Schrödinger-like ($\hbar \to T$) *differential* eigenvalue equation for $\Psi(\phi) \equiv \exp[-\tfrac{1}{2}\beta\ell A\omega_0^2 V(\phi)]\Phi(\phi)$:

$$\hat{H}(\phi)\Psi_n(\phi) = \varepsilon_n\Psi_n(\phi)$$

$$\hat{H}(\phi) = -(2m^*)^{-1}d^2/d\phi^2 + V(\phi) + V_0 \quad , \qquad (13)$$

with $m^* = A^2\omega_0^2c_0^2\beta^2 \propto [\beta E_K^{(0)}]^2$ and $V_0 = (2\beta\omega_0^2\ell A)^{-1}\ell n(Ac_0^2\beta/2\pi\ell)$. In terms of experimental applicability we note [8] that effects on eigenvalues from using (13) instead of the true TIO are at most a few percent unless $d \lesssim 2\ell$. The two lowest eigenvalues deduced from (13) are shown as a function of T in [Ref.19, Fig.3]. Note that for periodic potentials $V(\phi)$ as in SG we have a band-structure problem [(13) is the Mathieu equation for SG], with eigenvalues $\varepsilon_{n,k}$ labeled by a "band" index n (= 0,1,2, ----) and "wave vector" k in the first Brillouin zone ($-\tfrac{1}{2} < k \leq \tfrac{1}{2}$). The corresponding eigenfunctions have Bloch form [2]. The above-mentioned figure in [19] shows the lowest two characteristic eigenvalues (top and bottom of the first band); these correspond most closely to the lowest two levels of ϕ-four-like models. ε_0 and ε_1 converge rapidly for $k_BT \lesssim 0.2\,E_K^{(0)}$ (i.e., $m^* \gg 1$). Indeed, in this regime, they can be usefully analyzed [2,6] as tunnel-split single-well oscillator levels, $\varepsilon_0 = E_0 - t_0$; $E_0 = \tfrac{1}{2}m^{*-\frac{1}{2}} + O\,[\beta E_K^{(0)}]^{-2}$. The tunneling term can be extracted by standard WKB schemes, and for class A is found to have a "universal" form giving [2]

$$F_t \equiv -A\omega_0^2 t_0 = -\eta A\omega_0^2\pi^{-\frac{1}{2}}\left[\beta E_K^{(0)}\right]^{-\frac{1}{2}}\exp\left[-\beta E_K^{(0)}\right] \quad . \qquad (14)$$

η is a numerical coefficient depending on the particular $V(\phi)$, e.g., $\eta(SG) = 16\sqrt{2}$, $\eta(\phi^4) = 2\sqrt{2/3}$. Equation (14) includes a correction factor to WKB $[(\pi/e)^{\frac{1}{2}}]$ common to class A (at asymptotic low T) [20]. It should seem physically plausible that E_0 is related to harmonic (and anharmonic) phonon modes, and t_0 to kink configurations. This suggestion [6] is now verified for all class A (Sect.4).

We will not dwell here on numerical solutions of (13) (or the TIO) for nonasymptotic (high- or low-) T regimes. These are straightforward and have been abundant since at least [16] for class A and many of the spin Hamiltonians C. Renewed interest has been stimulated by speculations on certain magnetic chains (see STEINER). In this context information from the TIO on *correlation* functions [see (12)] is important. In particular, certain interesting correlations in planar (classical) *anti*ferromagnets are dominated by kink-*sensitive* functions for which a simple Ising-like kink phenomenology can be constructed. (Solitons enter through eigen*value* tunnel splitting.) By contrast, corresponding properties in corresponding *ferro*magnets are determined from kink-*insensitive* functions, which are dominated by anharmonicity, with kinks only entering weakly through a careful eigenfunction study. Phenomenology is then more difficult. We do not have space to develop these interesting topics here [1,17].

3.2 Statics: Other Models

Many of the models in 2A-C have been studied via the TIO. Tunnel splitting ($\longrightarrow$ *kink* solitons) is not found in the Toda lattice [21,22]. *Periodic* nearest-neighbor coupling in A is possibly interesting in view of planar approximations to easy-plane magnetic chains. This is considered in detail in [11]. The SG approximation $[\cos(\phi_{n+1} - \phi_n) \rightarrow 1 - \frac{1}{2}(\phi_{n+1} - \phi_n)^2]$ is generally found to be reasonable (for static properties) in the displacive regime at low T, with one essential provision. This is that the TIO or (13) are restricted to the range $(-\pi,\pi)$, not $(-\infty, +\infty)$. Thus, even in a SG approximation, we must only use 2π-periodic Mathieu functions, not the whole band structure. This has obvious consequences, particularly for correlation functions since certain previously finite matrix elements in the TIO expression for $C_g(x)$ (below (12)) may be excluded [11].

TIO applications to coupled scalar field Hamiltonians are also numerous (e.g., [9,10]). In some cases analytic attention to tunnel-splitting features can be profitable; see, for instance, [10] for a complex-order parameter model. Spin Hamiltonians are also extensively documented [14]. We report one example here for later use, the continuum uniaxial Heisenberg ferromagnetic [c.f., (6-9)] [14]. Using the TIO formalism, the free-energy density is [14]

$$F = \beta^{-1}\ln(\beta J/2\pi) - J + \varepsilon_0 \quad , \tag{15}$$

where ε_0 is the lowest eigenvalue of a (two-variable) "hindered rotator" transfer matrix eigenvalue equation, which can be transformed into the single-variable form

$$d^2\psi(u)/du^2 + V(u;E)\psi(u) = 0$$

$$V(u;E) = -\tfrac{1}{4}(\beta E_W)^2[(E/A) + \tanh^2 u]\,\mathrm{sech}^2 u \quad . \tag{16}$$

The eigenvalue problem (16) differs from the Schrödinger form (13), but for $|E| < A$ (sufficient for ε_0) a very similar *tunneling* problem emerges. With the same low-T WKB approach, we find [14] $\varepsilon_0 = E_0 - t_0$ with $E_0 = -A[1 - 4(\beta E_W)^{-1} + O(\beta E_W)^{-2}]$ and

$$t_0 \xrightarrow{T\to 0} 4(e/\pi)d^{-1}\beta(\beta E_W)\exp(-\beta E_W) \quad . \tag{17}$$

The factor (e/π) is cancelled by the same WKB corrections which removed the factor $(e/\pi)^{\frac{1}{2}}$ for class A. The important difference from the *one*-component class A is the prefactor βE_W in (17); for class A this appears universally as $[\beta E_K^{(0)}]^{\frac{1}{2}}$ [c.f., (14)]. This universal feature of class A can be understood very physically in terms of the phase-shifting effect of kinks on the extended harmonic excitations and the related zero-frequency kink translation mode (Sect.4). The spin model signature βE_W can be understood equally physically, by recognizing *two* degrees of freedom for fluctuations, and *two* symmetries—spin-rotation about the easy-axis as well as translation [see (7)].

4. Nonlinear-Mode Phenomenology

4.1 Low-T Statics: Noninteracting Kinks

Several *equivalent* schemes have been described in the last 20 years (e.g., [5,23,24]) which evaluate various statistical properties of models exhibiting *kink* solitons by explicitly recognizing those nonlinear configurations or "collective coordinates". Basically, these schemes are all attempts to build a "phenomenological" representation of a partition function around nonperturbative steepest descent or saddle-point trajectories. Examples from class A have been popular, but most schemes are more general (below). Typically, an exact phenomenology is possible at low k_BT (<< kink energy) because kink-kink interactions can be neglected and we only need to analyze the effect of a *single* kink on the *harmonic* modes. One then finds that "zero-frequency" modes, reflecting continuous symmetries, play a central role. We begin by summarizing the approach of CURRIE, et al. [2] since this has been applied in generality to class A, and leads to some more sophisticated ideas in Sect.5. Complete details can be found in [2]. Extended harmonic wave forms ("phonons") are modified in the neighborhood (scale d) of a kink and most importantly their *density of states is changed*. This nonlinear feature is also central to perturbation theories and to the treatment of quantum or critical fluctuations about nonlinear configurations. We can handle the problem relatively easily because of the assumption of a simple isolated kink. Analyzing *linear* oscillations about a kink is well known from stability theory [2,24]. The spectrum of small oscillations, $\chi(x,t) = f(x)e^{-i\omega t}$, comprises both bound (i.e., spatially localized) and continuum (i.e., extended) states. Since the general continuum Hamiltonian (2) is *translationally* invariant there must be a "zero-frequency" ($\omega = 0$) ("translation" or Goldstone) mode describing rigid kink translations. In linear order this corresponds to the lowest member of the bound-state spectrum: $\omega_{b,1}^2 = 0$, $f_{b,1}(x) \propto d\phi_K^{(0)}(x)/dx$. In addition there may, depending on the potential $V(\phi)$, be $(N_b - 1)$ finite frequency bound states ($0 < \omega_{b,n} < \omega_0$; $n = 2, \ldots, N_b$) describing harmonic kink-shape oscillations, e.g., for SG or double quadratic (class A) $N_b = 1$, for ϕ-four, $N_b = 2$. The remaining continuum spectrum (with wave vector k) has dispersion $\omega_k^2 = \omega_0^2 + c_0^2k^2$, unchanged by the kink presence (4). Only the asymptotic behavior of $f_k(x)$ is important below: the most general form is [2] $f_k(x) \xrightarrow{x\to\pm\infty} A_k \exp\{i[kx \pm \tfrac{1}{2}\Delta(k)]\} + B_k \exp\{-i[kx \pm \tfrac{1}{2}\Delta(k)]\}$. Here $\Delta(k)$ is a *phase shift* [depending on $V(\phi)$]. In fact SG and ϕ-four belong to a class [2] with *reflectionless* scattering ($B_k = 0$). This is of *no* physical consequence in the present context but implies some mathematical simplicity which has made these examples popular demonstrations. The parallels with other scattering problems should already be evident and it will not be surprising that only *asymptotic* scattering data is important (i.e., the phase shift). This contains information about the bound states and also about *conservation* of states. These are not well-posed questions in a continuum, and we also want to make contact with the discrete TIO results (Sect.3), so we consider a large system $L(N = L/\ell)$ and impose, e.g., periodic boundary conditions on $\{f_k(x)\}$. (We will use the *continuum* $f_k(x)$, $\Delta(k)$, which is strictly inconsistent, but has been validated for our purposes to leading order in ℓ/d [25].

It *is* essential to use the discrete phonon dispersion.) We see that the phonon density of states is changed by the presence of a kink, $\rho(k) = \rho_0 + \Delta\rho(k)$, $\rho_0 = L/2\pi$, $\Delta\rho(k) = (2\pi)^{-1}d\Delta(k)/dk$. Conservation of states is assured by a form of Friedel sum rule, $P\int dk\Delta\rho(k) = -\pi^{-1}\Delta(0+) = -N_b$. $P \equiv$ principal value. We can view this as "trapping" of phonon states by the kink—a precise mechanism for *sharing degrees of freedom in a nonlinear system*.

Procedural steps to a kink gas phenomenology are now straightforward. We will *assume* that kinks (and antikinks) form an ideal gas (at low T), but we *cannot* assume an independent phonon spectrum: the available phonon phase space is unavoidably dependent on the kink presence (and velocity, via a "Lorentz boost"). This *precludes* partition-function factorization as truly independent excitations. The phonon free-energy density associated with ρ_0 is $F_0 = (k_BT/2\pi)\int_{-\pi/\ell}^{\pi/\ell} dk\,\ell n(\beta\hbar\omega_k)$, i.e.,

$$F_0 \xrightarrow{d \gg \ell} k_BT[\ell^{-1}\ell n(\hbar\omega_0\beta d/\ell) + (2d)^{-1}] \quad , \tag{18}$$

which is *precisely* identified with harmonic oscillator pieces in the TIO formalism, $O[k_BT/E_K^{(0)}]$: $F_0 \equiv A\omega_0^2E_0 - k_BTL^{-1}\ell nZ_\phi + A\omega_0^2V_0$ [see (10), (13)]. Assuming independent kinks (i.e., additive phase shifts), the *change* in phonon free-energy density from $\Delta\rho$ is (for N_K kinks with velocities $\{v_i\}$ and $N_{\bar{K}}$ antikink with velocities $\{\bar{v}_i\}$) $\Delta F(\{v_i,\bar{v}_i\}) = \sum_i\Delta F(v_i) + \sum_i\Delta F(\bar{v}_i)$, with the single-kink contribution

$$L\Delta F(v) = k_BTP\int_{-\pi/\ell}^{\pi/\ell} dk\Delta\rho(k;v)\ell n(\beta\hbar\omega_k) + k_BT\sum_{n=2}^{N_b}\ell n(\beta\gamma\hbar\omega_{b,n})$$

$$\xrightarrow{\ell \ll d} k_BT\sum_{n=2}^{N_B}\ell n(\beta\gamma\hbar\omega_{b,n}) - k_BTN_b\ell n(\beta\hbar\omega_0)$$

$$+ \frac{k_BT}{2\pi}P\int_{-\infty}^{\infty} dk\,\frac{d\Delta(k;v)}{dk}\,\ell n(1 + k^2d^2)^{\frac{1}{2}} \quad . \tag{19}$$

We have kept track of "relativistic" dependencies (γ) in (19); N.B. $\Delta(k;v) = \Delta\{\gamma[k - v\omega_0c_0^{-2}(1 + k^2d^2)^{-\frac{1}{2}}];0\}$. In both (18) and (19) we have used a high-T limit [$k_BT \gg \hbar\omega_0$ but $\ll E_K^{(0)}$] of the quantum harmonic oscillator free-energy expression to be consistent with the *classical* TIO. For use in Sect.5, consider (19) for the SG case. Using the well-known phase-shift formula (below), a simple contour integration yields $L\Delta F(v) = -k_BT\ell n[\beta\hbar\omega_0(1 + \gamma)]$. As in quantization schemes [3], it is convenient to associate $\Delta F(v)$ with the *kink* (as a kink "self-energy"). A grand canonical partition function G can then be constructed for *ideal gases* of kinks and antikinks but with *renormalized energies*

$$E_K^*(\gamma) \equiv \gamma E_K^{(0)} + L\Delta F(\gamma) \quad . \tag{20}$$

We have effectively already integrated over phonon degrees of freedom. Thus, with the independent kink assumption, $G(T,L,\mu_K,\mu_{\bar{K}}) = \exp(-\beta F_0)G_K G_{\bar{K}}$ with

$$G_K(T,L,\mu_K) = \sum_{N_K=0}^{\infty} e^{\beta\mu_K N_K} Z_K(N_K) \quad , \quad \text{etc.}$$

$$Z_K(N_K) = (N_K!)^{-1}\left[h^{-1}\int_0^L dq_K \int_{-\infty}^{\infty} dp_K \; e^{-\beta E_K^*(p_K)}\right]^{N_K} \quad . \tag{21}$$

q_K is a kink position coordinate, p_K its momentum (class A), and μ_K the chemical potential. For periodic boundary conditions we set $\mu_K = \mu_{\bar{K}} = 0$ after any thermodynamic manipulations. Standard thermodynamic formula lead [2] to explicit expressions for, e.g., free energy, specific heat, internal energy, kink density $[n_K^T = n_K + n_{\bar{K}} = 2(\beta L)^{-1}(\partial \ln G_K/d\mu_K)(\mu_K = 0)]$. Indeed the momentum integral in (21) can be performed exactly giving (see also [26])

$$n_K^T(T) = 2(\pi d)^{-1}\left\{\beta E_K^{(0)}[K_0(\beta E_K^{(0)}) + K_1(\beta E_K^{(0)})] + K_1(\beta E_K^{(0)})\right\} \tag{22}$$

$$\xrightarrow{\beta E_K^{(0)} >> 1} 2(2/\pi)^{\frac{1}{2}} d^{-1}(\beta E_K^{(0)})^{\frac{1}{2}}\left[1 + \frac{5}{8}(\beta E_K^{(0)})^{-1} + O(\beta E_K^{(0)})^{-2}\right] e^{-\beta E_K^{(0)}} \quad . \tag{23}$$

(K_0, K_1 modified Bessel functions) and $F = F_0 - k_B T n_K^T$. This last result agrees *exactly* with the TIO result [see (14)] to $O(\beta E_K^{(0)})^0$ in the parentheses of (23).

We could have achieved this low-T result *universally* for class A by working with the $\Delta(k;0)$ only and expanding the momentum integral to Gaussian order (i.e., $\gamma E_K = E_K^{(0)} + \frac{1}{2}M_K v^2$). The universal T dependence found by TIO (Sect.3.1) then follows easily [1]. Indeed a general *low-T* n_K^T formula can be obtained [7] which does not require *explicit* knowledge of the kink waveform or small oscillations about it (i.e., asymptotic phase shifts and kink vibration modes; there is a general cancellation [7]). These are only needed implicitly via an integral involving $V(\phi)$ and the location of its degenerate minima. This result can also be derived within other collective-mode formalisms (c.f., Sect.4.2). Of course, information on the perturbed "phonons" requires case-by-case small-oscillation study. Having appreciated the significance of E_K^*, it is a simple textbook ideal-gas calculation to introduce kink chemical potentials and handle phenomenology for varying boundary conditions, multikink species (class A), and topological restrictions (e.g., kink-antikink ordering in ϕ-four). Agreement with TIO results has been obtained [26].

4.2 Equivalent Low-T Approaches

Our conclusion from Sect.3.1 is that static phenomenology for class A *at low T* is complete and of little further interest. However, it is important to emphasize the equivalence of all competing "collective-coordinate" phenomenologies [5,23,24]. An example is probably best. Class A examples can be found in several references [5,24], so we present instead a more recent example [14] with *two* zero-frequency modes: the Ising symmetry ferromagnetic (class C, Sect.3.2). Of the several formalisms available we choose a collective coordinate scheme for evaluating the partition function in a steepest descent approximation, which emphasizes the very close connections between treatments of "inhomogeneous" states in different problems [14]—metastable state decay, disordered systems, quantum tunneling, nuclear physics, nucleation theory (see BÜTTIKER), quantization schemes (see MAKI). The method discussed at length by LANGER [23] evaluates the classical partition function in a path integral representation by a steepest descent calculation with Gaussian corrections. To this end a normal-mode ("linear stability") analysis must be performed about (nonlinear) solutions to the governing equations of motion, as in Sect.4.1. The original functional integration can then be replaced by integrations over the normal-mode amplitudes. However, continuous symmetries have to be treated specially, since large fluctuations must be accommodated properly, e.g., rigid translations [x_0 in (7)]. We therefore exclude zero-frequency eigenmodes from the normal-mode set and integrate separately over the corresponding collective coordinate (e.g., x_0), introducing appropriate Jacobians of transformation and ensuring orthogonality w.r.t zero eigenfunctions. Modifications to the fluctuation (e.g., spin-wave) spectrum due to nonlinear excitations are explicit via effects on the density of states, as in Sect.4.1. Here we will make use of a slight generalization of the technique to O(n) symmetry spin models [14], with the further modification that our single-site anisotropy model has only O(2) symmetry. Thus *two* continuous symmetries [rotation (ϕ_0) and translation (x_0)] must be treated specially.

First consider the single-wall sector partition function, Z_1. Following [23] we find $(S \equiv 1)$

$$Z_1 = \exp[N\beta(A + J) - \beta E_W] \int [d\psi][d\psi^*][dS_\xi]\exp(-\beta \int \psi^* M\psi dx) \tag{24a}$$

$$= \exp[N\beta(A + J) - \beta E_W](\det \mathcal{J})[(2\pi)^N/\det(\beta M)] \tag{24b}$$

with $\det \mathcal{J} \equiv \det \mathcal{J}_\phi \cdot \det \mathcal{J}_{x_0}$ (below). To derive (24) we have used the Ising symmetry and transformed to a local coordinate frame rotating with the wall (W) profile (7): $S^x = S_\zeta \cos\theta_W + S_\xi \sin\theta_W$, $S^y = S_\eta$, $S^z = -S_\zeta \sin\theta_W + S_\xi \cos\theta_W$. To Gaussian fluctuation order we find $\int dx\, \mathcal{H}[\vec{S}(x)] \cong -(A + J)N + E_W + \int \psi^* M\psi dx$, where $\sqrt{2}\, S_\zeta = (\psi + \psi^*)$, $\sqrt{2}\, S_\eta = -i(\psi - \psi^*)(|\psi|^2 << 1)$ and the stability operator $M = -Jd^2/dx^2 + 2A \cos 2\theta_W(x)$. This stability operator is familiar from the stability analysis for SG kinks [2] (c.f., Sect.4.1). The eigenspectrum comprises a "zero-frequency" bound state, f_0, describing rigid translations (x_0), and a continuum of "scattering states", f_k: $M(x)f(x) = Ef(x)$; $E_0 = 0$, $f_0 = (2d)^{-\frac{1}{2}}\operatorname{sech}(x/d)$; $E_k = 2A(1 + k^2d^2)$, $f_k = (2\pi)^{-\frac{1}{2}}(1 + k^2d^2)^{-\frac{1}{2}} \exp(ikx) \cdot [kd + i \tanh(x/d)]$. As in Sect.3.1, the density of scattering states is $\rho(k) = \rho_0(k) + (2\pi)^{-1}d\Delta(k)/dk$, where $\rho_0 = L/2\pi$ and the phase shift $\Delta(k) = \pi k/|k| - 2\tan^{-1}(kd)$. The last term in (24b), representing the scattering state contributions, can then be evaluated after straightforward integrations (c.f. [23]): $(2\pi)^N/\det(\beta M) = \exp[-\int dE\rho(E)\ln(\beta E/2\pi)] = [4A\beta/\pi] \cdot \{\exp[-N\ln(\beta J/2\pi) + N(2A/J)^{\frac{1}{2}}]\}$. The second term is the contribution from free spin waves (ρ_0) (using a discrete dispersion and taking the limit $d >> \ell$ as in Sect.4.1) and the first term describes the effect of the *change* in the density of states $\Delta\rho$. The second term in (24b) describes contributions from the wall symmetry modes. As in [23], $\det \mathcal{J}_{x_0} = \int_{-\infty}^{\infty} dx_0 \{\int_{-\infty}^{\infty} dx\, |\partial\vec{S}_W/\partial x_0|^2\}^{\frac{1}{2}} = N(2d)^{\frac{1}{2}}$. The coordinate transformations above were designed to suppress the rotational symmetry about the easy-axis. However, the full stability *matrix* for the wall contains a second zero-frequency mode $\propto|\vec{S}_W^\perp|$ (from $\partial\vec{S}_W/\partial\phi_0$) $[\vec{S}^\perp \equiv (S^x, S^y)]$, reflecting the rotational symmetry in spin space [14]: $\det \mathcal{J}_{\phi_0} = \int_0^{2\pi} d\phi_0 \{\int_{-\infty}^{\infty} |\vec{S}_W^\perp|^2 dx\}^{\frac{1}{2}} = 2\pi\{\int_{-\infty}^{\infty} dx\, \operatorname{sech}^2(x/d)\}^{\frac{1}{2}} = 2\pi(2d)^{\frac{1}{2}}$. Gathering results, we have [suppressing the trivial ground-state energy $N(A + J)$] $Z_1 = \exp[-N(2A/J)^{\frac{1}{2}} - N\ln(\beta J/2\pi) - \beta E_W] \cdot (4\pi N) \cdot (4A\beta/\pi)$. At low T $(<< E_W/k_B)$ (i.e., low wall density), this single wall sector calculation can be extended to the multiwall regime [14]. The total partition function Z_T is obtained from a partial exponentiation of Z_1, with associated free energy per spin $\tilde{F} \equiv -\beta^{-1}N^{-1}\ln Z_T$:

$$\tilde{F} = 4Ak_BT/E_W + k_BT\ell n(\beta J/2\pi) - 16A\exp(-\beta E_W) \quad . \tag{25}$$

Equation (25) must be compared with the transfer integral result (15), (17)

$$F = 4Ak_BT/E_W + k_BT\ell n(\beta J/2\pi) - 16(e/\pi)A\exp(-\beta E_W) \quad . \tag{26}$$

We see that F and $\tilde{F}$ agree exactly except for the purely numerical factor (e/π), which can be removed by improved evaluation of tunneling integrals, as remarked in Sect.3.2. Most importantly, we see that the dominant low-T dependence is precisely reproduced *if* we account for all zero-frequency modes and associated scattering state phase shifts.

4.3 Higher T

Since we wish to emphasize rigorous mode phenomenology, it is important to stress how *little* is known beyond asymptotic low T (except at trivial high T, where a soliton basis is a poor starting point), both for statics and even more for dynamics. TIO numerics are only consistency checks for plausible phenomenologies [17]. MD simulations (SCHNEIDER) provide very visual information on collision channels for strongly nonlinear modes and motivate approximate phenomenologies. In many cases there is little doubt about the fundamental role of kinks or other nonlinear modes, but treatment of mode interactions is ad-hoc or perturbational. Of course such phenomenologies can be valuable theoretical and experimental guides if they are not used beyond their regimes of validity. For example, estimates [19] of kink-antikink density from appropriate correlation lengths within TIO show how n_K^T is decreased w.r.t. the low-T expression (23). Very little is known about virial kink-kink interactions. Expansion of the in-well oscillation contribution E_0 (see Sect.3.1) in powers of $k_BT/E_K^{(0)}$ can be identified exactly term-by-term as an anharmonic "phonon" expansion [2]. However, this does not recognize the intrinsic spatial and temporal coherence of modes such as "breathers", even though these are very apparent in SG-like and ϕ-four-like model simulations, and can be described in anharmonic expansions. The same problem arises if perturbation expansions are made (as in collective-coordinate schemes) about an N-kink configuration [24]. A related problem occurs in all collective-coordinate schemes [24] where kinetic and configurational partition function components are separated (as in TIO). It is then difficult to assign parts of $Z_{\dot{\phi}}$ to individual nonlinear modes. For low-T kinks this is possible; for breathers and similar pulse or envelope solitons it is a problem. In principle, *exactly* integrable Hamiltonian systems provide a means of overcoming this difficulty and extending rigorous phenomenology to arbitrary T. In practice, difficulties remain, but we consider such systems a major soliton SM cutting edge.

5. Integrable Hamiltonian Systems

The beautiful mathematics developed in the last 10 years for integrable Hamiltonians has been introduced by McLAUGHLIN. For SM the major interest is that canonical transformations are prescribed to (generalized) "action-angle" variables or natural "nonlinear normal modes", e.g., in the sense that they form a complete orthogonal set. (The Hamiltonian separability is a generalization of that found in *linear* perturbation order for a kink, Sect.4.1). Nonlinear modes necessarily interact, but in these special cases only through generalized *position* coordinates, i.e., asymptotic phase shifts, again generalizing Sect.4.1. Despite normal-mode energy separability, the phase space for *extended* modes *is* affected by the local mode environment; it is then natural to try to generalize the calculation of Sect.4.1 by discretizing the system and attempting to keep track of mode sharing. We will discuss below the problems with this approach for modes whose density can be large, but first we illustrate these difficulties for SG.

The inverse scattering transform (IST) allows us to perform a canonical transformation from $\{\phi, \phi_x, \phi_t\}$ to variables effectively labeling "position", "velocity", "frequency" of component nonlinear modes. For SG, there is i) a continuous spectrum (extended modes or "radiation") parametrized by "momentum" $p = \hbar k(-\infty < p < \infty)$; $\rho(p)$ density of states) $(0 \leqq \rho(p) < \infty)$ and $\zeta(p)$ $(0 \leqq \zeta(p) < 2\pi)$ constitute action-angle variables as in harmonic theory with similar Poisson brackets $\{\zeta(p), \rho(p')\} = \delta(p - p')$; ii) a finite number of kink-soliton variables p_j (asymptotic "momentum") and q_j (initial "position") (see A), $(-\infty < p_j, q_j < \infty)$ and $\{p_j, q_j\} = \delta_{j,k}$; iii) a finite number of breather coordinates $p_j^B, q_j^B(-\infty < p_j^B, q_j^B < \infty)$, $\alpha_j^B(0 \leqq \alpha_j^B < \pi/2)$, $\tilde{\beta}_j^B(0 \leqq \tilde{\beta}_j^B < 4\pi E_{K_B}^{(0)}\omega_0^{-1})$ α_j^B is related to the internal frequency of the j-th breather [$\cos \alpha_j^B = \omega_j^B/\omega_0$; see (5)] and $\tilde{\beta}_j^B$ is related to the phase angle of the breather oscillation. The Hamiltonian *separates* as [27] (c.f., A)

$$H = \int_{-\infty}^{\infty} dx\, \mathcal{H}(x) = \int_{-\infty}^{\infty} dp\, \rho(p)(p^2c_0^2 + \hbar^2\omega_0^2)^{\frac{1}{2}} + \sum_{j=1}^{N_S} \left[p_j^2c_0^2 + E_K^{(0)2}\right]^{\frac{1}{2}} + \sum_{j=1}^{N_B} \left[p_j^{B^2} c_0^2 + E_j^2(\alpha_j^B)\right]^{\frac{1}{2}} \tag{27}$$

with N_s kink solitons (and anti-solitons) and N_B breathers. The final remarkable feature is that all mode interactions are pairwise additive. Thus asymptotic phase shifts can be taken in pairs and the low-T additivity *assumption* in Sect.4.1,2 should extent rigorously to arbitrary T for kinks and breathers. Explicit expressions for linear mode phase shifts from an isolated kink, $\Delta(k;v)$, (Sect.4.1,2), or breather, $\Delta_B(k;v_B,\omega_B)$ (below), are available [27].

These circumstances make the conceptual generalization of the low-T partition-function calculation immediate. We discretize on a finite length L, *assume* (27) still holds, and *assume* continuum phase shifts. Indeed the calculation for kinks was given for an arbitrary kink velocity, so that (22) is unchanged. Following a parallel procedure for breathers we need $\Delta F_B(v_B,\omega_B)$, the change in extended ("phonon") mode free-energy density due to a single breather, $L\Delta F_B(v_B,\omega_B) = k_BT\ P\int_{-\pi/\ell}^{\pi/\ell} dk\Delta\rho_B(k;v_B,\omega_B)\ell n(\beta\hbar\omega_k)$, i.e.,

$$L\Delta F_B(v_B,\omega_B) \xrightarrow{\ell << d} -\ 2k_BT\ \ell n(\beta\hbar\omega_0)$$

$$+\frac{k_BT}{2\pi} P\int_{-\infty}^{\infty} dk \frac{d\Delta_B}{dk}(k;v_B,\omega_B)\ell n(1+k^2d^2)^{\frac{1}{2}} \qquad (28)$$

$$= -\,2k_BT\ell n(\beta\hbar\omega_0) + \frac{k_BT}{4\pi}\int_{-\infty}^{\infty} dk\frac{d\bar{\Delta}_B}{dk}(k;0,\omega_B)\ell n(1+\gamma^{-2}k^2d^2) \quad ,$$

where we have used a Lorentz-boosted phase-shift (c.f., Sect.4.1); $P\int dk(d\Delta_B/dk) = -4\pi$ (corresponding to the removal of *two* linear modes per breather—a breather can be viewed as a kink-antikink bound state); and $d\bar{\Delta}_B/dk$ is even in k - $d\bar{\Delta}_B/dk$ is the part of $d\Delta_B/dk$ analytic on the real k axis, $d\bar{\Delta}_B(k,0,\omega_B)/dk = -4d(1-\omega_B^2/\omega_0^2)^{\frac{1}{2}}(1+k^2d^2-\omega_B^2/\omega_0^2)^{-1}$, to be compared with the corresponding kink expression $d\bar{\Delta}(k,0)/dk = -2d(1+k^2d^2)^{-1}$. (The breather result → twice the kink result as $\omega_B \to 0$). Evaluating (28), we define a *renormalized* breather energy $[\eta \equiv (1-\omega_B^2/\omega_0^2)^{\frac{1}{2}} = \sin\alpha_B]$

$$E_B^*(\gamma_B,\omega_B) = 2\gamma_B E_K^{(0)}\eta - 2k_BT\ell n[\beta\hbar\omega_0(\gamma+\eta)] \qquad (29)$$

to be compared with (20), especially as $\omega_B \to 0$. Using asymptotic phase-shift additivity and Hamiltonian separability, we can now extend (21) by including an extra grand canonical partition function for breathers

$$G_B(T,L,\mu_B) = \sum_{N_B=0}^{\infty} \exp(\beta\mu_B N_B)Z_B(N_B) \quad \text{with} \quad Z_B(N_B) = (N_B!)^{-1}Z_B^{N_B} \quad \text{and}$$

$$Z_B = h^{-2} \int_{-\infty}^{\infty} dp_B \int_0^L dq_B \int_0^{\pi/2} d\alpha_B \int_0^{4\pi E_K^{(0)}/\omega_0} d\beta \exp[-\beta E_B^*(p_B,\alpha_B)] \;, \quad \text{giving}$$

$$Z_B = \frac{4L}{\pi d}\left(\beta E_K^{(0)}\right)^2 \int_0^1 d\eta \int_1^{\infty} d\gamma \, \frac{\eta\gamma(\eta+\gamma)^2 \, e^{-2\beta E_K^{(0)}\eta\gamma}}{(\gamma^2-1)^{\frac{1}{2}}(1-\eta^2)^{\frac{1}{2}}} \;. \tag{30}$$

(Note how the phase-shift effects cancel powers of h in (21) and (30), a feature lost if we use *pure* ideal relativistic gases.) We presented this calculation to illustrate some characteristic problems (below). Note already the renormalization of E_B (29) even as $\omega_B \to \omega_0$, the harmonic limit. This limit ($\omega_B \to \omega_0$; $\eta \to 0$) is the source of many related difficulties; (30) can be evaluated tediously in terms of modified Bessel functions, with low-T expansion

$$Z_B \xrightarrow{\beta E_K^{(0)} \gg 1} \frac{3L}{2d}\left\{1 + \frac{2}{3}\left(\beta E_K^{(0)}\right)^{-1} + O\left(\beta E_K^{(0)}\right)^{-2} + \frac{4}{3\pi}\int_0^1 d\eta \, \frac{K_1(2\beta E_K^{(0)}\eta)}{\eta(1-\eta^2)^{\frac{1}{2}}}\right\} \;. \tag{31}$$

The reader can check the several unhappy features in (31). For instance, the breather density $n_B(T) = Z_B/L$ (assuming $\mu_B = 0$), and the free-energy density is $F = F_0 - k_B T n_K - k_B T n_B$, to be compared with the TIO result for SG[2], $F = F_0 - k_B T n_K^T - (k_B T/4d)(\beta E_K^{(0)})^{-1} + O[(k_B T)^3/E_K^{(0)2}]$. Even more indicative is the explicit *divergence* of the remaining integral (for any finite T) in (31) *as* $\omega_B \to \omega_0$ [dominantly from $\omega_B/\omega_0 \gtrsim 1 - \frac{1}{8}(k_B T/E_K^{(0)})^2$]. These problems arise from our lack of control over mode conservation and taking a consistent thermodynamic limit (below). They are quite generic to gapless *envelope* solitons. We have, for instance, found the *same* difficulties in analyzing *classical* SM for the cubic Schrödinger equation or the isotropic continuum ferromagnet (2C), for which mode sharing from asymptotic phase shifts can also be established (see also [28]). Slightly more control may be possible for *pulse* solitons as in the Toda lattice, even though they also have a gapless energy spectrum. For instance YOSHIDA [22] has used *infinite* chain action-angle variables (for "solitons" and "ripples") on a finite chain and imposed periodicity and mode conservation by hand (restricting ripple wave vectors). In this way he finds agreement with TIO results at *low* T. Here again, however, controlled use of phase-shift information is more appropriate (work in progress) as well as integrability structure for the *finite* discrete chain (below). For *kink* solitons (i.e., finite energy gap) we saw

(Sect.4.1) that exact agreement with TIO could be obtained at *low* T. In view of SG integrability, it might be thought that result (22) would give an exact kink description at all T. In fact comparing with the predicted [19] n_K^T from correlation lengths in TIO, we find that (22) is a slightly *worse* prediction than the asymptotic formula (23). This seems to argue for the importance of nontrivial mode interactions on a discrete SG chain with periodic boundary conditions. [We note in passing, however, that it is advantageous [29] to use E_K^* (20) to construct an "ideal gas" velocity distribution, as used in phenomenological structure factor models].

We now briefly summarize the difficulties of using available IST mathematics to construct a classical SM (see also [2]). They are all consequences of a gapless spectrum or of high T.

i) We have implicitly assumed above that there is no special distinction of multiple occupations of the same "soliton state" (specified by positions of poles and pairs of poles in the eigenvalue plane in inverse scattering theory). In general this is untrue which will be important for high soliton densities (from pulses, envelopes or, at high T, kinks).

ii) The possibility of arbitrarily low-energy breathers and correspondingly large densities, and spatial extensions) means that consistent mode counting is more sensitive than for a low density of local kinks. The canonical action-angle variables and phase shifts used above were strictly valid for a continuum system with decaying boundary conditions. The use of these data for a finite discrete system with periodic boundary conditions (as used for TIO) is clearly uncontrolled as is the process of taking the thermodynamic limit. (Finite gap, localized kink excitations were insensitive to these concerns.) One can try to enforce mode conservation and soliton size cutoff in approximate ways, but the most desirable route is to construct SM with IST data for a finite, discrete, integrable system with, e.g., periodic boundary conditions, and *finally* take the thermodynamic limit. This is ongoing work; IST status for such problems is summarized by McLAUGHLIN. The Toda lattice is especially attractive as well as some discrete spin models related to discrete generalizations of nonlinear Schrödinger.

iii) *Classical* SM was implied in TIO by the separation $Z = Z_{\phi} Z_{\phi}$ (Sect.3.1). It was implemented in our nonlinear phenomenology by the use of high-T ($k_B T \gg \hbar\omega_0$) harmonic phonon free-energy expressions [e.g., (18)]. There is some inconsistency therefore in our use of a purely *classical* breather spectrum, especially for low-energy (extended) breathers with $E_B \lesssim \hbar\omega_0$. (This problem is intrinsic to the dual particle-oscillator character of breathers and related solitons.) In particular the *quantum* breather spectrum is *discrete*.

This is often not serious for applications but *is* for low-energy breathers since the *lowest* energy breather is physically equivalent to the quantum harmonic oscillator, i.e., Klein-Gordon quantum (c.f., the classical spectra, (class A). Indeed any breather may be viewed as a multimagnon (or phonon, etc.) bound state; breathers and solitons exhaust the quantum SG spectrum. While it is a technically challenging problem to construct a truly classical SM (for arbitrary T), this was not really achieved even for kinks (Sect.4.1) and physical concern necessarily centers on *quantum* SM from which a high-T (classical) limit can be extracted if desired. It is therefore our opinion that a major effort should now be devoted to quantum SM of "soliton" systems. Quantization schemes for nonlinear objects will be reviewed separately by MAKI; some useful progress has been made in constructing a quantum SM for ϕ-four, SG and double quadratic, etc. (class A) using nonlinear generalizations (analogous to classical collective-coordinate manipulations, Sect.4) of conventional *perturbative* functional integral formalisms. An alternative approach is to concentrate on fully *integrable quantum* Hamiltonians (including SG), particularly motivated by connections with *Bethe ansatz* literature [30]; see also BONNER, FOWLER, McLAUGHLIN. We do not have space to describe this philosophy here; details will be published elsewhere. The basic ingredients are [30] i) the unification of the majority of soluble many-body models as integrable quantum Hamiltonians with explicit connections to a generalized Bethe ansatz; ii) physical understanding of elementary "soliton" excitations from integrable *classical* Hamiltonian predecessors [12]; iii) use of quantum SM from Bethe ansatz literature (e.g., for the spin-½ Ising-XY-Heisenberg model, with connections to both SG-like and NLS-like integrable Hamiltonians).

6. Outlook

We hope that the central role of solitons in large areas of 1-D physics will be apparent from other talks at this Meeting, e.g., AXE, BAK, BONNER, FOWLER, RICE, SCHRIEFFER, STEINER, as well as contributions to the soliton session. As far as statistical properties are concerned, we suggest among important future areas i) exploitation of quantum integrability, the Bethe ansatz and soluble model equivalences; ii) study of impurity effects on soliton SM (classical and quantum); iii) study of turbulent transitions and fully developed turbulence in driven, damped soliton systems with thermal noise (having potential lessons for both solid-state and fluid turbulence theory). These are all on-going projects.

References

1. A.R. Bishop, "Statistical Mechanics of Nonlinear Dispersive Systems", in *Solitons and Condensed Matter Physics*, ed. by A.R. Bishop, T. Schneider, Springer Series in Solid-State Sciences, Vol.8 (Springer, Berlin, Heidelberg, New York 1978) p.85
2. J.F. Currie et al., Phys. Rev. B22 (1980); A.R. Bishop et al., Physica D1, 1 (1980); A.R. Bishop et al., Adv. Phys., in prep.
3. e.g., R. Jackiw, Rev. Mod. Phys. 49, 681 (1977)
4. S.F. Edwards and A. Lenard, J. Math. Phys. 3, 773 (1962)
5. A. Seeger and P. Schiller, in *Physical Acoustics*, ed. W.P. Mason (Academic, New York 1966), Vol.3
6. J.A. Krumhansl and J.R. Schrieffer, Phys. Rev. B11, 3535 (1975)
7. R.M. DeLeonardis and S.E. Trullinger, preprint (1980); and refs.
8. A.R. Bishop, unpublished (1975)
9. J.F. Currie et al., Phys. Rev. A20, 2213 (1979)
10. S.E. Trullinger and R.M. DeLeonardis, preprint (1980); K.R. Subbaswamy and S.E. Trullinger, preprint (1980)
11. A. Patkós and P. Rujàn, Z. Phys. B33, 163 (1979); P.S. Riseborough and S.E. Trullinger, Phys. Rev. B (1980)
12. e.g., L.A. Takhtajan, Phys. Lett. 64A, 235 (1977); A.E. Borovik, Pisma JETP 28, 629 (1978); E.K. Sklyanin, preprint (1980)
13. K.A. Long and A.R. Bishop, J. Phys. A12, 1325 (1979)
14. A.R. Bishop, et al., J. Phys. C (Lett.) 13 (1970); and in prep.
15. V.E. Zhakarov and L.A. Takhtadzhan, Teor. i Mat. Fiz. 38, 26 (1979)
16. D.J. Scalapino et al., Phys. Rev. B6, 3409 (1972); and refs.
17. A.R. Bishop, J. Phys. A (1981)
18. e.g., N. Gupta and B. Sutherland, Phys. Rev. A14, 1790 (1976); J.F. Currie et al., Phys. Rev. A16, 796 (1977)
19. A.R. Bishop, Solid State Comm. 30, 37 (1979); see also A.R. Bishop and J.A. Krumhansl, Phys. Rev. B12, 2824 (1975)
20. S. Goldstein, Proc. Roy. Soc. (Edinburgh) T49, 210 (1929); R.M. DeLeonardis and S.E. Trullinger, Phys. Rev. B20, 2603 (1979)
21. e.g., H. Büttner and F.G. Maertens, Solid State Comm. 29, 663 (1979)
22. F. Yoshida, private communication (1980); preprint (1980)
23. e.g., R. Landauer and J.A. Swanson, Phys. Rev. 121, 1668 (1961); J.S. Langer, Annal. Phys. (NY) 14, 108 (1967)
24. e.g., G.F. Mazenko and P.S. Sahni, Phys. Rev. B18, 6139 (1978); W. Apel et al., Z. Phys. B34, 1983 (1979); S. Coleman, Phys. Rev. D15, 2929 (1977)
25. N. Theodorakopoulos, J. Phys. A12, L211 (1979)
26. S.E. Trullinger, Phys. Rev. B22 (1980)
27. J.F. Currie, Phys. Rev. A16, 1692 (1977); V.E. Korepin and L.D. Faddeev, Phys. Rep. C42, 1 (1978)
28. H.C. Fogedby, Physica Scripta (in press)
29. S.E. Trullinger and A.R. Bishop, Solid State Comm. (1981)
30. A.R. Bishop, Z. Phys. B37, 357 (1980); and refs.

The Quantum Inverse Scattering Method and Applications to Spin Chains

M. Fowler

Department of Physics, University of Virginia
Charlottesville, VA 22901, USA

1. Introduction

Over the past twelve years, the development of the Inverse Scattering Method (ISM) has provided much insight into the dynamics of certain classical nonlinear one dimensional systems [1]. Within the last two years or so, the method has been applied to quantum systems by Fadeev, Sklyanin, Kulish and Takhtadzhyan [2 - 4] in Leningrad, Thacker, Wilkinson and Creamer [5 - 8] at Fermilab and Honerkamp [9] and coworkers in Freiburg. It turns out to be closely related to two other techniques which have been developed for one dimensional systems - the Bethe Ansatz (BA), first applied to spin chains [10,11], and the transfer matrix methods of Onsager [12], Baxter [13] and others for lattice models.

In this review we shall concentrate on two particular systems : the field described by the nonlinear Schrödinger equation (NLS) [2,5], and the Heisenberg-Ising (HI) [14] spin one-half chain. In section two, we review the application of ISM techniques to the classical NLS field [15,16], yielding a set of action-angle type variables for the system. The parallel quantum analysis is carried out in section three - the quantized NLS field is a gas of bosons with δ-function interactions. The eigenstates of this system were shown to have BA form by Lieb and Liniger [17]. Creation operators for BA excitations turn out to be given by the quantum analogs of certain of the action-angle variables. Thus the quantum ISM gives - for the first time - a second quantized formulation of the Bethe Ansatz states. Correlation functions for this system can (in principle) be computed using a quantum version of the Gelfand-Levitan method, due to Thacker and coworkers [6]. Section three closes with a brief introduction to the elegant techniques of Fadeev et al. [2] for finding the commutation relations among the action-angle type variables.

In section four, the application of the method to the HI spin chain is considered. We emphasize the close mathematical analogy between the action-angle type variables for the HI spin chain and those for the quantum NLS. The appropriate mathematical formulation for HI emerges from a study of Baxter's analysis of the eight-vertex model [13]. A quantum Gelfand-Levitan analysis to find formal expressions for the correlation functions in the spin chain system has not yet been carried out, but it appears likely that this will be done in the near future. These spin-spin correlation functions are of great current interest, because neutron scattering experiments on systems well approximated by the HI Hamiltonian are being done at the present time, with good enough resolution to give significant information about the correlation functions [18].

For one simple case of the HI model - the isotropic XY spin one-half chain- the correlation functions have already been found by a Jordan-Wigner transformation [19]. As is discussed in section four, it can be shown explicitly that for this case the BA excitation creation operator, given by the quantum ISM, is identical to the Jordan-Wigner excitation creation operator [20], so the methods are equivalent for the isotropic XY model. This might serve as a guide toward the more general quantum Gelfand-Levitan analysis for the HI system.

2. The Classical Inverse Scattering Method

The classical ISM has two steps [1]. In the first, or "direct" step, a wave scattering problem is solved where the classical field being analyzed appears as a potential. The scattering amplitudes are then functions of the classical field and it turns out for NLS, Sine-Gordon and some others that they are, remarkably, action-angle variables having a simple time dependence. In the second, or "inverse" step of the method, the classical field (potential) at a later time is expressed in terms of the (time developed) scattering amplitude variables, and hence the variation of the classical field in time can be found.

To exhibit the basic ideas of the method, which we shall generalize to the quantum case in the next section, we briefly outline the steps for NLS, the classical field given by the Hamiltonian

$$H = \int dx \left(\frac{\partial \phi^*}{\partial x}\frac{\partial \phi}{\partial x} + c|\phi|^4\right) . \tag{2.1}$$

The equation of motion of the field is

$$i\frac{\partial \phi}{\partial t} = \{H,\phi\} = -\frac{\partial^2 \phi}{\partial x^2} + 2c|\phi|^2\phi \tag{2.2}$$

where the Poisson Bracket is defined by

$$\{\alpha,\beta\} = i\int dx\left[\frac{\delta\alpha}{\delta\phi(x)}\frac{\delta\beta}{\delta\phi^*(x)} - \frac{\delta\alpha}{\delta\phi^*(x)}\frac{\delta\beta}{\delta\phi(x)}\right] . \tag{2.3}$$

Zakharov and Shabat [15] noticed that (2.2) could be written in the form

$$i\dot{L} = [L,A] \tag{2.4}$$

where L,A are ϕ dependent linear differential operators having a 2×2 matrix structure.

Explicitly,
$$L = i\begin{pmatrix} 1+p & 0 \\ 0 & 1-p \end{pmatrix}\frac{\partial}{\partial x} + \begin{pmatrix} 0 & \phi^* \\ \phi & 0 \end{pmatrix} , \tag{2.5}$$

$$A = -p\begin{pmatrix} 1 & 0 \\ 0 & 1 \end{pmatrix}\frac{\partial^2}{\partial x^2} + \begin{pmatrix} \frac{|\phi|^2}{1+p} & i\phi_x^* \\ -i\phi_x & -\frac{|\phi|^2}{1-p} \end{pmatrix} ,$$

$$c = \frac{1}{p^2-1} . \tag{2.5}$$

Studying the structure of the operator L above, we notice that it has the form of a Dirac scattering equation - its eigenfunctions ψ_i, defined by $L\psi_i = \lambda_i\psi_i$, are two component spinors and ϕ,ϕ^* play the role of a potential. Furthermore, equation [2.4] implies that as ϕ changes in time, it must change in such a way that the eigenvalues λ_i are invariant (although the eigenfunctions do of course change).

To study these eigenfunctions further, it is convenient to factor out a plane wave term and rescale (see Zakharov and Shabat [15] for details) to give the equation $L\psi = \lambda\psi$ in the form (following Thacker and Wilkinson [5])

$$(i\frac{\partial}{\partial x} + \frac{1}{2}\xi)\psi_1 = -\sqrt{c}\,\psi_2\,\phi$$

$$(i\frac{\partial}{\partial x} - \frac{1}{2}\xi)\psi_2 = \sqrt{c}\,\phi^*\,\psi_1 . \tag{2.6}$$

The standard analysis of such quantum mechanical scattering problems is based on the Jost functions, which are scattering eigenstates having a particular plane wave behavior at large distances in one direction. Here they are

$$\psi(x,\xi) = \begin{pmatrix} \psi_1 \\ \psi_2 \end{pmatrix} \underset{x\to-\infty}{\sim} e^{\frac{i\xi x}{2}}\begin{pmatrix} 1 \\ 0 \end{pmatrix}$$

$$\chi(x,\xi) = \begin{pmatrix} \chi_1 \\ \chi_2 \end{pmatrix} \underset{x\to+\infty}{\sim} e^{-\frac{i\xi x}{2}}\begin{pmatrix} 0 \\ 1 \end{pmatrix} \tag{2.7}$$

Equation (2.6) is also satisfied by $\tilde{\psi} = \begin{pmatrix} \psi_2^* \\ \psi_1^* \end{pmatrix}$ and ψ, $\tilde{\psi}$ form a complete set of solutions. The scattering coefficient $a(\xi)$, $b(\xi)$ are defined by

$$\psi(x,\xi) = a(\xi)\tilde{\chi}(\xi,x) + b(\xi)\chi(\xi,x) \tag{2.8}$$

so $a(\xi)$ is the Wronskian $\psi_1\chi_2 - \psi_2\chi_1$, independent of x. Using (2.5) at large distances, where ϕ is zero, it is not difficult to show, following Zakharov and Shabat, that $a(\xi)$, $b(\xi)$ change in a simple way as ϕ develops in time. In fact

$$a(\xi,t) = a(\xi,0), \qquad b(\xi,t) = b(\xi,0)e^{-i\xi^2 t} . \tag{2.9}$$

A standard method in quantum scattering theory - the Gelfand-Levitan method - determined the potential uniquely if the scattering amplitudes are known. (We ignore complications caused by possible bound states, which can however be handled quite easily). This method can be used to find $\phi(x,t)$, given $a(\xi,t)$, $b(\xi,t)$.

Summarizing, in the classical ISM, equation (2.2) is solved as follows :

(i) The initial values of ϕ are inserted in (2.6), and the scattering coefficients $a(\xi)$, $b(\xi)$ are found.

(ii) Their time development is given by (2.9).

(iii) The Gelfand-Levitan analysis is applied to the set of scattering data $a(\xi,t)$, $b(\xi,t)$ to find $\phi(x,t)$.

We now turn to a more complicated (but quite illuminating) approach to the problem [16]. The scattering amplitudes $a(\xi)$, $b(\xi)$ are functionals of the classical field ϕ, that is, they are dynamical variables. They can be expressed as series in ϕ, ϕ^* by iterative solution of the Dirac scattering equation (2.6) - and hence their time development can be found by evaluating their Poisson Brackets with the Hamiltonian. The results are, of course

$$\{H,a(\xi)\} = 0 , \qquad \{H,b(\xi)\} = i\xi^2 b(\xi) . \tag{2.10}$$

Thus $a(\xi)$, $b(\xi)$ are dynamical variables of the action-angle type. Poisson Brackets of various pairs of these variables can be evaluated by the same technique, giving

$$\{a(\xi),a(\xi')\} = \{a(\xi),a^*(\xi')\} = \{b(\xi),b(\xi')\} = 0$$

$$\{a(\xi),b(\xi')\} = \frac{c}{\xi - \xi' - i\varepsilon}\, a(\xi)b(\xi') \tag{2.11}$$

$$\{b(\xi),b^*(\xi')\} = 2\pi i a^*(\xi)a(\xi')\delta(\xi - \xi') .$$

The constancy of the function $a(\xi)$ in time is equivalent to an infinite set of conservation laws for the dynamical system.

3. The Quantum Inverse Scattering Method

Over the last two years or so, Fadeev **et al.**[2] and Thacker **et al.**[5] have extended the ISM to quantum systems. Here we outline the new technique for the simplest case - the NLS discussed in the last section - following Thacker and coworkers. Later in the section, the elegant formulation developed by Fadeev and coworkers is discussed.

If the classical NLS Hamiltonian (2.1) is replaced by its quantum equivalent

$$H = \int dx\,(\frac{\partial\phi^+}{\partial x}\frac{\partial\phi}{\partial x} + c\phi^+\phi^+\phi\phi) \tag{3.1}$$

and the field variables have canonical commutation relations

$$[\phi(x), \phi^+(y)] = \delta(x-y) \qquad (3.2)$$

we have a gas of Bose particles with δ-function interactions. The eigenstates of this system were found some years ago by Lieb and Liniger [17] using the Bethe Ansatz (BA). Here we shall set up a scattering analysis parallel to that in section two, leading to quantum operators $a(\xi)$, $b(\xi)$, and discuss the relationship between these operators and the BA analysis.

Rewriting the equation (2.6),

$$\begin{aligned} (i\frac{\partial}{\partial x} + \frac{1}{2}\xi)\psi_1 &= -\sqrt{c}\,\psi_2\,\phi \\ (i\frac{\partial}{\partial x} - \frac{1}{2}\xi)\psi_2 &= \sqrt{c}\,\phi^+\,\psi_1 \end{aligned} \qquad (2.6)$$

the ordering of the terms on the right hand side is now important, because ψ_1, ψ_2 depend on ϕ, ϕ^+ and hence do not in general commute with ϕ, ϕ^+. The ordering chosen above leads to normally ordered expressions for $a(\xi)$, $b(\xi')$ on iteration, giving the desired result that $a(\xi) \equiv 1$, $b(\xi) \equiv 0$ on the vacuum state.

Following Thacker and Wilkinson [5], we find

$$\begin{aligned} a(\xi) &= 1 + c\int dx_1 dy_1 \theta(x_1<y_1) e^{i\xi(x_1-y_1)} \phi^+(x_1)\phi(y_1) + \ldots \\ b(\xi) &= \int dx_1 e^{i\xi x_1}\phi^+(x_1) + \\ &\quad + c\int dx_1 dx_2 dy_1 \theta(x_1<y_1<x_2) e^{i\xi(x_1+x_2-y_1)} \phi^+(x_1)\phi^+(x_2)\phi(y_1) + \ldots . \end{aligned} \qquad (3.3)$$

The commutation relations among the $a(\xi)$, $b(\xi')$ can be found from (3.2), (3.3) and are

$$\begin{aligned} &[a(\xi), a(\xi')] = [a(\xi), a^+(\xi')] = [b(\xi), b(\xi')] = 0 \\ &[a(\xi), b(\xi')] = -\frac{ic}{\xi-\xi'-i\varepsilon}\, b(\xi')a(\xi) \\ &[H, a(\xi)] = [H, b(\xi)] = \xi^2 b(\xi) \quad . \end{aligned} \qquad (3.4)$$

Notice that $b(\xi_1)b(\xi_2)\ldots b(\xi_n)|0\rangle$ is an exact eigenstate of H with eigenvalue $\xi_1^2+\xi_2^2+\ldots+\xi_n^2$. But we already know all the eigenstates of H - they are given by the Bethe Ansatz [17]. We can confirm by direct computation using the form of $b(\xi)$ in (3.3) that this state is just the BA state with pseudomomenta $\xi_1, \ldots, \xi_n$. (Actually there are simpler ways of proving this than direct computation, which has only been carried through for a few particles). Hence we have derived a <u>second quantized</u> formulation of the Bethe Ansatz. Since $[H, a(\xi)] = 0$, the BA eigenstates of H are also eigenstates of the continuum of operators $\{a(\xi)\}$.

The operator

$$R^+(\xi) = \frac{i}{\sqrt{c}}\, b(\xi) a^{-1}(\xi) \tag{3.5}$$

creates correctly normalized BA eigenstates [2,5]. It satisfies the commutation relations

$$R(\xi)R^+(\xi') = S(\xi,\xi')R^+(\xi')R(\xi) + 2\pi\delta(\xi - \xi') \tag{3.6}$$

$$\text{where} \quad S(\xi,\xi') = \frac{\xi - \xi' - ic}{\xi - \xi' + ic} \quad \text{(S-matrix).}$$

Creamer et al. [6] completed the quantum ISM for this system by a Felfand-Levitan analysis to find

$$\phi(x) = \int \frac{d\xi_1}{2\pi} R(\xi_1) e^{i\xi_1 x} + c\int \frac{d\xi_1}{2\pi}\frac{d\xi_2}{2\pi}\frac{d\xi_3}{2\pi} \frac{R^+(\xi_2)R(\xi_1)R(\xi_3)e^{i(\xi_1-\xi_2+\xi_3)}}{(\xi_2 - \xi_1 + i\varepsilon)(\xi_3 - \xi_2 + i\varepsilon)} + \ldots \tag{3.7}$$

This could in principle be used to find correlation functions, which are already known in part from the work of Vaidya and Tracy [21]. For the particular case $c = \infty$, the series exponentiates [8] to give a Jordan-Wigner form

$$\phi(x) = e^{i\pi\int_x^\infty \tilde{R}^+(y)\tilde{R}(y)dy}\, \tilde{R}(x) \tag{3.8}$$

where $\tilde{R}(x)$ is the Fourier transform of $R(\xi)$.

A different and extremely elegant approach to the quantum ISM for NLS was developed by Fadeev and coworkers [2]. Writing the Dirac scattering equation (2.6) in the form

$$\left(\frac{d}{dx} + Q(x,\xi)\right) G(x,\xi) = 0 \tag{3.9}$$

with $G = \begin{pmatrix} I & 0 \\ 0 & I \end{pmatrix}$ at $-\infty$ to give the Jost functions (with the $\phi \equiv 0$ behavior factored out), we have the formal solution

$$G(x,\xi) = \overleftarrow{\exp}\left\{- \int_{-\infty}^{x} Q(y,\xi)dy\right\}$$

where $\leftarrow$ denotes x-ordering of the operators. The limit of $G(x,\xi)$ at $x = +\infty$ is called the monodromy matrix $T(\xi)$, its elements being $a(\xi)$, $b(\xi)$ and their Hermitian conjugates. To avoid ultraviolet difficulties, the problem is formulated on a lattice, with new canonical variables ϕ_n, ϕ_n^+ corresponding to suitably defined averages of $\phi(x)$, $\phi^+(x)$ over an nth step of length δ. The monodromy matrix then becomes a product of local matrices,

$$T(\xi) = \overleftarrow{\prod_n} L_n(\xi) \tag{3.10}$$

where $$L_n(\xi) = \begin{pmatrix} (1 - \frac{i\xi}{2}\delta)I_n & -i\sqrt{c}\,\phi_n^+ \\ i\sqrt{c}\,\phi_n & (1 + \frac{i\xi}{2}\delta)I_n \end{pmatrix} . \qquad (3.11)$$

It is evident from (3.10), (3.11), that the elements $a(\xi)$, etc., of $T(\xi)$ can be expressed as sums of products of many operators ϕ_n, $\phi_{n'}^+$, etc. The problem is to find the commutation relations for these $a(\xi)$, $b(\xi')$. Fadeev solves this problem using a trick of Baxter's [13]. To show how it works, we take the case of a product of a finite number N of local matrices, denoting the monodromy matrix $T_N(\xi)$.

It is not difficult to show that for the local matrices $Ln(\xi)$, $Ln(\xi) \otimes Ln(\xi')$ and $Ln(\xi') \otimes Ln(\xi)$ are mathematically similar - that is, there exists a 4×4 matrix $R(\xi,\xi')$ with nonoperator elements such that

$$R(\xi,\xi')(Ln(\xi) \otimes Ln(\xi')) = (Ln(\xi') \otimes Ln(\xi))R(\xi,\xi') . \qquad (3.12)$$

Note that the direct product $\otimes$ is defined by

$$\begin{pmatrix} M_{11} & M_{12} \\ M_{21} & M_{22} \end{pmatrix} \otimes \begin{pmatrix} N_{11} & N_{12} \\ N_{21} & N_{22} \end{pmatrix} = \begin{pmatrix} M_{11}N_{11} & M_{11}N_{12} & M_{12}N_{11} & M_{12}N_{12} \\ M_{11}N_{21} & M_{11}N_{22} & M_{12}N_{21} & M_{12}N_{22} \\ M_{21}N_{11} & M_{21}N_{12} & M_{22}N_{11} & M_{22}N_{12} \\ M_{21}N_{21} & M_{21}N_{22} & M_{21}N_{21} & M_{22}N_{22} \end{pmatrix}$$

Hence (3.12) is just a long way of writing the (simple) commutation relations among the elements of $Ln(\xi)$, $Ln(\xi')$ - these determine the elements of $R(\xi,\xi')$. The fact that the commutation relations can be written in the form (3.12) reflects some very special properties of the matrices $Ln(\xi)$. Putting (3.12) together with

$$\overset{\longleftarrow}{\prod_{n=1}^{N}} Ln(\xi) \otimes \overset{\longleftarrow}{\prod_{n'=1}^{N}} Ln'(\xi') = \overset{\longleftarrow}{\prod_{n=1}^{N}} Ln(\xi) \otimes Ln(\xi') \qquad (3.13)$$

which follows from the fact that operators belonging to different sites commute, we find

$$R(\xi,\xi')(T_N(\xi) \otimes T_N(\xi')) = (T_N(\xi') \otimes T_N(\xi))R(\xi,\xi') ,$$

from which the commutation relations of $a(\xi)$, $b(\xi)$, etc. follow immediately ! The limiting processes necessary on the infinite line change the commutation relations somewhat, but the method is straightforward, and is analyzed in detail by Fadeev [2].

4. The Heisenberg-Ising Spin Chain

In this section we consider the application of the quantum ISM to the Heisenberg-Ising (HI) spin chain, having the Hamiltonian

$$H = -\sum(\sigma_i^x\sigma_{i+1}^x + \sigma_i^y\sigma_{i+1}^y + \Delta\sigma_i^z\sigma_{i+1}^z) . \qquad (4.1)$$

The eigenstates of this system are already well known - they are given by Bethe's Ansatz [11,14], and are very similar to the Bose gas eigenstates discussed in section three. The main differences are that the single excitation (magnon) dispersion curve has the cosine form imposed by the lattice, and the magnon-magnon phase shift is not simply a function of the difference of the pseudomomenta. We mention here that a standard procedure for evaluating the energy levels is to transform from the pseudomenta k_i to variables α_i in which the phase shift <u>is</u> a function of $\alpha_i - \alpha_j$, whereupon Fourier convolution methods can be used [11].

Even though the form of the eigenstates has been written down, their mathematical complexity is such that it has not proved possible so far to evaluate matrix elements of spin operators, so the spin-spin correlation functions of the system are not known, except in certain limits. However, as mentioned in the introduction, these functions are currently being measured in neutron scattering experiments [18] on materials well represented by the HI spin chain Hamiltonian. Hence it seems worth-while to try to set up a quantum ISM scheme, parallel to that discussed in section three, to find expressions for the correlation functions. Some progress in this direction has already been made - the scattering data operators analogous to $a(\xi)$ and to $b(\xi)$, the creation operators for the BA excitations, have been found. The local spin operator has not at this time been expressed as a series in BA creation operators (corresponding to (3.7)) although this should not be too difficult. It must be added however, that even when this has been done, the evaluation of correlation functions will still be rather tricky because the $R^+(\xi)$ have nontrivial commutation relations among themselves.

We turn now to a discussion of the first step in the quantum ISM for the spin chain. The construction of operators analogous to the scattering amplitudes $a(\xi)$, $b(\xi)$ in the quantum NLS. As we have already stated, the eigenstates of both systems have the same BA mathematical structure, with some rather minor differences of detail. Since the $b(\xi)$ are the creation operators for individual BA excitations, and since all the BA states are exact eigenstates of all the $a(\xi)$, it is difficult to imagine that the monodromy matrix for the spin chain system can be very different from that of the NLS given in (3.10) and (3.11), especially since the latter is already in lattice form. In fact, as pointed out by Fadeev and coworkers [2], just such mathematical objects for spin chains have already appeared in the literature ! Baxter [13], in his solution of the eight-vertex model, introduced a one-parameter set of commuting transfer matrices, all of which commuted with a spin chain Hamiltonian. We shall see that this set of operators is equivalent to the set $\{a(\xi) + a^+(\xi)\}$.

To show how Baxter's work leads to a monodromy matrix for the spin chain, it is necessary to recapitulate his construction of the eight-vertex model transfer matrix [13]. The eight-vertex model has a square lattice and directed bonds with an even number pointing into each vertex. The vertex energy depends on the configuration of bonds at that vertex, but is invariant with respect to reversing all four bonds. This implies that there are four distinct possible vertex energies, in general. Labelling the north, south, east and west bonds at a vertex by μ', μ, λ', λ (each variable being $+1$ or -1) Baxter wrote the weight at a single vertex in the form

$$R(\mu,\mu'|\lambda,\lambda') = \sum_{j=1}^{4} W_j\, \sigma^j_{\mu\mu'}\, \sigma^j_{\lambda\lambda'} \tag{4.2}$$

where σ^1, σ^2, σ^3 are the Pauli matrices, σ^4 is the unit matrix, and the W_j's are simple linear combinations of the weights $e^{-\beta\varepsilon_j}$, $\{\varepsilon_j\}$ being the set of possible vertex energies. The free energy is found by multiplying together one R for each vertex in the system and summing over all λ's and μ's. In the transfer matrix method, the sum is done one row at a time. To sum over all the λ's in one row, R is written as a 2×2 matrix with rows and columns labelled by $\lambda = +,-$ $\lambda' = +,-$. Then matrix multiplication of $R_1 \ldots R_N$ along a row automatically sums over the λ's. The product matrix

$$T_N = R_1 \ldots R_N = \begin{pmatrix} A_N & B_N \\ C_N & D_N \end{pmatrix} \tag{4.3}$$

is still a function of the set of μ's and μ''s for the N sites. In the present analysis, we need only consider (4.2) for $w_1 = w_2$ (the more general case turns out to correspond to completely anisotropic spin chains). With this equality, the matrix R_n in (4.3) has the form [7] (using (4.2))

$$R_n = \begin{pmatrix} w_3\sigma_n^3 + w_4\sigma_n^4 & 2w_1\sigma_n^- \\ 2w_1\sigma_n^+ & -w_3\sigma_n^3 + w_4\sigma_n^4 \end{pmatrix} . \tag{4.4}$$

Thus the individual elements of R_n are 2×2 matrices with their rows and columns labelled by $\mu = +,-$ $\mu' = +,-$. That is to say, the elements of T_N operate on a Hilbert space $\mathcal{H}$ which is a direct product of spin one-half spaces corresponding to each site n along the row.

$$\mathcal{H} = \bigotimes_{n=1}^{N} \mathcal{H}_n \tag{4.5}$$

This is the Hilbert space of a spin one-half chain.

From Baxter's work, we know that for a Heisenberg-Ising spin chain with given $\Delta = -\cos\mu$ (for $|\Delta| \leq 1$) the HI Hamiltonian commutes with the set of operators $\mathrm{Tr}T_N(\xi)$, where $T_N(\xi)$ is defined by (4.3), (4.4) with

$$\begin{aligned} w_1 &= w_2 = \sinh i\mu \\ w_4 + w_3 &= 2\sinh\tfrac{1}{2}(\xi + i\mu) \\ w_4 - w_3 &= 2\sinh\tfrac{1}{2}(\xi - i\mu), \quad -\infty < \xi < \infty . \end{aligned} \tag{4.6}$$

Studying (4.3), (4.4) we realize that T_N has a mathematical structure very close to that of the monodromy matrix $T_N(\xi)$ defined by (3.10), (3.11) for the quantum NLS problem. Furthermore, the operators $\mathrm{Tr}T_N(\xi)$, which would correspond to $a(\xi) + a^+(\xi)$, are known to be diagonal on all the BA eigenstates of the Hamiltonian. Hence it is tempting to conjecture that $T_N(\xi)$ <u>is</u> the monodromy matrix for the spin chain problem, the $B_N(\xi)$ being creation operators for BA excitations, and so forth. In fact, following Fadeev [2], it is not difficult to prove this is the case, using the method - Baxter's trick again - outlined at the end of section three for the NLS problem. It is elementary to establish that $R_n(\xi)\otimes R_n(\xi')$ and $R_n(\xi')\otimes R_n(\xi)$ are mathematically similar, and it follows from (3.12) for the R_n's that the same similarity relation holds for $T_N(\xi)\otimes T_N(\xi')$ and $T_N(\xi')\otimes T_N(\xi)$, from which the commutation relations for the $A_N(\xi)$, $B_N(\xi)$, etc., can be found. As mentioned in section three, the rather different commutation relations for an infinite system can be found by a suitable limiting procedure.

It turns out that the commutation relations for finite N are considerably more complicated than those for $N \to \infty$. This is because boundary conditions are imposed on the finite system, and the algebra must reflect the constraints on the pseudomomenta. In fact, the algebra must lead to the BA equations for spectral densities [2]. For the infinite case with no boundary conditions, the algebra is simpler but of course cannot give us, for example, the spectral densities. A further unsatisfactory feature of the infinite case is that the limiting procedure mentioned above is done assuming that the system is in a state with all spins up at great distances. However, this is not the physical ground state for $|\Delta| \lesssim 1$.

To illustrate the parallel with the quantum NLS, we give the $N \to \infty$ commutation relations [3] :

$$[A(\xi), A(\xi')] = [A(\xi), A^+(\xi')] = [B(\xi), B(\xi')] = 0$$

$$A(\xi)B^+(\xi') = B^+(\xi')A(\xi) \frac{\sinh\frac{1}{2}(\xi - \xi' + 2i\mu)}{\sinh\frac{1}{2}(\xi - \xi')} . \tag{4.7}$$

If $\quad R^+(\xi) = B(\xi)A^{-1}(\xi)$,

$$R(\xi)R^+(\xi') = \frac{\sinh\frac{1}{2}(\xi - \xi' + 2i\mu)}{\sinh\frac{1}{2}(\xi - \xi' - 2i\mu)} R^+(\xi')R(\xi) + 2\pi\delta(\xi - \xi') .$$

We also find :

$$[H, A(\xi)] = 0, \qquad [H, B(\xi)] = \varepsilon(\xi)B(\xi) \tag{4.8}$$

where $\varepsilon(\xi)$ is the magnon energy. At this point, it is evident that products of $B(\xi)$ operators acting on the vacuum give states which are eigenstates of H and $A(\xi)$. It can be verified directly that there are the BA states, thus establishing that $T(\xi)$ is the monodromy matrix for the spin chain system.

The coefficient appearing in the R, R^+ commutation relation above is the magnon-magnon S-matrix. It depends on $\xi - \xi'$. The ξ's are not in fact the magnon pseudomomenta, but are equivalent to the parameters α introduced by Yang and Yang [11].

The magnon pseudomomenta are given by

$$p = -i \ln \left\{ - \frac{\sinh(\frac{\xi - i\mu}{2})}{\sinh(\frac{\xi + i\mu}{2})} \right\} \tag{4.9}$$

with $-\infty < \xi < \infty$ corresponding to $-(\pi - \mu) < p < \pi - \mu$. (For p outside this range, $\mathrm{Im}\xi \neq 0$).

It is interesting to write out the monodromy matrix T as a function of the variable p. Removing a factor $e^{-ip/2}$, the local matrix becomes

$$R_n = \begin{pmatrix} \alpha_n & \beta_n \\ \gamma_n & \delta_n \end{pmatrix}$$

where $\alpha = \begin{pmatrix} e^{ip} & 0 \\ 0 & 1 \end{pmatrix} \qquad \beta = e^{ip/2}\sqrt{2(\cos p + \cos\mu)}\begin{pmatrix} 0 & 0 \\ 1 & 0 \end{pmatrix}$

$$\gamma = e^{ip/2}\sqrt{2(\cos p + \cos\mu)}\begin{pmatrix} 0 & 1 \\ 0 & 0 \end{pmatrix} \qquad \delta = \begin{pmatrix} 1 & 0 \\ 0 & e^{ip} \end{pmatrix} . \tag{4.10}$$

Expanding out the matrix product first for finite N [20],

$$\begin{aligned} A_N &= \alpha_1\alpha_2 \ldots \alpha_N + \alpha\alpha\beta\delta\delta\gamma\alpha \ldots \alpha + \ldots \\ B_N &= \alpha\alpha\alpha\beta\delta\delta \ldots \delta + \ldots . \end{aligned} \tag{4.11}$$

The elements are sums over all products which have the form of runs of α's, δ's with β, γ signalling the changes from α to δ and δ to α respectively. The α, δ runs may have zero length.

Using a solid line to denote an α run, a dotted line for a δ run, (4.11) can be represented by :

$$\begin{aligned} A &= \text{———} + \text{———}\ \sigma'^{-} \ldots\ldots \sigma'^{+}\ \text{———} + \text{ etc.} \\ B &= \text{———}\ \sigma'^{-} \ldots\ldots + \text{———}\ \sigma'^{-} \ldots \sigma'^{+}\ \text{———}\ \sigma'^{-} \ldots\ldots + \text{ etc.} \end{aligned} \tag{4.12}$$

where $\sigma'^{\pm} = e^{ip/2}\sqrt{2(\cos p + \cos\mu)}\ \sigma^{\pm}$.

These expansions are analogous to (3.3). We can also check immediately that B(p) applied to the vacuum gives the appropriate pseudomagnon state. It is amusing to note that for the special case $p = 0$, $\cos\mu = -1/2$ the expansion (4.12) for A was first written down many years ago by Lieb [22].

For the special case $\Delta = 0$, as mentioned in the introduction, the quantum ISM is equivalent to the Jordan-Wigner transformation. This can be seen very directly in the present formulation. Writing the Jordan-Wigner creation operator (with an all spins up vacuum)

$$S^+(p) = \sum e^{ipn}\, \sigma_n^- \prod_{n'<n} \sigma_{n'}^z , \tag{4.13}$$

we need to prove that

$$S^+(p)A(p) = B(p) . \tag{4.14}$$

Here S^+ lowers a single spin, A, B can raise and lower many spins. It is not difficult to verify (4.14) by applying the operators to successively more complicated states. The main nontrivial point is that one must include terms in which S^+ lowers a spin raised by A, so different orders in the expansion of A correspond to the same term in B.

Finally, we note that for any Δ such that $|\Delta| \lesssim 1$, if (4.11) is regarded as a series expansion in orders of β, γ it is rapidly convergent for low energy excitations. In the ground state of the antiferromagnet [11], there is a Fermi sea of magnons filling the pseudomomentum states $|p| < \pi - \mu$. Low energy excitations involve states with $|p|$ close to $\pi - \mu$, so $\cos p - \cos\mu$ is small. This turns out to reflect the fact that a magnon close to the Fermi surface phase shifts almost all the other magnon by the same amount. However, two low energy magnons in the same energy range exhibit rather complicated scattering behavior.

Acknowledgments

This work was began at the Aspen Physics Institute, and completed when the author was a guest of Prof. Hans Beck at the University of Neuchâtel. The author wishes to thank Prof. Beck and Dr. Marcos Puga for many useful discussions concerning this work, and Dr. Puga for a critical reading of the manuscript.

References

1. See the review article by D.W. McLaughlin in this volume
2. L.D. Fadeev: Steklov Mathematical Institute Report P-2-79, Leningrad 1979. For earlier work see E. Sklyanin and L.D. Fadeev: Sov. Phys. Dokl. 23, 902 (1978), E. Sklyanin ibid. 24, 107 (1979)
3. P.P. Kulish and E.K. Sklyanin: Phys. Lett. 70A, 461 (1979)
4. E.K. Sklyanin, L.A. Takhtadzhyan and L.D. Fadeev: Teor. Math. Fiz. 40, 194 (1979)
5. H.B. Thacker and D. Wilkinson: Phys. Rev. D 19, 3660 (1979)
6. D.B. Creamer, H.B. Thacker and D. Wilkinson: Phys. Rev. D 21, 1523 (1980)
7. H.B. Thacker: J. Math. Phys. 21, 1115 (1980)
8. D.B. Creamer, H.B. Thacker and D. Wilkinson: Fermilab Pub 80/17 THY, Jan 1980
9. J. Honerkamp, P. Weber and A. Wiesler: Nucl. Phys. B 152, 266 (1979)
10. H.A. Bethe: Z. Physik 71, 205 (1931)
11. C.N. Yang and C.P. Yang: Phys. Rev. 150, 321, 327 (1966)
12. L. Onsager: Phys. Rev. 65, 117 (1944)
13. R.J. Baxter: Ann. Phys. (NY) 70, 193 (1972)
14. J. des Cloizeaux and M. Gandin: J. Math. Phys. 7, 1384 (1966); M. Fowler: Phys. Rev. B 17, 2989 (1978)
15. V.E. Zakharov and A.B. Shabat: Sov. Phys. JETP 34, 62 (1972)
16. V.E. Zakharov and S.V. Manakov: Theor. Math. Phys. 19, 551 (1975)
17. E.H. Lieb and W. Liniger: Phys. Rev. 130, 1605 (1963); E.H. Lieb: Phys. Rev. 130, 1616 (1963)
18. S.K. Satija et al. : Phys. Rev. Lett. 44, 1548 (1980); W.J.L. Buyers et al. : Sol. State Comm. 33, 857 (1980)
19. B.M. McCoy, E. Barouch and D.B. Abraham: Phys. Rev. A 4, 2331 (1971)
20. M. Fowler: Phys. Lett. B 94, 189 (1980)
21. H.G. Vaidya and C.A. Tracy: Phys. Rev. Lett. 42, 3 (1979)
22. E.H. Lieb: Phys. Rev. Lett. 18, 692 (1967)

Bethe Ansatz, Connection Between One-Dimensional Models and Their Classification [1]

E. Berkcan[2] and L.N. Cooper

Department of Physics, Brown University
Providence, RI 02912, USA

The relationship between one dimensional (ID) theories in field theory and Solid State Physics as well as one and two dimensional (2D) models in statistical mechanics has proved rather useful for the study of quasi-one dimensional systems. As for the exact diagonalization (of the Hamiltonian or the transfer matrix)of these models the Bethe Ansatz [1](BA) has proved to be a very powerful method. In addition to the well known solutions [2,3] this method was recently used to diagonalize the massive Thirring [4,5] and the chiral-invariant Gross-Neveu [7,8] models and the Kondo problem [9].

We will use the expression for the periodic boundary conditions (PBC's) that results from the BA method to relate various models and to suggest a classification scheme [10].

The BA approach is inherently connected with the quantum inverse scattering method. Since this Ansatz provides the exact eigenstates, one can calculate the S-Matrix of the theory [11]. This is a more powerful approach than the method of factorization equations [12] since one does not skip over ranges of coupling constants where there are anamolies [6]. Also factorization explicitly follows from the former approach. In addition to the exact spectrum the BA method yields an infinite number of conserved charges [13].

Given the Hamiltonian H of a theory and a conserved charge N, the Bethe Ansatz method consists of diagonalizing H and N simultaneously. For a typical δ-function potential [14], the Hamiltonian is

$$H = (-i)^n \sum_j \gamma_j \frac{\partial^n}{\partial x_j^n} + \kappa \sum_{i \neq j} \delta(x_i - x_j) f(\gamma_i, \gamma_j), \tag{1}$$

where γ_j refers to the generalized "chirality" [7,9] of the particles, n = 1 or 2, and no assumption is made about the symmetry of the wave function [15]

$$\psi = \sum_{P \in S_N} \xi_P (Q) e^{i \sum_j k_{Pj} x_{Qj}}. \tag{2}$$

[1] Supported in part by Brown University's Material Research Laboratory founded by the National Science Foundation.

[2] Address after October 1, 1980: Laboratoire de Physique Théorique de l'Ecole Normale Supérieure, 24, rue Lhomond, 75231 Paris Cedex 05, France.

Here $X_{Qi} \varepsilon Q \equiv \{0 \leq X_{Q1} \leq --- \leq X_{QN} \leq L\}$. The diagonalization condition can be written as

$$\xi_{P'} = Y_{ij}^{ab} \xi_{P}, \tag{3}$$

where Y_{ij}^{ab} depends on the boundary Qa = Q'b, Q'a = Qb, b = a + 1, on the momenta coupled to these coordinates Pa = P'b = i, Pb = P'a = j, the internal symmetry and the "chiralities". This gives the form of the wave function. Upon imposing periodic boundary conditions (PBC's) we obtain an eigenvalue equation for the identity coefficient ξ_I,

$$e^{ik_j L} \xi_I = X_j \xi_I, \tag{4}$$

where X_j is obtained by repeated use of Eq'n. (3). This last equation is solved by a generalized Bethe Ansatz.

Eq. (4) is the fundamental equation and contains all the physics of the problem; the diagonalization condition and the symmetry of the wave function are built into the PBC's and X_j is a function of the two-body phase shifts only. Finding a set of unequal momentum labels $\{k_i\}$ that satisfy (4) is equivalent to solving the problem by quantum inverse scattering method and constructing an infinite set of conserved charges [13]. All the effects of the interaction are contained in the PBC's and the physically allowed states follow from these coupled equations. These equations have to be supplemented by the expression for the energy $\Sigma_i \gamma_i k_i^n$, and the density $\Sigma_i 1/L$. The physical properties are obtained in the thermodynamic limit. (For a δ-function potential, the two-body phase shift has the form $\theta = 2 \tan^{-1} \lambda$, where λ depends on n, the γ_i (thus momentum) and the coupling constant.)

To illustrate the power and the usefullness of the expression for the PBC's consider the Spin - 1/2 x y z chain,

$$H = -1/2 \Sigma \{J_x S_x S_x' + J_y S_y S_y' + J_z S_z S_z'\}, \tag{5}$$

where the primes denote the nearest neighbor spins. For simplicity take $J_x = J_y$ and $\Delta = J_z/J_x$ [16].

In this case, the PBC's can be written as

$$\psi_{n_1 n_2 --- n_m} = \psi_{n_2 --- n_m n_1 + N}, \tag{6}$$

where $n_1 n_2$, --- n_m are the locations of the m down spins, N is the total number of sites and m/Na is the eigenvalue of the charge $\frac{1}{aN} \Sigma S_Z$. Here a is the lattice constant. Using the two body phase shift [3]

$$\theta(p,q) = 2 \tan^{-1} \{\frac{\Delta \sin\{a(p-q)/2\}}{\cos\{a(p+q)/2\} - \Delta \cos\{a(p-q)/2\}}\}, \tag{7}$$

(6) takes the form

$$e^{ip_j aN} = (-1)^{m-1} \exp\{-i \sum_{\ell=1}^{m} \theta(p_j, p_\ell)\} . \tag{8}$$

The energy is given by

$$E = -\frac{\Delta}{2} N + \Sigma_j (2\Delta - 2 \cos p_j a). \tag{9}$$

We will now take a particular continuum limit and show that the above theory yields the Non-Linear Schrödinger equation (Bose Gas ≡ BG) [2]. For $\Delta = +1 - \kappa a/2$ we identify aN with the length L of the system and noting that m is the eigenvalue of the conserved charge, (8) yields the PBC's of the Non-Linear Schrödinger equation in the continuum limit $a \to 0$,

$$(-1)^{m-1} e^{-iLp_j} = e^{i\Sigma_\ell \theta'(p_\ell - p_j)}, \tag{10}$$

where $\theta'(p,q) = -2 \tan^{-1} \frac{1}{\kappa}(p-q)$ and κ is the coupling constant of the Bose Gas. The energy of the Bose Gas is given by

$$\lim_{a\to 0} (E - E_0)/a^2 = \sum_{j=1}^{m} k_j^2. \tag{11}$$

One also obtains from the magnetization the density

$$\rho = \frac{m}{aN} = \frac{m}{L} = \frac{1}{L} \sum_j^m 1. \tag{12}$$

E_0 in (11) is chosen so as to obtain a finite result in the continuum limit. This establishes the Bose gas as the continuum limit of the xyz model. This result is expected classically [17,18] and was recently pointed out in the context of the inverse scattering method [19].

The study of the PBC's allow us to draw general and interesting conclusions. We already pointed out the Bose gas as a particular continuum limit of the xyz model. It is well known that taking another continuum limit of the xyz model one obtains the massive Thirring model [20] and that the xyz chain is related to the transfer matrix of the 8V model [21]. These considerations lead us to conjecture,that for a given internal symmetry, there is only one fundamental model and only one Bethe Ansatz. The others are obtained from this model either by lattice transformations or by taking continuum limits. For the case we are discussing the model is the generalized Baxter model [22] and the BA is the generalized Bethe Ansatz that diagonalizes the transfer matrix.

We expect that this structure can be generalized to include models with higher internal symmetry. The purpose is to construct 2D lattice models that will be related in the above fashion to (1+1) dimensional models like the non-linear σ-model, the Gross-Neveu or the SU(N) massive Thirring models. We believe there are generalized models that occupy the slot corresponding to the Baxter model. These points are under current investigation.

In the thermodynamic limit the PBC's are replaced by an inhomogeneous Freedholm Integral equation of the second kind

$$p(x) + \lambda \int K(x,y)\, p(y)\, dy = g(x). \tag{13}$$

Given the Fredholm integral equations the "gauge" equivalence of exactly integrable field theoretical models is then realized as a mapping of the kernel and/or the inhomogeneous parts of these integral equations [10]. Such gauge equivalences have been found in the classical theories [18]. This generalization together with the Bethe Ansatz thus allows classification of the (1+1) dimensional (or similarity 2 space dimensional) models. We expect that this classification will give some hint as to the construction of (1+1) dimensional models with particle production.

References

1. H.A. Bethe, Z. Physik 71(1931)205.
2. E.H. Lieb and W. Liniger, Phys.Rev.130(1963)1605;E.H.Lieb.ibid 130, (1963)1616. E.Brezin and J.Sinn-Justin,C.R. Acad. Sci.(Paris)B263,(1966)670.
3. C.N.Yang and C.P.Yang, Phys.Rev. 150(1966)321;150(1966)327.
4. F.A. Berezin and V.N. Sushko,Sov. Phys. JETP 21 (1965)865.
5. H.Bergnoff and H.B. Thacker,Phys.Rev.Lett.42(1979)135;Phys.Rev.D19(1979) 3666.
6. V.E. Korepin, KFKI preprint.
7. N.Andrei and J.H. Lowenstein,Phys.Rev.Lett.43(1979)1698.
8. A.A. Belavin, Phys.Lett.87B(1979)117.
9. N.Andrei, Phys.Rev.Lett.45(1980)379.
10. E. Berkcan, U. Mohanty and L. N Cooper, to be published.
11. F.A. Berezin and V. N. Sushko, Ref. 4 C.N. Yang, Phys. Rev. 168(1968) 1920. E. Berkcan, unpublished. V.E. Korepin,Ref.6. N.Andrei and J.H. Lowenstein,NYU preprint.
12. See, for example,M.Karowski,Phys.Rep.C49(1979)222 and the references therein.
13. E.Berkcan, L.N Cooper, to be published.
14. Since we have local relativistic theories in mind,we restrict ourselves to δ-function potentials.
15. C.N.Yang, Phys.Rev.Lett. 19(1967)1312.
16. For $J_x \neq J_y$ $(\neq J_z)$ this problem is solved by an Ansatz [21] which will henceforth be referred to as the generalized Bethe Ansatz. This is justified by the fact that in a particular continuum limit the former reduces to the Bethe Ansatz that solves the massive Thirring model [4,5].
17. M. Lakshmanan, Phys. Lett. 61A(1977)53.
18. V.E.Zakharov and L.A. Takhtadzhyan, Teor.Mat.Fiz 38(1979)26.
19. P.P. Kulish and E.K. Sklyanin,Phys.Lett. 70A(1979)461. After this work has been completed it has been brought to our attention that J.H.H.Perk, Scriptie, Univ. of Amsterdam 1974 (unpublished) has also obtained the same result.
20. A Luther, Phys. Rev. B14(1979)2153.
21. R.J. Baxter,Phys.Rev.Lett.26(1971)832;26(1971)834. J.D.Johnson, S. Krinsky and B.M. McCoy, Phys.Rev. A8(1973)2526.
22. R.J. Baxter, Ann.Israel Phys. Soc. 2(1)(1978)37.

Quantum Statistics of Solitons

Kazumi Maki[1]

Department of Physics, University of Southern California
Los Angeles, CA 90007, USA[2], and

IBM Zurich Research Laboratory
CH-8803 Rüschlikon, Switzerland

1. Introduction

In spite of the remarkable success of classical statistic mechanics [1,2] in describing correctly the crossover behavior of one-dimensional systems with degenerate vacuum from the high-temperature topologically disordered phase to the low-temperature ordered phase, the classical theory becomes inadequate in the low-temperature region where the energy associated with fluctuation quanta becomes comparable to the temperature. Furthermore, the question of renormalizations of the boson mass and the soliton energy and bound states of fluctuations (e.g., the breathers in the sine-Gordon system) are only handled in the framework of the quantum statistics.

In order to deal with the quantum-field theory of nonlinear systems in 1 + 1 dimensions, two methods are available: a) BETHE's Ansatz [3], which gives the exact results but applies only for limited classes of nonlinear systems [4-6] and, b) method of functional integral [7,8], which is applicable to any nonlinear system, although the actual evaluation relies on the perturbation theory, and the results are useful in general only in the weak-coupling limit. Here, we shall review the work done by TAKAYAMA and myself [9,10] on the sine-Gordon system and the ϕ^4 system in the framework of the functional integral technique. In the following, we shall consider the system described by

$$\mathscr{H}_{sG} = \frac{1}{2} \int dz \left\{\pi(z)^2 + \left(\frac{\partial\phi}{\partial z}\right)^2 - \frac{2m^{*2}}{g^2} \cos [g\phi]\right\} \tag{1}$$

and

$$\mathscr{H}_{\phi^4} = \frac{1}{2} \int dz \left\{\pi(z)^2 + \left(\frac{\partial\phi}{\partial z}\right)^2 - \frac{1}{2} (m^*)^2 \phi^2 + \frac{1}{4} \lambda\phi^4\right\} , \tag{2}$$

where $\pi(z) = \frac{\partial\phi}{\partial t}$.

First, we shall consider the sine-Gordon system to illustrate our method.

---

[1]Supported by the National Science Foundation, under Grant Number DMR76-21032.
[2]Permanent Address.

2. Mass Renormalization

The quantum systems in 1 + 1 dimensions defined by the Hamiltonians (1) and (2) contain a divergent integral associated with the simple loop diagram of the fundamental boson of the systems. These divergences are completely eliminated from the theory by the mass renormalization [11]. In the case of the sine-Gordon system given in (1), we can rewrite it as

$$ {}_{sG} = \frac{1}{2} \int dz \left\{ \pi(z)^2 + \left(\frac{\partial \phi}{\partial z}\right)^2 - \frac{2m^2}{g^2} N[\cos(g\phi)] \right\}, \tag{3} $$

where N is the normal ordering operator defined by [11]

$$ V(\phi) = N\left[\exp\left(\frac{1}{2} D \frac{\partial^2}{\partial \phi^2}\right) V(\phi)\right], \tag{4} $$

where

$$ D \equiv \langle \phi^2 \rangle \cong T \sum_{\nu} \int_{-\Lambda}^{\Lambda} \frac{dk}{2} (\omega_\nu^{\,2} + \omega_k^2)^{-1} \tag{5} $$

and

$$ \omega_k = (k^2 + m^2)^{1/2} \tag{6} $$

and $\omega_\nu = 2\pi T\nu$ is the Matsubara frequency, m the renormalized mass of the linear quanta (say, the magnon in the magnetic system), and Λ is the cutoff momentum. Note that we use the unit system $\hbar = k_B = C_0 = 1$ throughout the text, where C_0 is the limiting velocity of the linear quanta of the system. Here, we have generalized the definition of the normal ordering N at T = 0 K [11] to finite temperatures [9]. Making use of relation (4), the renormalized mass m is determined self-consistently by

$$ m^2 = (m^*)^2 \exp [- \frac{1}{2} g^2 D]. \tag{7} $$

First, let us consider the T = 0 K case. Eq. (5) then reduces to

$$ D_0 = \frac{1}{2\pi} \ln \left(\frac{2\Lambda}{m_0}\right), \tag{8} $$

where m_0 is the renormalized mass at T = 0 K. Substituting (8) into (7), we obtain

$$ m_0 = m^* \left(\frac{m^*}{2\Lambda}\right)^{\hat{g}^2/8\pi}, \tag{9} $$

where

$$ \hat{g}^2 = g^2 / \left(1 - \frac{g^2}{8\pi}\right). \tag{10} $$

The renormalized mass decreases as the cutoff momentum increases.

In a discrete system, where the dispersion of the linear quanta is given by

$$ \omega_k = [m^2 + 4a^{-2} \sin^2 (\frac{1}{2} ak)]^{1/2} \tag{11} $$

rather than by (6), the lattice constant a provides a natural cutoff constant of the system. Substituting (11) into (5), we have at T = 0 K,

$$D_0 = \frac{1}{2\pi} \ln\left(\frac{8}{m_0 a}\right) + 0\,(m_0 a)^2, \tag{12}$$

where we assumed that $m_0 a << 1$. Here, the integral is made for $|k| \le \pi/a$. In this case, the renormalized mass is given by

$$m_0 = m^* \left(\frac{m^* a}{8}\right)^{\hat{g}^2/8\pi}. \tag{13}$$

Some of the quasi-one-dimensional planar magnetic systems like $CsNiF_3$ and TMMC in a magnetic field in the spin plane and in low temperatures are known to be described in terms of the sine-Gordon equation [12,13]. The renormalized masses in these systems are estimated to be 10 to 20% less than the corresponding bare masses of the system [14].

At finite temperatures, D defined in (5) is written as

$$D = \frac{1}{2\pi} \ln\left(\frac{\Lambda}{2m}\right) + f_0(\beta m), \tag{14}$$

where

$$f_0(z) = \frac{1}{\pi} \sum_{n=1}^{\infty} K_0(nz), \tag{15}$$

$\beta = T^{-1}$ and $K_0(z)$ being the modified Bessel function. There are two temperature regions of particular interest. For $T \sim m_0$, the renormalized mass is still practically given by (13), as the temperature-dependent correction is exponentially small ($\sim e^{-\beta m_0}$). On the other hand, for $T >> m$ the renormalized mass is given by

$$m \cong m_0 \left[1 + \left(\frac{4\pi}{\gamma_0 \beta m_0}\right)^2\right]^{g^2/16\pi} \exp\left[-\frac{g^2}{8} (\beta m)^{-1}\right], \tag{16}$$

where $\gamma_0 = 1.781....$ the Euler constant. Here, I have made use of the asymptotic behavior of $f_0(z)$ for $z << 1$,

$$f_0(z) = \frac{1}{2} z^{-1} + \frac{1}{2\pi} \ln\left(\frac{\gamma_0 z}{4\pi}\right) - \frac{\zeta(3)}{2(2\pi)^3} z^2 + \ldots \tag{17}$$

If I neglect the coefficient in the square bracket of (16), Eq. (14) reproduces a few terms of the classical result [2]

$$m = m^* \left[1 - 2\left(\frac{T}{E_s}\right) - \frac{1}{2}\left(\frac{T}{E_s}\right)^3 - \frac{3}{2}\left(\frac{T}{E_s}\right)^4 ..\right]^{1/2} \tag{18}$$

if m^* is replaced by m_0. Therefore, the factor in the square bracket is the quantum correction to m. Since the static magnetization of $CsNiF_3$ at low temperatures should be proportional to m^2, in principle the temperature dependence of the renormalized mass should be accessible experimentally. Furthermore, the renormalized mass at T = 0 K, m_0, also appears in the expression of the renormalized soliton energy E_s, as we shall see later.

3. Thermodynamics

Making use of the functional integral technique, we calculate the partition function of the s-G system at low temperatures. We shall first decompose the total functional space into sectors in terms of number of solitons [9]. Then in the temperature region $T \ll E_s$, the partition function Z of the system is dominated by z_0 (that of the soliton free sector) and Z_1 (that of the one-soliton sector). Indeed, the thermodynamic potential of the system per unit length Ω is given by

$$\Omega \; (\equiv - L^{-1}\beta^{-1}\ell n Z) = \Omega_0 - 2\beta^{-1} \bar{n}_s, \tag{19}$$

where

$$\Omega_0 = - L^{-1}\beta^{-1}\ell n Z_0 \tag{20}$$

and

$$\bar{n}_s = L^{-1} Z_1 / Z_0, \tag{21}$$

$\bar{n}_s$ being the total soliton density, and L the total length of the system. Eq. (19) agrees with the expression deduced from the ideal gas phenomenology of solitons [15,16].

The thermodynamic potential for the soliton free sector Ω_0 is easily obtained by the perturbation calculation as [9]

$$\Omega_0 = \text{const.} + \beta^{-1} \int_{-\infty}^{\infty} \frac{dk}{2\pi} \ell n \, (1 - e^{-\beta\omega_k}) + \frac{g^2}{8} m^2 f_0{}^2 (\beta m) + Og^4. \tag{22}$$

Furthermore, the total soliton density $\bar{n}_s$ is given in terms of the density of the soliton with velocity v as [9]

$$\bar{n}_s = \int_{-\infty}^{\infty} \frac{dp}{2\pi} n_s(v), \tag{23}$$

where $p \equiv (E_s\gamma v)$ is the momentum of the soliton, E_s the soliton energy, and $\gamma = (1 - v^2)^{-1/2}$. Here, $n_s(v)$ is calculated as follows [9]: $n_s(v)$ is expressed in terms of the partition function associated with the classical soliton with the velocity v by

$$n_s(v) = Z_1(v)/Z_0, \tag{24}$$

where $Z_1(v)$ is obtained as

$$Z_1(v) = \int \mathscr{D}(\hat{\phi}) \exp \left[- \int_0^{\beta} \mathscr{H}_{sG}(\phi) d\tau\right] \tag{25}$$

Here, we have substituted in (25)

$$\phi = \phi_{c\ell} \, (z, t) + \hat{\phi} \tag{26}$$

with

$$\phi_{c\ell} \, (z, t) = 4g^{-1} \tan^{-1} [e^{m\gamma(z-vt)}] \tag{27}$$

which is the classical soliton solution with velocity v. Eq. (25) is then evaluated within the Gaussian approximation, and we obtain [9]

$$n_s(v) = e^{-\beta\Omega_s(v)}, \tag{28}$$

where

$$\Omega_s(v) = \left(\frac{8m}{g^2} + 2mf_0\right)\gamma + 2Tf_2(v) \tag{29}$$

and

$$f_2(v) = -\frac{m}{2\pi}\gamma^{-1}\int_{-\infty}^{\infty} dk\left[\omega_k(\omega_k + vk)\right]^{-1}\ell n(1 - e^{-\beta\omega_k}). \tag{30}$$

At T = 0 K, $\Omega_s(v)$ reduces to the energy of the soliton with velocity v,

$$\Omega_s(v) = E_s\gamma, \tag{31}$$

where

$$E_s = \frac{8m_0}{g^2} - \frac{m_0}{\pi} \tag{32}$$

which was first obtained by DASHEN et al. [8]. The soliton energy obtained here differs from the classical soliton $E_s^* = 8\, m^* g^{-2}$ in two respects: first, the bare mass m^* has to be replaced by the renormalized mass m_0, and second, g^2 has to be replaced by $\hat{g}^2$ defined in (10). Both corrections reduce the soliton energy from the classical value E_s^*.

For $m_0 << T << E_s$ on the other hand, $\Omega_s(v)$ is approximately given by [9]

$$\Omega_s(v) \cong E_s\gamma - T\,\ell n\,[(1+\gamma)\,\beta m] - \frac{2}{3}\frac{\zeta(3)}{(2\pi)^3}\beta^{-1}(\beta m)^3 + \dots, \tag{33}$$

where E_s is the same as given by (32). This implies that the soliton energy is independent of temperature at least for $T << E_s$. From (31) and (32), we obtain

$$n_s(v) = \begin{cases} e^{-\beta E_s\gamma} & \text{for } T \sim m_0 \quad (34) \\ (1+\gamma)\,\beta m e^{-\beta E_s\gamma} & \text{for } E_s > T >> m_0, \quad (35) \end{cases}$$

and

$$\bar{n}_s = \begin{cases} \dfrac{E_s}{\pi} K_1(\beta E_s) \simeq \left(\dfrac{E_s}{2\pi\beta}\right)^{1/2} e^{-\beta E_s} & \text{for } T \sim m_0 \quad (36) \\ \dfrac{\beta m E_s}{\pi} [K_1(\beta E_s) - K_1'(\beta E_s)] \\ \simeq 2m \left(\dfrac{\beta E_s}{2\pi}\right)^{1/2} \left\{1 + \dfrac{5}{8}(\beta E_s)^{-1} + \dfrac{21}{128}(\beta E_s)^{-2}\right\} e^{-\beta E_s} \\ & \text{for } T_s > T >> m_0, \quad (37) \end{cases}$$

where $K_n(z)$ are the modified Bessel functions. Eq. (34) implies that the soliton behaves like an elementary particle for $T \sim m_0$, which justifies TRULLINGER's conjecture [17]. In the temperature region $m_0 << T << E_s$ on the other hand, the soliton behaves like a particle with internal structure [i.e. the second and the third term in (33)]. In particular, the first term in (37) agrees with the transfer-matrix technique result by GUPTA and SUTHERLAND [2], if we replace E_s^* and m^* in the classical theory by the renormalized ones E_s and m, respectively. The other terms in (37) are the relativistic corrections.

Then from Ω_0 and $\bar{n}_s$, the thermodynamic potential Ω is calculated by (19). The specific heat for $m_0 << T << E_s$ is calculated, for example, as [9]

$$C_s/L = - T \frac{\partial^2 \Omega}{\partial T^2} \tag{38}$$

$$= T\left\{\frac{g^2}{16} + \frac{g^4}{128}\left(\frac{T}{m_0}\right) + 2\left(\frac{2}{\pi}\right)^{1/2} (\beta m)\,(\beta E_s)^{5/2}\, e^{-\beta E_s} - \frac{g^2}{16\pi}\left(\frac{m_0}{T}\right) + \frac{g^4}{64\pi}\left[\ell n\left(\frac{\gamma_0 \beta m_0}{4\pi}\right) + 1\right] + \frac{g^6}{2^9\pi}\left(\frac{T}{m_0}\right) \ell n\left(\frac{\gamma_0 \beta m_0}{4\pi}\right)\right\}. \tag{39}$$

It is easy to see that the first three terms are consistent with the transfer-matrix technique result [2], while the last three terms are the lowest-order quantum corrections. In general, the thermodynamic quantities depend on two parameters, g^2 and βm, while the classical results depend only on $T/E_s (\equiv g^2T/7m)$. Therefore, in the temperature region $m_0 << T << E_s$, the quantum corrections are expanded in powers of g^2 and βm after the renormalization of m^* and E_s^* has been made. At low temperatures ($T \sim m_0$), the classical result breaks down completely. In this temperature region, the thermodynamics of the sine-Gordon system are controlled by bosonic excitations (the breathers).

4. Breathers in the sine-Gordon System

The classical sine-Gordon equation contains breathers as a nonlinear solution, which is given by [18,19]

$$\phi^B_{c\ell}(z, t) = 4\, g^{-1}\tan^{-1}\left\{\frac{(1-\omega^2)^{1/2}}{\omega}\, \frac{\sin[m\omega t]}{\cosh[m(1-\omega^2)^{1/2}Z]}\right\} \tag{40}$$

and

$$0 < \omega < 1. \tag{41}$$

This solution may be considered as a bound state of soliton and antisoliton. In a remarkable paper, DASHEN et al. [8] have shown within the semi-classical approximation that the breather becomes a class of bosons in the quantum-field theory of the sine-Gordon system with discrete mass spectrum

$$M_n = 2E_s \sin\left(\frac{n\hat{g}^2}{16}\right) \tag{42}$$

with

$$n = 1, 2, 3, \ldots, n_0 < 8\pi\hat{g}^{-2}, \tag{43}$$

and $\hat{g}^2$ having been defined in (10). Furthermore, they have shown that there are no other bosonic excitations at T = 0 K. Later, the above mass spectrum was rederived by LUTHER [20], who exploits the relation between the bound-state spectrum of the XYZ chain and that of the sine-Gordon system. In the low-temperature limit $T \ll m$, the thermodynamics is dominated by breathers. For example, the specific heat is given by

$$C_s/L \cong \beta^2 \sum_{n=1}^{n_0} \int_{-\infty}^{\infty} \frac{dk}{2\pi} E_n^2(k)\, e^{-\beta E_n(k)}$$

$$= \beta \sum_{n=1}^{n_0} M_n^2 \left(\frac{\beta M_n}{2\pi}\right)^{1/2} [1 + \frac{11}{8}(\beta M_n)^{-1} + \frac{113}{128}(\beta M_n)^{-2}]\, e^{-\beta M_n}, \tag{44}$$

where

$$E_n(k) = (M_n^2 + k^2)^{1/2} \tag{45}$$

and M_n has been defined by (42). Here, we have made use of the fact that the quantized breathers behave like elementary particles at low temperatures [9]. In (44) we have neglected the multi-breather terms, since they will be suppressed by the interaction between breathers.

In the weak-coupling limit, a simpler picture of breathers emerges. As DASHEN et al. [8] have shown, the n = 1 breather (the first breather) is nothing but the renormalized linear mode. Furthermore, in the same limit, the n-th breather is the bound state of n first breathers. Indeed, in the weak-coupling limit, (42) is approximated as

$$M_1 \cong m_0\left[1 - \frac{1}{6}\left(\frac{\hat{g}^2}{16}\right)^2\right] \tag{46}$$

and

$$M_n \cong nM_1 \left[1 - \frac{n^2-1}{6} \left(\frac{\hat{g}^2}{16} \right)^2 \right], \quad \text{for } n > 1, \tag{47}$$

where the latter exhibits the bound-state spectrum of the interacting bosons with the δ-function potential in the 1 + 1 dimensions [21,22].

In the intermediate temperature region ($m_0 << T << E_s$), the breather spectrum is no longer given by (42). Although the general expression for the breather mass spectrum is not known, in the weak-coupling limit ($g^2 \lesssim 1$) it is given by [23]

$$M_1 = m\left\{1 - \frac{1}{6}\left[\frac{\hat{g}^2}{16}\coth\left(\frac{1}{2}\beta m\right)\right]^2\right\}, \tag{48}$$

$$M_n = nM_1\left\{1 - \frac{n^2-1}{6}\left[\frac{\hat{g}^2}{16}\coth\left(\frac{1}{2}\beta m\right)\right]^2\right\} \quad \text{for } n > 1. \tag{49}$$

In the limit where the temperature tends to zero, the above expressions reduce to (46) and (47), respectively. In sharp contrast to the soliton energy E_s, which is independent of temperature, the breather mass decreases as the temperature increases. These decreases have two origins; first, m is reduced by the thermal fluctuations (16), and second, the interaction energy between two bosons increases with the thermal boson number. The first three breather masses M_1, M_2 and M_3 are evaluated numerically for $\hat{g}^2 = 1$, and shown as functions of temperature in Fig.1.

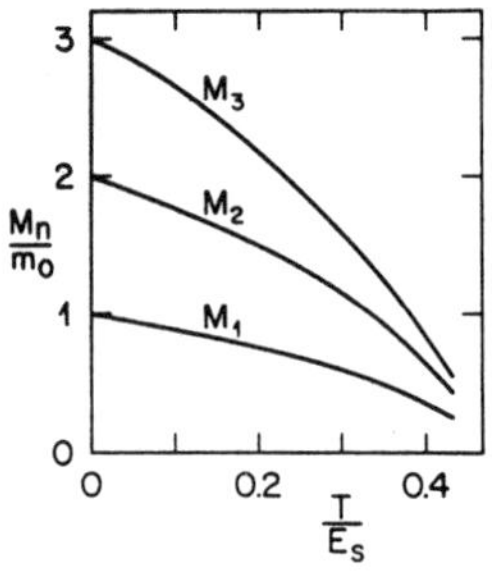

Fig.1 Three breather masses M_1, M_2, and M_3 are shown as functions of the reduced temperature T/E_s, where E_s is the soliton energy

Unfortunately, our perturbation scheme to calculate the breather masses breaks down around $T/E_s \simeq 1/3$. Therefore, we cannot say anything about the breather spectrum for $T/E_s \geq 1/3$. The complete determination of the breather spectrum for $T \neq 0$ K is one of the most urgent questions in the sine-Gordon system. From the experimental point of view, it is rather difficult to extract the breather-related properties from the thermodynamic quantitities. The dynamical correlation functions, on the other hand, will reveal the clear signatures of breathers.

5. Correlation Functions

Dynamical correlation functions of the one-dimensional systems are extremely valuable, as they contain detailed information about the excitation spectrum of the systems. In particular, great interest in the sine-Gordon system was stimulated by observation of the sine-Gordon solitons in $CsNiF_3$ in a magnetic field through inelastic neutron scattering by KJEMS and STEINER [24]. Similarly, a new theoretical interest has been generated [13,25,26] by the observation of a large central peak in the quasi-one-dimensional planar antiferromagnet TMMC in a magnetic field through both elastic and inelastic neutron-scattering experiments by BOUCHER et al. [27].

In the framework of quantum statistics, we can calculate the correlation function for $T \ll E_s$ by the generalized perturbation theory. In this temperature region, the correlation functions are decomposed into two terms: the boson term and the soliton term [10]. Furthermore, in the case of the sine-Gordon system, the boson term is dominated either by the single-boson term or by the two-boson term, depending on the ϕ parity of the observable involved in the correlation function [10]. As far as the soliton term is concerned, the quantum-statistical analysis yields much the same results as those obtained by the configuration average within the classical theory [12,28], except that m, E_s, and $n_s(v)$ have to be replaced by those calculated within the framework of quantum statistics. The single-boson term describes the creation or the annihilation of the $n = 1$ breather (i.e. the renormalized boson), while the two-boson term contains a structure associated with the $n = 2$ breather, which would not be easily obtained within the classical theory. To see the $n = 2$ breather contribution, let us consider the cos-cos correlation function defined by

$$\chi_{cc}(z, t) = \langle \cos[g\phi(z, t)] \cos[g\phi(0, 0)] \rangle . \tag{50}$$

The imaginary part of the Fourier transform of the above correlation function is given by [10]

$$\mathrm{Im}\chi_{cc}(\omega, k) = \left(\frac{m}{m^*}\right)^4 \Big\{ (1 - 16 \frac{\bar{n}_s}{m}) \frac{g^4}{2} I_2(\omega, k) + \left[\frac{2k}{m^2} \operatorname{cosech}\left(\frac{\pi k}{2m}\right)\right]^2 \frac{E_s}{2k} n_s(\omega/k) \Big\} , \tag{51}$$

where the first term is the two-boson term, and the second term is the soliton term. Here $n_s(\omega/k)$ is $n_s(v)$ defined in (34) and (35) with $v = \omega/k$. The two-boson term $I_2(\omega, k)$ depends on temperature. In the temperature region $m_0 \ll T \ll E_s$, we have

$$\begin{aligned} I_2(k, \omega) \simeq{} & \frac{|\omega| T}{k(2M)^3} \Theta(-s)\, (1 - Z)^{1/2} \{[1 - \eta(1 - Z)]^2 + \eta^2(1 - Z)\}^{-1} \\ & + \frac{g^2\pi}{128M} (\coth \tfrac{1}{2} \beta M)^2 \left[\delta\left(\omega - E_2(k)\right) - \delta\left(\omega + E_2(k)\right)\right] \\ & + T\, \omega^{-2}\, \Theta(s - 4M^2)\, (s - 4M^2)^{1/2} \times \\ & \Big\{ (s - 4M^2) \left[1 + \eta \left(\frac{2M}{\omega}\right)^2\right]^2 + (2\eta M)^2 \left(\frac{2M}{\omega}\right)^4 \Big\}^{-1} , \end{aligned} \tag{52}$$

where

$$Z = (\omega/k)^2, \; s = \omega^2 - k^2, \; E_2(k) = (M_2^2 + k^2)^{1/2},$$

$$M = M_1 \text{ and } \eta = \frac{T}{E_s} . \tag{53}$$

Now, the two-boson term consists of three different contributions; the central peak due to the thermal bosons (the n = 1 breathers) for $|\omega| < k$, the n = 2 breather structure at $\omega = \pm E_2(k)$, and the two-boson continuum for $|\omega| \geq (4M^2 + k^2)^{1/2}$. We have calculated the last term of (52) numerically for k = 0.2 M, and $\eta = 0.2$ and shown in Fig.2. For comparison,

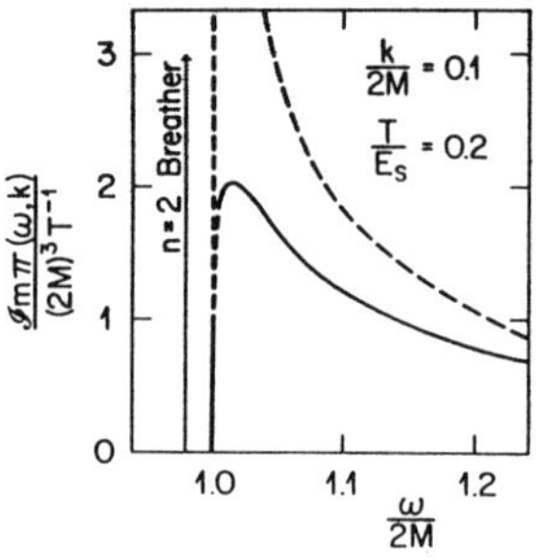

Fig.2 The imaginary part of the two-boson correlation function near the two-boson threshold is shown as a function of $\omega/2M$. The broken curve is that for the noninteracting bosons

in the same figure, we have also plotted with the broken curve the one for the case of noninteracting bosons. In the sine-Gordon system, these bosons (i.e. the n = 1 breathers) interact attractively and give rise to a bound state (the n = 2 breather). Due to the appearance of the bound state, a part of the intensity associated with the two-boson continuum is shifted to the bound state. If we integrate over the two-boson continuum of $S_{cc}(\omega, k)$ for small k, where the dynamical-structure factor $S_{cc}(\omega, k)$ is related to the correlation function $\chi_{cc}(\omega, k)$ by

$$S_{cc}(\omega, k) = 2(1 - e^{-\beta\omega})^{-1} \mathrm{Im}\chi_{cc}(\omega, k), \tag{54}$$

we can estimate the relative intensity of the n = 2 breather to the two-boson continuum as $4T/E_s$. Though expression (52) is valid only for $T/E_s \lesssim 1/3$, the contribution of the n = 2 breather to χ_{cc} is already substantial even when $T/E_s = 0.1$. On the other hand, the position of the n = 2 breather peak is only slightly below the threshold of the two-boson continuum

$$\Delta\omega = (4M^2 + k^2)^{1/2} - (M_2{}^2 + k^2)^{1/2}$$
$$\cong M\left(\frac{T}{E_s}\right)^2 . \tag{55}$$

Therefore, we believe that the study of the dynamical-structure factors in the vicinity of the two-boson threshold will throw light onto the existence of the n = 2 breather. In this respect, inelastic neutron-scattering experiments on $CsNiF_3$ and TMMC in a magnetic field are of great interest, as those systems appear to be described by the sine-Gordon equation at low enough temperatures.

6. The ϕ^4 System

A similar analysis has been carried out for the system described by (2). The renormalized mass, the soliton density $n_S(v)$, and the total soliton density has been obtained [10]. In the temperature region $m_0 << T << E_S$, we find complete agreement with the classical thermodynamics if the classical values for E_S and m are replaced by the renormalized ones, and if the higher-order terms in λm_0^{-2} are neglected. Also, I should like to point out that the ϕ^4 system possesses a breather spectrum similar to that in the sine-Gordon system [8,23] in the weak coupling limit. In particular we expect that the n = 2 breather will give rise to a similar structure near the two-boson threshold in $\chi_{\phi^2\phi^2}(\omega, k)$, as in the case of the sine-Gordon system.

7. Concluding Remarks

We shall summarize the present analysis as follows.

1) In the weak-coupling limit, the method of the functional integral is extremely useful. We can determine the thermodynamic as well as the dynamic properties of the system.

2) In the temperature region $m_0 << T << E_S$, where m_0 is the mass of the fundamental boson, and E_S is the soliton energy, the present results are compared with the classical statistical ones. The present results agree with the classical ones, if the classical values of m and E_S are replaced by the renormalized ones, and if the higher-order terms in the coupling constant are neglected.

3) The renormalized soliton energy is independent of temperature for $T << E_S$, while the renormalized boson mass decreases as the temperature is increased.

4) In both the s-G and the ϕ^4 systems, the existence of the quantized breathers has been established. The breather mass spectrum in the weak-coupling limit and for $T << E_S$ is determined.

5) The n = 2 breather gives rise to a structure in the vicinity of the two-boson threshold, which should be accessible by inelastic neutron-scattering experiments.

6) In the case of the strong-coupling limit, the only method available appears to be Bethe's Ansatz. A further study of the sine-Gordon system in the framework of Bethe's Ansatz is extremely desirable.

References

1. D.J. Scalapino, M. Sears, R.A. Ferrell: Phys. Rev. B 6, 3409 (1972)
2. N. Gupta, B. Sutherland: Phys. Rev. A 14, 1790 (1976)
3. H.A. Bethe: Z. Phys. 71, 205 (1931)
4. E. Lieb, W. Liniger: Phys. Rev. 103, 1605 (1963)
5. C.N. Yang, C.P. Yang: J. Math. Phys. (NY) 7, 1115 (1969)
6. H. Bergkoff, H.B. Thacker: Phys. Rev. Lett. 42, 135 (1979); Phys. Rev. D 19, 3666 (1979)

7. R.F. Dashen, B. Hasslacher, A. Neveu: Phys. Rev. D 10, 4114 (1974); ibid 10, 4130 (1974)
8. R.F. Dashen, B. Hasslacher, A. Neveu: Phys. Rev. D 11, 3424 (1975)
9. K. Maki, H. Takayama: Phys. Rev. B 20, 3223 (1979); ibid 20, 5002 (1979)
10. H. Takayama, K. Maki: Phys. Rev. B 20, 5009 (1979); ibid 21, 4558 (1980)
11. S. Coleman: Phys. Rev. D 15, 2929 (1977)
12. H.J. Mikeska: J. Phys. C 11, L29 (1978)
13. K. Maki: J. Low Temp. Phys. 41, 327 (1980)
14. K. Maki: *Solitons in Low Temperature Physics*, in Prog. of Low Temperature Physics, ed. by D.F. Brewer (North Holland, Amsterdam) to be published
15. J.A. Krumhansl, J.R. Schrieffer: Phys. Rev. B 11, 3535 (1975)
16. J.F. Currie, J.A. Krumhansl, A.R. Bishop, S.E. Trullinger: Phys. Rev. B (1980), in press
17. S.E. Trullinger: Solid State Commun. 29, 27 (1979)
18. A. Seeger, H. Donth, A. Kochendörfer: Z. Phys. 134, 173 (1953)
19. J.K. Perring, T.H.R. Skyrme: Nucl. Phys. 31, 550 (1962)
20. A. Luther: Phys. Rev. B 14, 2153 (1976)
21. J.B. McGuire: J. Math. Phys. (NY) 5, 622 (1964)
22. C.N. Yang: Phys. Rev. 168, 1920 (1968)
23. K. Maki, H. Takayama: Phys. Rev. B (1980) in press
24. J.K. Kjems, M. Steiner: Phys. Rev. Lett. 41, 1137 (1978)
25. K.M. Leung, D.W. Hone, D.L. Mills, P.S. Riseborough, S.E. Trullinger: Phys. Rev. B 21, 4017 (1980)
26. H.J. Mikeska: J. Phys. C. 13, 2913 (1980)
27. J.P. Boucher, L.P. Regnault, J. Rossat-Mignod, J.P. Renard, J. Bouillot, W.G. Stirling: Solid State Commun. 33, 171 (1980)
28. K. Kawasaki, Prog. Theor. Phys. 55, 2029 (1976)

Classical Statistical Mechanics of Soliton-Bearing Systems

T. Schneider and E. Stoll

IBM Zurich Research Laboratory
CH-8803 Rüschlikon, Switzerland

1. Introduction

We review the status of classical statistical mechanics for some 1-D soliton or solitary wave-bearing systems. Much of the recent activity is attributable to the motivating impact of molecular-dynamics (MD) results on discrete ϕ^4- [1-4] and sine-Gordon (sG) systems [5,6] and the Toda lattice [7], which revealed the need for including soliton features in a statistical description, in particular, of dynamic properties. In keeping with these model systems, we discuss the evidence for kink, breather (envelope-soliton) and pulse-soliton effects, including the associated new excitation branches. Primary attention is given to: a) thermodynamic properties and static form factors, and b) dynamic form factors (DFF) and displacement patterns.

2. Models and Definitions

The sG and ϕ^4 chains belong to a general class of Hamiltonians of the form

$$\mathcal{H} = \sum_{\ell} Da\left[\frac{\dot{y}_{\ell}^{2}}{2} + \omega_0^2\, V(y_{\ell}) + \frac{c_0^2}{2a^2}\,(y_{\ell+1}-y_{\ell})^2\right], \tag{1}$$

where y_ℓ is a scalar displacement on a one-dimensional reference lattice of points (labeled by ℓ) with lattice constant a. The parameter $D = Ma$ sets the energy scale, where M denotes the mass of the particles. The single-site potentials $V(y_\ell)$ are given by

$$V(y) = 1 - \cos y \qquad : \text{ sG chain}, \tag{2}$$

$$V(y) = \frac{b}{4}\,(y^2-1)^2 - \frac{1}{4b} \qquad : \ \phi^4 \text{ chain}. \tag{3}$$

For the ϕ^4 chain, it is sometimes more convenient to use the parameters

$$\omega_0^2 = \Delta, \quad b\,\omega_0^2 = B. \tag{4}$$

In the continuum limit, where $\phi_x = \partial\phi/\partial x$ replaces $(\dot{y}_{\ell+1}-y_{\ell})/a$, (1) reduces to

$$H = D \int dx \left[\frac{1}{2}\dot{\phi}^2 + \omega_0^2\, V(\phi) + \frac{c_0^2}{2}\,\phi_x^{\,2}\right]. \tag{5}$$

With $V(\phi)$ given by (2) and (3), these continuum models correspond to the sG and ϕ^4 systems. In this limit, the sG system is exactly integrable [8]. Examples for soliton solutions are the kink and breather solitons. Because the ϕ^4 system is not exactly integrable, the ϕ^4 kink is a solitary wave only. Some kink properties are listed in Table 1.

Table 1 Kink properties of the sG and ϕ^4 systems $\zeta = x - vt$, $\gamma_0 = (1-(v/c)^2)^{-\frac{1}{2}}$, $d = c_0/\omega_0$. v denotes the velocity of the kinks and E_K^0 their rest energy

V	$1 - \cos\phi$	$\frac{b}{4}\left(\phi^2 - \frac{1}{b}\right)^2 - \frac{1}{4b}$
ϕ_K	$4\tan^{-1}\exp \pm \frac{\gamma\zeta}{d}$	$\frac{1}{\sqrt{6}} \tanh \pm \frac{\gamma\zeta}{d} \frac{6}{2}$
E_K^0	$8\, D\omega_0\, c_0$	$2\frac{\sqrt{2}}{3}\, d\omega_0\, c_0 \frac{1}{b}$

The breather solution reads [9]

$$\phi_B(y,t) = 4\tan^{-1}\left\{\left(\frac{\omega_0^2}{\Omega^2} - 1\right)^{1/2} \sin\gamma\,\Omega\left(t - \frac{vy}{c_0^2}\right) \cdot \operatorname{sech}\frac{\omega_0}{c_0}(y-vt)\gamma\left(1 - \frac{\Omega^2}{\omega_0^2}\right)^{1/2}\right\} \tag{6}$$

with rest energy

$$E_B = 2\,E_K^0\left(1 - \Omega^2/\omega_0^2\right)^{1/2}, \tag{7}$$

Ω being the oscillation frequency of the breather. The appearance of the relativistic term γ in the kink and breather solutions is a consequence of the Lorentz invariance of Hamiltonian (1).

The Toda lattice, proposed more than a decade ago [10], is a 1-D string of equal masses with an exponential spring between pairs of nearest neighbors. It is a discrete counterpart of numerous continuous systems admitting soliton solutions. The Hamiltonian, after appropriate scaling is [11]

$$H = \sum_\ell \left[\frac{\dot{x}_\ell^2}{2} + e_\ell + r_\ell - 1\right], \tag{8}$$

where

$$e_\ell = \exp-(x_{\ell+1} - x_\ell) = \exp - r_\ell. \tag{9}$$

The pulse-soliton solution reads [11]

$$e_\ell = 1 + \sinh^2\alpha \; \mathrm{sech}^2(\alpha\ell \pm t\sinh\alpha) \tag{10}$$

or

$$T_\ell = \frac{\dot{x}_\ell^2}{2} = \frac{1}{2}\sinh^2\alpha \left[\tanh(\alpha\ell \pm \alpha v t) - \tanh(\alpha(\ell+1) \pm \alpha v t)\right]^2 , \tag{11}$$

α being an arbitrary parameter. Velocity and energy of the pulse soliton are, respectively,

$$v = \frac{\sinh\alpha}{\alpha} , \quad E_S = 2\left[\cosh\alpha|\sinh\alpha| - |\alpha|\right] . \tag{12}$$

Another important solution is the cnoidal wave [11]

$$e_\ell = 1 + \sum_{n=-\infty}^{+\infty} \beta^2 \mathrm{sech}^2\left[\alpha(\ell-\lambda n) - \beta t)\right] - 2\beta\nu , \tag{13}$$

corresponding to an infinite sequence of solitons and troughs, with

$$\alpha = \pi \frac{K(\kappa)}{\lambda K(\sqrt{1-\kappa^2})} , \quad \beta = \alpha\nu\lambda . \tag{14}$$

The wavelength λ and frequency ν are related by the dispersion relation

$$2K\nu = \left(\frac{1}{\mathrm{sn}^2 \frac{2K}{\lambda}} - 1 + \frac{E}{K}\right)^{-1/2} , \tag{15}$$

K and E being the complete elliptic integrals of the first and second kind, and sn representing the Jacobian elliptic function. In the limit $\kappa \to 1$, α = finite, $\lambda \to \infty$, the periodic wave (13) reduces to the soliton (10) in an infinite lattice.

3. Thermodynamic Properties and Static Form Factors

The calculation of thermodynamic properties and static form factors can essentially be reduced to the solution of a transfer-integral (TI) equation. For the Toda lattice, an exact treatment is possible [7], while for the sG and ϕ^4 chains one has to rely on numerical solutions [12] or an approximate replacement of the TI equation by a pseudo-Schrödinger (PS) equation [12,13]. The PS approximation turns out to be valid only for [12]

$$\frac{d}{a} = \frac{c_0}{\omega_0 a} >> 1, \quad \frac{k_B T}{Mc_0^2} << 1. \qquad (16)$$

The first condition defines a strong coupling regime (stiff chains), including the continuum limit ($a \to 0$), where the soliton or solitary-wave properties hold. In the limit $T \to \infty$, one can perform a high-temperature expansion by noting that at $T = \infty$, the sG chain becomes equivalent to the harmonic chain, while the ϕ^4 chain represents independent quartic oscillators [12]. Consequently, in this limit solitons become irrelevant. For intermediate temperatures, one is left with a numerical solution of the TI equation [12]. These points are illustrated in Fig.1 showing the temperature dependence of the specific heat at constant length C_L for the sG chain, with parameters listed in Table 2. Note that $C_L/k_B = 1$ is just the phonon contribution. The identification of soliton-sensitive properties is restricted, therefore, to low temperatures where the PS approximation holds [12-14]. In the sG case, the PS equation reduces to the Mathieu equation which is well documented [13]. The resulting low-temperature behavior of some thermodynamic properties and static form factors are listed in Table 3 [12]. The static form factors are defined by

$$S_{AA}(q) = \left\langle \left| \frac{1}{\sqrt{N}} \sum_\ell (A_\ell - \langle A_\ell \rangle)\, e^{iqa\ell} \right|^2 \right\rangle . \qquad (17)$$

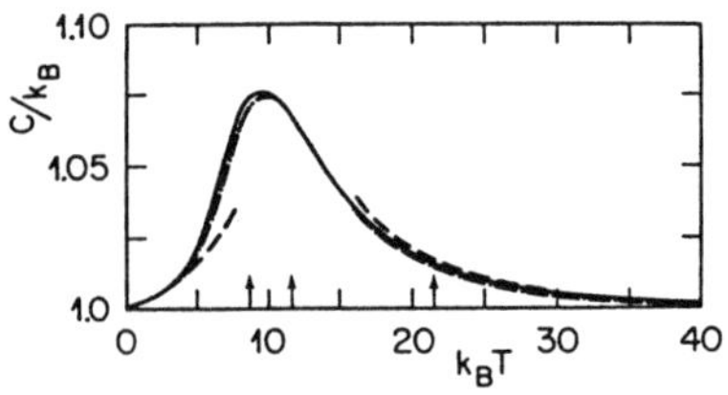

Fig. 1 Temperature dependence of the specific heat C_L of the sG chain for the parameters listed in Table 2. The full line represents the numerical solution of the TI equation; the dashed-dotted line, the numerical solution of the PS equation, and the dashed line asymptotic expansions (Table 3). The arrows mark the temperatures where MD calculations have been performed

Table 2 Choice of parameters for the sG and ϕ^4 chains, with $M = 1$, $a = 1$. The units are $[k_BT] = [\omega_0^2]$, $[C_0^2] = [\omega_0^2]$

Ma	ω_0^2	$d^2 = \frac{c_0^2}{\omega_0^2}$	B	E_K^0
1	1	29.22	-	43.244
1	0.05	10	1/3	0.0224

Table 3 Asymptotic low-temperature behavior of some thermodynamic properties and static form factors of the sG chain as obtained from the PS approximation. The subscripts denote A_ℓ: c = $\cos y_\ell$, s = $\sin y_\ell$, y = y_ℓ, c/2 = $\cos(y_\ell/2)$, s/2 = $\sin(y_\ell/2)$

$$c_k/k_B = 1 + 4Da\omega_0^2\left[\frac{k_BT}{(E_K^0)^2} + \frac{3}{2}\frac{(k_BT)^2}{(E_K^0)^3} + \ldots + O(\exp - \beta E_K^0)\right]$$

$$\langle\cos y\rangle = 1 - \frac{2}{\beta E_K^0} - \frac{1}{2(\beta E_K^0)^3} - \ldots + O(\exp - \beta E_k^0)$$

$$S_{cc}(q{=}0) = \frac{k_BT}{Da\omega_0^2}\left[\frac{1}{\beta E_K^0} + \frac{3}{2^2(\beta E_K^0)^3} + \ldots + O(\exp - \beta E_K^0)\right]$$

$$S_{ss}(q{=}0) = \frac{k_BT}{Da\omega_0^2}\left[1 - \frac{2}{\beta E_K^0} - \frac{1}{2^7(\beta E_K^0)^3} - \ldots + O(\exp - \beta E_K^0)\right]$$

$$\lim_{q=0} S_{yy}(q)\left(\frac{qa}{\pi}\right)^2 = \frac{32Da\omega_0^2}{k_BT}\left(\frac{1}{8}\beta E_K^0\right)^{-1/2}\exp - \beta E_K^0$$

$$S_{c/2\ c/2}(q{=}0) = \left(\beta Da\omega_0 c_0\right)^{1/2}\exp \beta E_K^0$$

$$S_{s/2\ s/2}(q{=}0) = \frac{2d}{\beta E_K^0 a} + \ldots + O(\exp - \beta E_K^0)$$

The anomalous exponential contribution is related to the tunnel splitting in the PS equation, and is identifiable as the kink contribution. It enters in leading order in $S_{yy}(q)$ and $S_{c/2\ c/2}(q)$ only. These are kink-sensitive properties. Otherwise, the kink contribution represents an exponentially small correction to the power laws, which can be fully accounted for by anharmonic perturbation theory.

Another important result is that the kink contribution to $S_{yy}(q)$ and $S_{c/2\ c/2}(q)$ is appreciable only for very small wave numbers. The absence of a clear-cut breather signature may be understood in terms of the fact that the breather solution is explicitly time dependent [see (6)], so that it does not enter the static classical properties. As Table 4 reveals, qualitatively similar results are obtained for the ϕ^4 chain [12]. To summarize: At low temperature, there are static form factors where the anomalous kink contribution enters in leading order for small wave numbers. The thermodynamic properties are dominated by the harmonic and anharmonic con-

Table 4 Asymptotic low-temperature behavior of some thermodynamic properties and static form factors of the ϕ^4 chain. The subscripts denote A_ℓ: $y = y_\ell$, $y^2 = y_\ell^2$

$$C_L/k_B = 1 + \frac{Da\omega_0^2}{b}\,\frac{8}{15}\,\frac{k_BT}{(E_K^0)^2} + \ldots \; 0(\exp - \beta E_K^0)$$

$$<Y_i^2> = \frac{\Delta}{B} - \frac{k_BT}{D\sqrt{2\Delta}\; c_0} - \ldots + 0(\exp - \beta E_K^0)$$

$$S_{yy}(q=0) = (\pi b)^{1/2}\,\frac{3}{2}\,(\beta E_K^0 b)^{1/2}\,(8\beta D\omega_0^2)^{-1/2}\,\exp\beta E_K^0$$

$$S_{y^2y^2}(q=0) = \frac{2k_BT}{Da\beta} + 2\left(\frac{k_BT}{Da\omega_0^2}\right)^2 + \ldots + 0(\exp - \beta E_K^0)$$

tributions, and the sG breather does not enter with a particular signature. At intermediate temperature, where the specific heat has a peak (Fig.1), the PS approach becomes questionable for parameter values of interest. Accordingly, an unambiguous identification of kink contributions is no longer possible. Some progress has been made in terms of a thermal renormalization of the kink energy [15]. At high temperatures, kinks become irrelevant [12].

For the Toda lattice, the thermodynamic properties and static form factors can be calculated exactly via the TI technique or for an isobaric canonical ensemble [7]. In fact, the TI equations can be solved by Laplace transform. As an example, we consider the specific heat at constant pressure, given by [7]

$$C_p|_{p=0} = \frac{1}{2} - \frac{2}{T} + \frac{1}{T}\frac{\partial\psi}{\partial\beta}\,, \quad \beta = \frac{1}{T}\;, \tag{18}$$

where ψ is the di-Gamma function. At low temperatures, one finds

$$C_p|_{p=0} = 1 + \frac{1}{6}T - \frac{1}{30}T^3 + \frac{1}{42}T^5 + 0(T^7)\;, \tag{19}$$

while for $T \to \infty$,

$$C_p|_{p=0} = \frac{3}{2} - \frac{1}{T} + \frac{\pi^2}{6T^2} + 0(T^{-3})\;. \tag{20}$$

The low-temperature behavior can be fully understood, term by term, by anharmonic perturbation theory [7]. Accordingly, the pulse solitons do not enter with a particular signature. They become apparent, however, for $T \to \infty$, by recognizing that $C_p|_{p=0} = 3/2$ is the specific heat of a hard-rod system. The Toda lattice, therefore, bridges two important limits: the harmonic chain ($T \to 0$) and the hard-rod system ($T \to \infty$) [7].

4. Dynamic Properties

Dynamic form factors (DFF) defined by

$$\hat{S}_{AA}(q,\omega) = \frac{S_{AA}(q,\omega)}{S_{AA}(q,t=0)} \quad , \tag{21}$$

where

$$S_{AA}(q,\omega) = \int_{-\infty}^{+\infty} dt e^{-i\omega t} S_{AA}(q,t) \tag{22}$$

$$S_{AA}(q,t) = \langle A(-q,t)\, A(q,0)\rangle \tag{23}$$

$$A(q,t) = \frac{1}{\sqrt{N}} \sum_{\ell} e^{iqa\ell}(A_{\ell} - \langle A_{\ell}\rangle) \tag{24}$$

represent the most revealing dynamic [2-7] [16-18] properties to unravel the excitation spectrum. $S_{AA}(q,t=0)$ is identical to the static form factor defined in (17). Even though progress has been made [3,7] [16-18], there is not yet any firmly based analytical technique to calculate these quantities for a system evolving according to Newton's equations. The only reliable technique presently available is molecular dynamics.

Next, we shall review the evidence for soliton excitation branches, associated with resonances, in particular DFF. These resonances are unexpected within the framework of the weakly interacting phonon-gas approximation. For DFF where the displacements enter linearly, this approach predicts Lorentzian peaks close to [7,18]

$$M\omega_A^2(q) = \omega_0^2 \langle \cos y\rangle + \frac{2c_0^2}{a^2}(1-\cos qa) \; : \; sG \tag{25}$$

$$M\omega_A^2(q) = -\omega_0^2 + 3b\omega_0^2\langle y_{\ell}^2\rangle + \frac{2c_0^2}{a^2}(1-\cos qa) \; : \; \phi^4 \tag{26}$$

$$\omega_A^2(q) = 2(1-\cos qa) \; : \; \text{Toda} \; . \tag{27}$$

For DFF where the displacements enter nonlinearly, such as $S_{CC}(q,\omega)$, ($A_{\ell} = \cos y_{\ell}$), $S_{TT}(q,\omega)$, ($A_{\ell} = T_{\ell} = \dot{x}_{\ell}^2/2$), in this approximation and lowest-order anharmonic perturbation theory, one expects two resonances [7,18]. A low-frequency peak due to the two-phonon difference process, and a high-frequency one close to $2\omega_A(q)$ stemming from the two-phonon sum process. For $q \to 0$, these approximate DFF exhibit a hydrodynamic singularity, signaling the possibility of a second-sound or heat-diffusion resonances [18].

Some key MD results for DFF are shown in Figs.2-5 [5-8]. Figure 2 reveals that in the sG chain and at low temperatures, $\hat{S}_{yy}(q,\omega)$ is exhausted by a low-frequency resonance. The occurrence of this peak is unexpected within the weakly interacting phonon-gas approach, accounting for the weak high-frequency

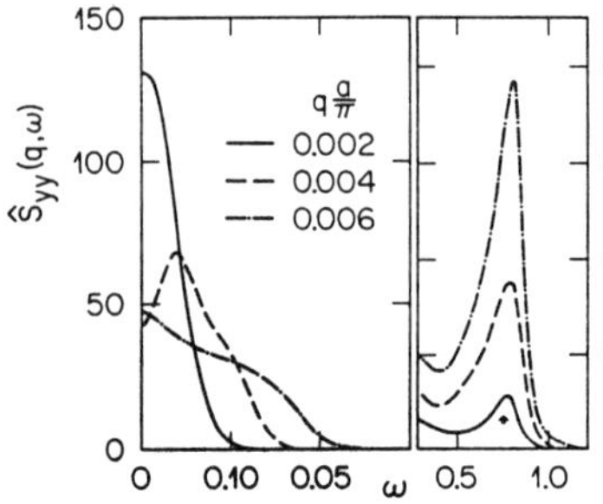

Fig.2 MD results for $S_{yy}(q,\omega)$ at $k_BT = 8.924$ ($E_K^0/k_BT = 4.85$) for the sG chain (Table 2) at small q values. The arrow marks the phonon frequency $\omega_A(q)$ (25)

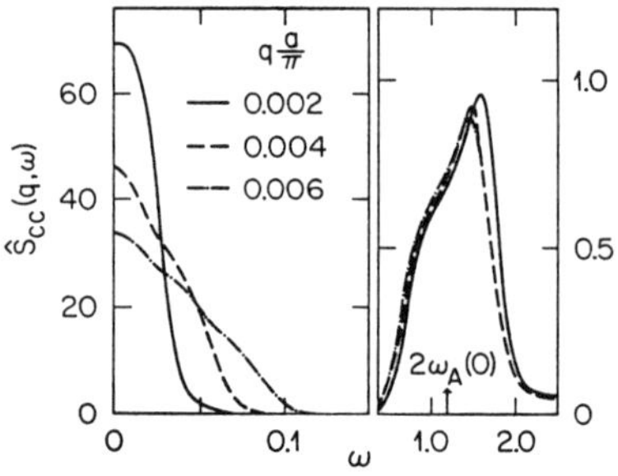

Fig.3 MD results for $S_{cc}(q,\omega)$ at $k_BT = 8.924$ for the sG chain (Table 2). The arrow marks $2\omega_A(q=0)$

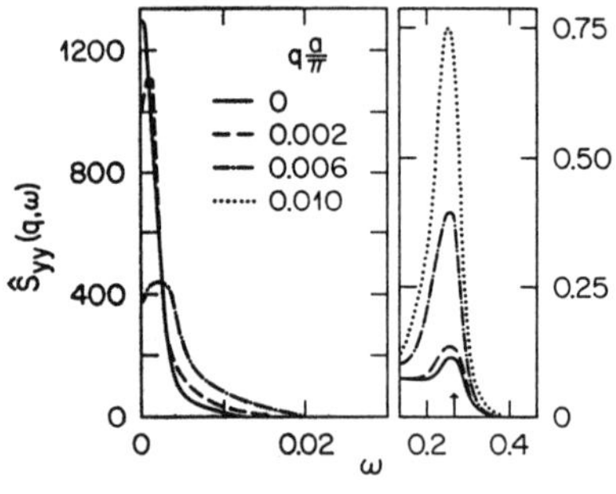

Fig.4 MD results for $S_{yy}(q,\omega)$ at $k_BT = 0.00448$ ($E_K^0/k_BT = 5$) for the ϕ^4 chain (Table 2) at some small q values. The arrow marks the phonon frequency $\omega_A(q)$ (26)

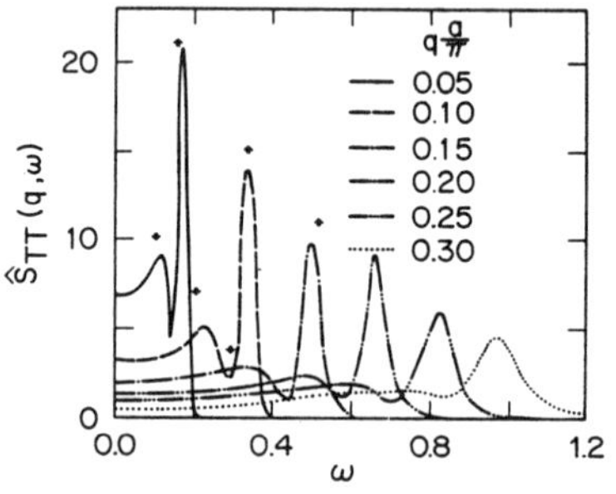

Fig.5 MD results for $S_{TT}(q,\omega)$ at $T = 0.25$ for the Toda chain. The arrows mark ↓ second-sound frequency estimate; ↑ peak position of the resonance predicted by the ideal pulse-soliton gas

phonon resonance close to $\omega_A(q)$. Recognizing that the associated static form factor $S_{yy}(q)$ is kink sensitive (Table 3), it is suggestive that the low-frequency resonance might be attributable to the kinks. Figure 3, showing the MD results for $\hat{S}_{cc}(q,\omega)$, does not, at least at first glance, exhibit unexpected features. In fact, the low-frequency resonance is consistent with the structure expected from the two-phonon difference processes, while the high-frequency peak might be attributed to the sum processes, leading to a resonance for $\omega \lesssim 2\omega_A(0)$. A quantitative comparison with the two-phonon approximation reveals excellent agreement with the low-frequency structure, but a sharp cut-off at $\omega \approx 2\omega_A(0)$. The observed cut-off is seen to be much closer to $2\Omega = 2\omega_0 \doteq 2$, ω_0 being the highest breather frequency [(7)] [18]. This evidence and the fact that $S_{cc}(q)$ is not kink sensitive, points to the possibility that the high-frequency resonance in $\hat{S}_{cc}(q,\omega)$ might be associated with breathers.

Let us then turn to Fig.4, showing the low-temperature behavior of $\hat{S}_{yy}(q,\omega)$ for the ϕ^4 chain [18]. As in the sG case (Fig.2), the new feature is the low-frequency resonance, exhausting the spectrum for small wave numbers. The weak high-frequency peak is consistent with the expected phonon resonance at $\omega_A(q)$ (26). Because the associated static form factor is kink sensitive (Table 4), one is naturally led to the conjecture that, in analogy to the sG case, the low-frequency resonance is attributable to the kinks.

Finally, we turn to Fig.5, showing $\hat{S}_{TT}(q,\omega)$ at $T = 0.25$ [7]. This value of T corresponds to a low temperature, because the static properties are well described by the anharmonic perturbation theory [(19)]. The low-frequency resonance is expected either from the two-phonon difference processes or in terms of second sound. The high-frequency peak close to $\omega_A(q)$ (27), however, cannot be accounted for within the weakly interacting phonon-gas approximation [7]. It constitutes clear-cut evidence for a soliton resonance in a temperature regime where the static properties give no indication for the relevance of pulse solitons.

So much about the direct evidence accumulated by MD on excitation branches in soliton-bearing systems, which are distinct from the familiar excitations in a weakly interacting phonon gas, namely phonons, second sound and heat diffusion. It is suggestive, therefore, that these excitation branches are subtle soliton features, having, with the exception of kinks, no particular signature in the static properties. Remarkably, we do not yet have any firmly based theory to illuminate this conjecture and to predict the consequences of soliton solutions on the excitation spectrum. Nevertheless, additional insight has been and will be obtained along several lines. Recent MD results aimed to determine the temperature-, wave-number-, frequency- and model-parameter dependence of the new excitation branches, indicate rather clearly that the low-frequency resonances seen in Figs.2 and 4 are kink excitation branches [18]. Moreover, the results are in reasonable agreement with the predictions of kink-gas approximations [16-18]. The breather problem appears to be considerably more subtle, due to the near coincidence with two-phonon resonances. This coincidence is absent for the high-frequency resonance appearing in $S_{TT}(q,\omega)$ of the Toda lattice. Accordingly, the identification in terms of a pulse-soliton excitation branch appears to be unambiguous [7].

In addition, and according to the Chinese proverb "A picture says more than a thousand words", snapshots or movies made from the time evolution of chains being in thermal equilibrium, represent the most efficient way of extracting and communicating unbiased information about the relevance of solitons in statistical mechanics. For the sG chain, we implemented a π detector, marking particles passing the maximum of the single-site potential (2) by a dot [6]. Figure 6 shows the time evolution of the π signals at

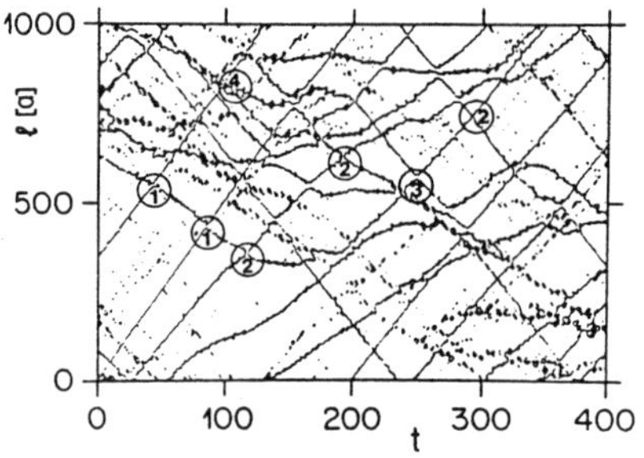

Fig.6 Time evolution of the π signals at $k_BT = 12.5$. The marked collisions are discussed in the text. ① kink-kink collision (0-2π-2π-4π); ② kink-antikink collision; ③ kink-antikink collision → breather creation; ④ breather decay into a kink and antikink.

$k_BT = 12.5$. An important feature is the line patterns, demonstrating the occurrence of propagating kinks and relativistic effects. The velocity is distributed around $v = \pm 4a$. This value agrees with the group velocity derived from the peak positions of the low-frequency peak in $\hat{S}_{yy}(q,\omega)$. Another important feature is the collisions. ① corresponds to a 0-2π - 2π - 4π kink-kink collision, while ② is a kink-antikink collision. These collisions appear to be rather ideal, because they are associated only with a phase shift (small shift of the trajectory) without changing the velocity. From (6), it is seen that large-amplitude breathers will appear in the form of bubbles. Figure 6 clearly demonstrates the occurrence of such large-amplitude breathers and their relevance as decay and creation channels for kinks. Examples are the kink-antikink collision marked by ③ leading to a breather creation, and the decay of a breather, marked by ④ into a kink and antikink. These creation and decay channels clearly reveal the limitations of the ideal kink-gas approximation, and the importance of lifetime effects, affecting the subtle scaling properties of $\hat{S}_{yy}(q,\omega)$ [18].

For the ϕ^4 chain, similar information may be obtained in terms of a sign detector, marking particles with a positive displacement by a black dot [18], and those with negative displacement by a white one. Propagating kinks will then be represented by a line separating black and white regions. Figure 7 shows the time evolution of the sign signals. The occurrence of line patterns clearly demonstrates the occurrence of thermalized kinks. Owing to topological reasons, a kink can be followed only by an antikink in the ϕ^4 chain [18]. Their propagation properties are distinctly different from those in the sG case (Fig.6). In fact, Fig.7 reveals a number of events where the ϕ^4 kinks reverse their velocity; an example is marked by ① . These processes lead to a nearly random-walk-type motion of the kinks. Examples of lifetime effects are marked by ② and ③ .

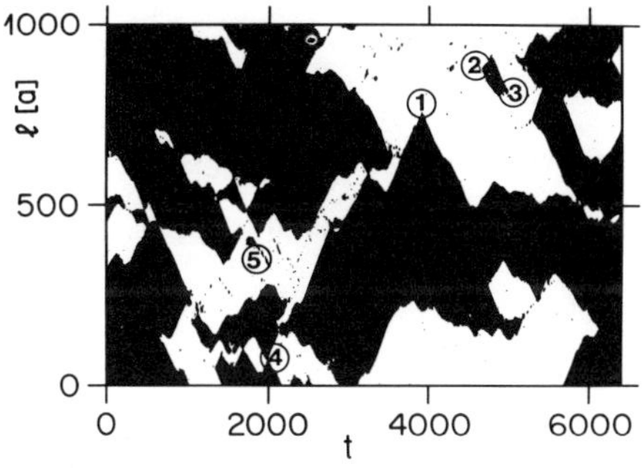

Fig.7 Time evolution of the sign signals at $k_BT = 0.00448$. ① kink reverses its velocity; ② creation of a kink-antikink pair; ③ kink-antikink collision → decay into "phonons"; ④ kink-antikink collision; ⑤ breather-like features

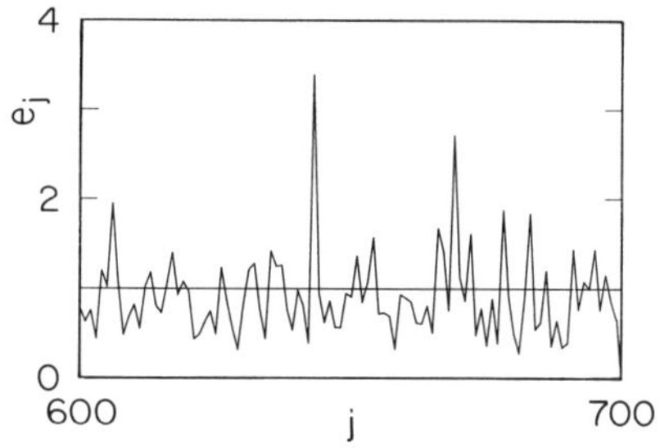

Fig.8 Snapshot of the displacements in terms of e_ℓ for a segment of the Toda lattice at T = 0.25

Finally, in Fig.8 we show a snapshot of the displacements of a segment of the Toda chain in terms of e_j (9). Adjacent points have been connected by straight lines. The total kinetic energy corresponds to T = 0.25. In view of the strong nonlinear characteristics seen in terms of narrow pulses followed by troughs, it is puzzling that at T = 0.25 the static properties are well described by the harmonic approximations (24). Given the evidence for the large number of high pulses, and recognizing that $\langle e_i \rangle|_{p=0} = 1$ [7], it becomes clear that the pulse must be followed by troughs. The occurrence of these troughs, which is not accounted for in the one-soliton solution, reveals the limitations of an ideal soliton-gas picture [7]. Such a displacement pattern is much better accounted for by the cnoidal wave (13), representing an infinite sequence of solitons, shifted downwards. In any case, these pulse-soliton features are an important example, where a particular signature in the static properties is missing, while the dynamics exhibit a soliton excitation branch.

Given this evidence for soliton sensitive properties and in particular for soliton excitation branches, it is natural to speculate on the observation of soliton features in real systems. This is largely uncharted territory but clearly of crucial importance. In this respect the reviewed and future MD work will certainly provide a solid basis for a quantitative comparison with approximate treatments and experimental results. As an example we refer to Ref. 19.

References

1. T. Schneider, E. Stoll: Phys. Rev. Lett. 31, 1254 (1973)
2. T. Schneider, E. Stoll: Phys. Rev. Lett. 35, 296 (1975)
3. S. Aubry: J. Chem. Phys. 64, 3392 (1976)
4. T.R. Koehler, A.R. Bishop, J.A. Krumhansl, J.R. Schrieffer: Solid State Commun. 17, 1515 (1975)
5. T. Schneider, E. Stoll: Phys. Rev. Lett. 41, 1429 (1978)

6. E. Stoll, T. Schneider, A.R. Bishop: Phys. Rev. Lett. 42, 937 (1979)
7. T. Schneider, E. Stoll: To be published
8. R.K. Bullough, R.K. Dodd: "Solitons in Mathematics: Brief History", in *Solitons and Condensed Matter Physics*, ed. by A.R. Bishop, T. Schneider, Springer Series in Solid-State Sciences, Vol. 8 (Springer, Berlin, Heidelberg, New York 1978) p.2
9. M.J. Ablowitz, D.J. Kaup, A.C. Newell, H. Segur: Phys. Rev. Lett. 30, 1262 (1973)
10. M. Toda: J. Phys. Soc. Japan 22, 431 (1967)
11. M. Toda: Prog. Theor. Phys. Suppl. 45, 174 (1970)
12. T. Schneider, E. Stoll: To be published
13. J.F. Currie, J.A. Krumhansl, A.R. Bishop, S.E. Trullinger: To be published
14. A.R. Bishop: To be published
15. A.R. Bishop: Solid State Commun. 30, 37 (1979)
16. K. Kawasaki, Prog. Theor. Phys. 55, 2029 (1976)
17. G.F. Mazenko, P.S. Sahni: Phys. Rev. B 18, 6139 (1978)
18. T. Schneider, E. Stoll: To be published
19. J.M. Loveluck, T. Schneider, E. Stoll, H.R. Jauslin: "A Comparison of Static and Dynamic Properties of One-Dimensional Magnets and Corresponding sG Systems", in this volume, p.157

Multistable Driven Systems

M. Büttiker and R. Landauer

IBM Thomas J. Watson Research Center
Yorktown Heights, NY 10598, USA

I. Classification of Phenomena

We are concerned with systems which are spread out in one spatial dimension, and where different spatial portions can be in different states of local stability, separated in space by transition regions. We will largely, but not exclusively, discuss systems where the differing states of local stability are not equally favored, and the transition regions thus have a preferred direction of motion. We will also emphasize questions which go beyond the deterministic kinetics of such systems, and treat fluctuations. We can go on to classify our subject in a few additional ways.

The number of competing states of local stability is variable. There are systems which have two competing states of local stability, with transition regions exemplified by a ferroelectric domain wall. In the absence of an external field, in the simplest case, there will be two equally favored competing states of stability. In the presence of an applied field we will have an unsymmetrical system, with a preferred state, toward which the system is driven. Alternatively, we can consider systems which have an unlimited sequence of states of local stability, typified by the Sine-Gordon chain. Such systems were studied in connection with dislocation theory, [1,2] at least as early as 1928. By the mid sixties the theory of such systems included their statistical mechanics [3]. Our discussion will emphasize the Sine-Gordon chain.

We shall invoke a distinction between "generalized Hamiltonian systems" and "open systems." Normally a Hamiltonian system is one which obeys conservative equations of motion. We shall use the expression "generalized Hamiltonian systems" to include such systems, even if they are supplemented by damping and by thermally induced fluctuations. Thus domain walls, as well as the kink in an edge dislocation [3], are generalized Hamiltonian systems. The contrasting systems, open systems, dissipate energy continually, even in the locally stable stationary states. A simple example of such a system is the ballast resistor, studied long ago, and in detail, by BUSCH [4], and also the object of a number of more recent discussions [5 – 8]. This is a wire carrying a *fixed current*, with a temperature dependent resistance, showing a particularly rapid increase in resistance over a narrow temperature range. The wire is cooled by a mechanism which gives heat transport roughly proportional to the wire's temperature elevation. Of the two locally stable states one corresponds to a small temperature elevation, which in turn gives a slightly elevated resistance and heat dissipation. The other state corresponds to a larger temperature elevation and thus to the appreciably higher resistance and heat dissipation needed to sustain the higher temperature. In the range of currents which permit bistability, i.e. allow both a low temperature and high temperature state, there will be only one value of current which permits a static transition region equivalent to phase equilibrium. At lower values of current the low temperature phase is favored, at higher currents the high temperature phase will grow. Similar phenomena occur in a material with a conductivity increasing with temperature kept under a constant transverse electric field [9]. If a bulk semiconductor exhibiting negative differential conductivity is driven away from equilibrium by an applied electric field [10 – 15], it is found that the system undergoes a nonequilibrium first order like phase transition to a solitary state (moving dipole domain). These phenomena have been studied from a semi-microscopic viewpoint in [14,15].

Another example of an open system consists of a linear string of active electronic oscillators. Each oscillator is coupled to its two neighbors so that neighboring oscillators tend to be in phase. At the same time the oscillators are synchronized by identical external signals. This external signal is not necessarily at the natural frequency of the oscillators. If the synchronization signal is large enough compared to the difference between the synchronization frequency and the natural frequency, the oscillators will be locked to the synchronization signal in their phase evolution. These phase-locked states are multistable states since an oscillator which gets ahead of its synchronization signal by 2π will once again be in a favored state. Thus we can have a chain of oscillators which has one region 2π ahead of an adjacent region, and a transition region in between. If the synchronization frequency and the natural oscillator frequency are identical then the transition region will have no reason to move. If, on the other hand, the oscillators have a natural frequency higher than the synchronization frequency, then the oscillators in the transition region which are not effectively locked to a synchronization signal, will tend to run at the higher natural frequency, thus causing the transition region to move, and the region with the more advanced phase will grow. Systems of this sort were described briefly in a recent paper by one of the authors [16] and in more detail in an earlier report [17]. The remainder of the present paper will be confined to generalized Hamiltonian systems. Note, however, that we are basically concerned with deterministic equations of motion, supplemented by sources of fluctuation which cause different ensemble members to behave differently. It is this description that counts, rather than the origin of these effects. Thus the oscillator chain described above has a qualitative resemblance to the Sine-Gordon chain commonly studied in generalized Hamiltonian systems. Proper choice of details can make that qualitative resemblance a detailed quantitative equivalence.

For the generalized Hamiltonian system, we can make a distinction between those where the fluctuations are characteristic of a single temperature, and more complex cases where, for example, different parts of configuration space are characterized by different temperatures. Ref. 16 discussed systems of that sort. In our subsequent discussion we will, however, stay with the simpler single temperature case.

The generalized Hamiltonian system, with damping, can be lightly damped, or heavily damped. *Some* of the subsequent discussion is specialized to the heavily damped case, where inertial effects are unimportant.

As a final point in this catalogue of situations we emphasize that moving transition regions, between local stability regimes, are not necessarily Sine-Gordon or ϕ^4 solitons, or any of their close relatives discussed above. Two recent review papers [18,19] discuss a long established field in which propagating shock waves induce phase transitions, i.e. transitions between two states of relative local stability. The recent soliton literature and soliton conferences do not recognize the existence of phase transitions induced by shock waves. A note by one of the authors discusses the relationship between shock waves and solitons in more detail [20].

II. Driven Kinks

Consider a chain of particles separated from each other in the x direction, with periodic boundary conditions in x, and subject to displacement in an orthogonal direction, θ. The particles are in a periodic potential $V_0 \cos\theta$, supplemented by an additional applied driving force potential, $-F\theta$, and also subject to thermal noise forces. Adjacent particles are tightly coupled, so that they remain close to each other in their θ position. This leads to the equation of motion [21],

$$\gamma \partial\theta/\partial t = -V_0 \sin\theta + F + \kappa \partial^2\theta/\partial x^2 + \zeta, \tag{2.1}$$

where κ is proportional to the coupling strength between neighboring particles. ζ is a random force with $<\zeta> = 0$, and $<\zeta(x,t)\zeta(x',t')> = 2\gamma kT\delta(t-t')\delta(x-x')$. We confine our attention to the case $V_0 \geq F \geq 0$. The case where $F > V_0$, and the potential minima of $V = V_0(1 - cos\theta) - F\theta$

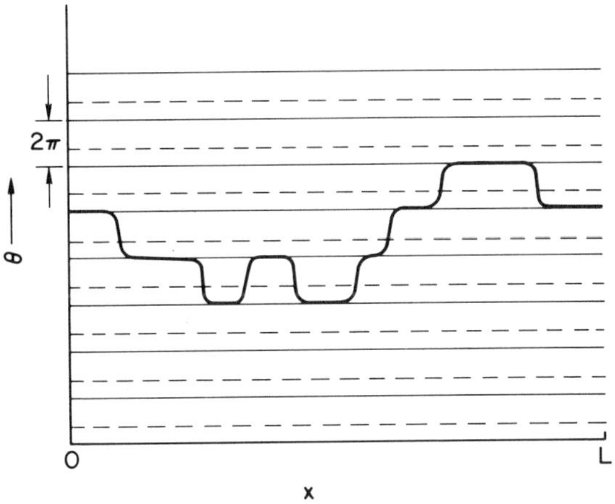

Fig.1 Typical configuration of the displacement field $\theta(x,t)$ at a given instant of time. Long segments of the chain lying in a valley of V (thin solid lines) are connected by kinks and antikinks which span the hills of V (thin broken lines).

are washed out, is uninteresting. In that case the particles can move down the potential gradient without stopping. The coupling will keep the particles moving together, within minor fluctuations generated by the noise. In the interesting case $F < V_0$ the chain is trapped in local potential minima. Long segments of the chain will lie in the potential minima. These segments of the chain are connected by kinks (transition from one valley to an adjacent one with $(\partial\theta/\partial x > 0)$ and antikinks $(\partial\theta/\partial x < 0)$. A typical chain configuration [3] is shown in Fig. 1. The steady state motion of the chain under the action of a field F is characterized by the density of kinks and their propagation velocity. Knowledge of these parameters permits, for instance, calculation of the mean angular velocity $<\partial\theta/\partial t>$ of the pendulums [21]. A calculation of this quantity was put forth by TRULLINGER et al. [22], by GUYER and MILLER [23], and LEE and TRULLINGER [24] using a perturbative approach. In contrast our approach is, in principle, nonperturbative. The relation of the two viewpoints has been discussed elsewhere [25].

In this section we will discuss the propagation velocity of the kinks. In the next section we will calculate the steady state density of kinks. To investigate the propagation of a kink under the action of a field F we return to the deterministic time evolution equation [21],

$$\gamma \partial\theta/\partial t = -V_0 \sin\theta + F + \kappa \partial^2\theta/\partial x^2. \tag{2.2}$$

For completeness, we will go beyond the purely viscous system (2.2) and also consider the chain with inertia [26 – 28],

$$I\partial^2\theta/\partial t^2 + \gamma\partial\theta/\partial t = -V_0 \sin\theta + F + \kappa\partial^2\theta/\partial x^2, \tag{2.3}$$

as well as the equation of motion describing a single damped particle in a periodic potential [29 – 31],

$$Id^2\theta/dt^2 + \gamma d\theta/dt = -V_0 \sin\theta + F. \tag{2.4}$$

The single particle system exhibits a critical value F_c such that for $F_c \leq F \leq V_0$ two stable states exist; in addition to the locked state in which a particle is trapped in one of the potential minima, there is a running state where the particle retains enough kinetic energy to pass all the potential hills. We will search for travelling waves (TW) of (2.2) and (2.3). The TW depend on x and t only through $t-x/u$. If we introduce the dimensionless time $\tau = (V_0/\gamma)(x/u - t)$ in (2.2) and (2.3) the TW's are found as solutions of

$$m\frac{d^2\theta}{d\tau^2} + \frac{d\theta}{d\tau} = -\frac{1}{V_0}\frac{dU}{d\theta} = -\sin\theta + F/V_0. \tag{2.5}$$

U is the original potential turned upside down, $U = -V$. For the purely viscous system (2.2), we find $m = m_1 = u_0^2/u^2$, where $u_0 = (\kappa V_0)^{1/2}/\gamma$. For the system with inertia, as in (2.3), we have $m = m_2 = (u_0^2 - \chi^2 u^2)/u^2$, where $\chi = (IV_0)^{1/2}/\gamma$. The single particle is also described by (2.5) with mass $m = m_3 = \chi^2$, when we introduce the time $\tau = (V_0/\gamma)t$ and the transformation $\theta \to \theta + \pi$ and $F \to -F$ in (2.4).

All three problems lead, therefore, to the investigation of the motion of a driven and damped particle in a periodic potential [26]. Every solution of (2.5) is a solution of (2.2)-(2.4) provided the masses are equal $m = m_1 = m_2 = m_3$. We search for solutions of (2.5) which describe a particle starting at a local maximum of U with zero velocity and reaching an adjacent maximum of U with zero velocity. Such a trajectory gives a solution of (2.2) and (2.3) which describes a transition of the chain from one valley of V (maximum of U) to an adjacent valley of V. Such a trajectory corresponds, therefore, to driven kinks of (2.2) and (2.3) and also to the limiting running state at the critical field F_c of the single particle problem (2.4). Eq. (2.5) depends only on the parameters m and F/V_0. To find such a state the "mass" m and F/V_0 must be related in a special way. We define $m = 1/\phi^2(F/V_0)$, where ϕ is a universal (parameter independent) function. From $m = m_1$, we find that a kink of the purely viscous system propagates with velocity

$$u = u_0\phi(F/V_0). \tag{2.6}$$

For a kink with inertia, $m = m_2$ yields the propagation velocity,

$$u = u_0\phi/[1 + \chi^2\phi^2]^{1/2}, \tag{2.7}$$

and $m = m_3$ defines the critical field F_c for the onset of the running state,

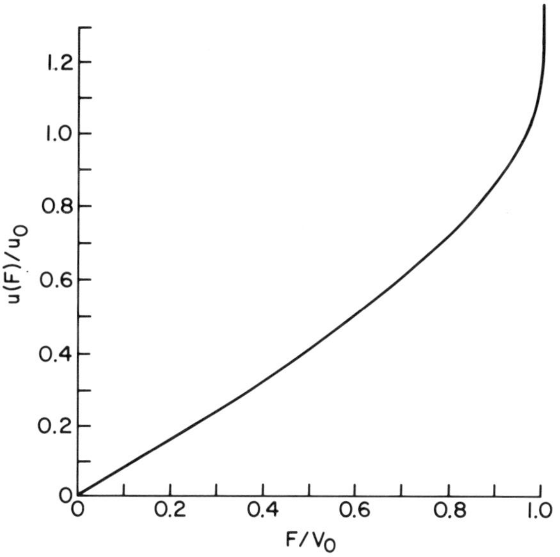

Fig.2 Propagation velocity u of a purely viscous kink. The curve shown defines the universal function $\phi = u/u_0$. At $F = V_0$ a driven kink can have any velocity $u \geq 1.19u_0$, where $u_0 = (\kappa V_0)^{1/2}/\gamma$.

$$\frac{1}{\chi} = \phi(F_c/V_0), \tag{2.8}$$

in the single particle problem [26].

The computation of ϕ thus becomes essential [21,32]. ϕ is linear at small fields [16,21,26,27,28,33]. The derivative of ϕ at the origin is $\pi/4$, giving rise to a kink mobility $\mu = (\pi/4\gamma)(\kappa/V_0)^{1/2}$. (The kink mobility is the ratio of kink velocity to F). As F grows, ϕ increases. As shown in Fig. 2, it reaches the vertical line $F = V_0$ at $\phi^* = 1.19$. (This actually makes ϕ^{-1}, rather than ϕ, a single-valued function, but we presume that our slight lack of precision will cause no confusion.) For a *single particle* [29] with mass $m < m^* = 1/(\phi^*)^2$ or $\chi < \chi^* = 1/1.19$, respectively, the motion is overdamped and the critical field F_c becomes equal to the stability limit of the locked state, i.e. $F_c = V_0$. Correspondingly, for $F = V_0$ the *kink* velocity is undetermined. For the purely viscous system every propagation velocity $u \geq u^* = 1.19u_0$ corresponds to a possible kink solution. The limiting kink with $u = \infty$ can be given analytically [26] ; the displacement θ becomes independent of x. In the case with inertia the limiting kink propagates with the velocity of sound $u_c = u_0/\chi$ and has a finite width [26,27].

The travelling wave solutions can be stable or unstable. We have shown that the driven kinks for $F < V_0$ are stable against small amplitude perturbations [25].

III. Nucleation Rate of Kink-Antikink Pairs

A chain lying initially in a single potential minimum will stay there. Thermal fluctuations are needed to cause motion of a part of the chain over an intervening barrier, into the adjacent *lower lying* potential valley. Once a big enough length of chain has been thrown into the next valley by fluctuations, it will tend to grow and bring a larger portion of the chain into the lower valley. The transferred chain is terminated by kinks of the type illustrated in Fig. 1. These require extra energy because they place particles away from the potential minima and also because $\partial\theta/\partial x$ is appreciable in these transition regions, requiring energy in the coupling terms. If the length of chain transferred into the lower lying valley is too short then the energy can be reduced by letting the transferred region collapse and eliminating the kink energy. Thus in the establishment of a region of transferred chain, long enough to grow, we must pass through a saddle point configuration, i.e. through the minimum possible barrier which will permit departure out of the initial potential valley. Thermal activation over such a barrier is familiar from nucleation theory, e.g. in drop formation in a supersaturated vapor [34], feroelectric domain reversal [35], or studies of phase slip in superconducting systems [36,37]. Methods for calculating such rates, from equations of motion of the form (2.1) including noise and damping, were formulated by LANDAUER and SWANSON [38], and have been described in detail by LANGER [39]. This approach yields a nucleation rate per unit length of the chain of the form $j = \Omega\, e^{-\Delta E/kT}$. ΔE is the energy of the saddle point configuration relative to the initial state in which the chain lies uniformly in one valley. All the subtleties of this approach relate only to the pre-exponential factor Ω. Evaluation of this factor requires a detailed characterization of all the normal modes near the initial state, and also about the saddle point configuration. The results for the overdamped Sine-Gordon chain have been presented in [21] and the detailed calculations in [25].

In the presence of a field, a nucleation event leads to a formation of a kink-antikink pair. Each member of the pair is driven with the respective velocity $\pm u(F)$, (see 2.6). Eventually the driven kink (antikink) will collide with an antikink (kink) leading to the annihilation of the kink (antikink). A steady-state density η, including kinks and antikinks, is maintained by a balance of this annihilation with the nucleation of new pairs. To find the kink lifetime, we proceed as discussed in [21]. The probability that a kink encounters an antikink in the next short time interval dt is simply the probability that there is an antikink in the range $2udt$, the range swept out by the relative motion, within that time. If we denote the density of antikinks by n, that probability is $2nudt$. The rate of destruction of m kinks per unit length and time is $2unm$ and this must be balanced by the nucleation. The balance equation in the steady state becomes,

$$j - \langle 2unm \rangle = j - 2un_0^2 = 0, \tag{3.1}$$

where j is the rate at which pairs of kinks and antikinks are generated by nucleation processes, as discussed above. For a chain with periodic boundary conditions, we have $\langle 2unm \rangle = 2un_0^2$, where $n_0 = \langle n \rangle = \langle m \rangle$ is the steady state density of kinks. A more fundamental justification for (3.1) is given elsewhere [40].

Each driven kink passing a location x along the chain reduces the displacement at x by 2π and each antikink advances the displacement at x by 2π. If j_K is the current density of the kinks and j_{AK} the current density of the antikinks, then the displacement of the chain is given by

$$\partial\theta/\partial t = -\,2\pi(j_K - j_{AK}). \tag{3.2}$$

In the average $\langle j_K \rangle = -um_0$ and $\langle j_{AK} \rangle = +\,un_0$ and hence we find,

$$\langle \partial\theta/\partial t \rangle = 4\pi u n_0. \tag{3.3}$$

The results from (3.3), using (3.1), have been presented in [21] and are discussed in detail in [25].

In (3.1) we have assumed that the driving field F is so strong that the diffusive motion of the kinks can be neglected compared to their deterministic drift $u(F)$. At very low fields this is not the case and (3.1) becomes invalid. However, $<\partial\theta/\partial t>$ is easily found from (3.3) directly. For small fields where the velocity is proportional to the field, we find [21],

$$<\partial\theta/\partial t> = [4\pi n_{eq}\,\mu]F = [(2\pi)^{3/2}\gamma^{-1}(E_0/kT)^{1/2}\exp(-E_0/kT)]F, \tag{3.4}$$

where $n_{eq} = (2V_0/\pi\kappa)^{1/2}(E_0/kT)^{1/2}\exp(-E_0/kT)$ is the equilibrium kink density [3,21,25,41] and $2E_0 = 16(\kappa V_0)^{1/2}$ is the activation energy for a kink-antikink pair at equilibrium. The method of TRULLINGER et al. [22], GUYER and MILLER [23], and LEE and TRULLINGER [24] fails to give the low field result of Eq. (3.4). A computer study by

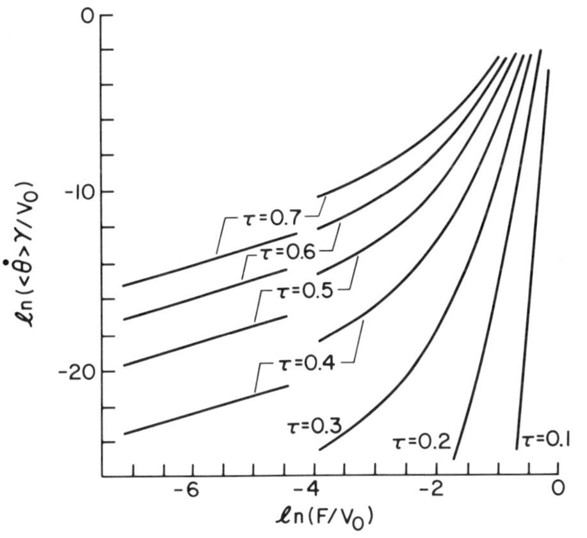

Fig.3 Velocity of the chain as a function of the field for different temperatures $\tau = kT/(\kappa V_0)^{1/2}$. The straight lines show the extrapolation of the result (3.4) to higher fields. The other curves are computational results.

SCHNEIDER and STOLL [42] of a short sample, not necessarily characteristic of the thermodynamic limit was best fit by a pre-exponential factor T^{-2}. Fig. 3 shows the extrapolation of (3.4) into the high field region together with the computational results from (3.1) and (3.3).

IV. Fluctuations in the Sine-Gordon Chain

We have shown in the preceding section that the dynamics of the Sine-Gordon chain is governed by the nucleation and recombination process of kink-antikink pairs. Based on this picture, we have investigated the steady state transport. In this section we will discuss the fluctuations away from the steady state, summarizing some of the key results reported elsewhere [40,43]. Here, we only discuss the case of an infinitely extended chain, whereas [40,43] also treat the finite chain. In Sec. IV.A. we will present results which are independent of the degree of damping of the chain, whereas in Sec. IV.B. we will return to the heavily damped case of the Sine-Gordon chain.

A. The Spread of the Chain in the Direction of Motion

Let us consider the spatial excursions along the chain. The density of kinks and antikinks is determined by the considerations of Sec. 3, but we additionally want to emphasize that for a given applied force F, and resulting density of kinks, n_0, all geometrical distributions of kinks and antikinks are equally likely. The kinks occur at random, just as in equilibrium. Pairs of kinks and antikinks are nucleated at random, and thus obviously uncorrelated with the position of other solitons. The pair then separates, and this leaves us with the possibility that there might be a correlation between the kink and antikink that were nucleated together. Ref. 40 shows, in fact, that if we do not try to make such an identification (i.e. identifying the pair that started together) then there is no correlation left, whatsoever, between kinks and antikinks aside from that imposed by their density.

If we have a Sine-Gordon chain at a particular position $\theta(x_0)$, at $x = x_0$, and move along the x axis, away from x_0, we can expect to see an uncorrelated sequence of jumps in θ, each of magnitude 2π, but of variable sign. This is then just like a diffusion process with the x axis replacing the usual time coordinate, and a diffusion constant,

$$D_\theta = \frac{1}{2} \bullet 2n_0(2\pi)^2, \tag{4.1}$$

where $2n_0$ takes the place of the usual frequency of jump factor. Therefore at a point x, with $(x - x_0) >> 1/n_0$, we can expect a probability distribution for $\delta\theta = \theta(x) - \theta(x_0)$ given by

$$p(\delta\theta) \propto \exp\,[-(\delta\theta)^2/4D_\theta(x - x_0)] = \exp\,[-G^2/2\mathcal{N}(x)], \tag{4.2}$$

where $\mathcal{N}(x) = 2n_0(x - x_0)$ is the expected number of kinks and antikinks between x_0 and x and $G = \delta\theta/2\pi$. The distribution (4.2) has a variance

$$<(\theta(x) - \theta(x_0))^2> = 2D_\theta(x - x_0). \tag{4.3}$$

Eqs. (4.1-4.3) are very general results which hold for the Hamiltonian and the damped system at equilibrium, as well as in the driven case [40,43].

B. The Dynamic Universality Class of $t^{1/2}$ Systems

We will now return to the purely viscous system described by (2.1) and discuss the dynamical behavior of the fluctuations away from the steady state. We will show that the overdamped Sine-Gordon soliton gas belongs to a class of one dimensional systems describing various physical situations but exhibiting a common long term behavior. These systems exhibit a constant of the motion $P = \int \rho dx$ and hence the associated density ρ and current j_ρ are related by a continuity equation,

$$\partial\rho/\partial t + \partial j_\rho/\partial x = 0. \tag{4.4}$$

The field ψ whose long term behavior is of interest is related to j_ρ and ρ by the self consistent equations,

$$\partial\psi/\partial t = -\Lambda j_\rho, \quad \partial\psi/\partial x = \Lambda\rho, \tag{4.5}$$

where Λ is a constant. Additionally we shall assume that for these systems j_ρ is given by a pure diffusion current,

$$j_\rho = -D\partial\rho/\partial x + \xi, \tag{4.6}$$

where the noise current ξ has the property $<\xi> = 0$ and $<\xi(x,t)\xi(x';t')> = C\delta(x-x')\delta(t-t')$. From (4.4, 4.5) and (4.6) one obtains

$$\partial\psi/\partial t - D\partial^2\psi/\partial x^2 = \Lambda\xi. \tag{4.7}$$

As can be shown [40,43], (4.7) yields, for the infinitely extended system, a mean square difference of the field amplitude at x which grows as $t^{1/2}$,

$$<(\psi(x,t)-\psi(x,0))^2> = \Lambda^2 C(t/\pi D)^{1/2}, \tag{4.8}$$

and a mean square difference of the field between x and 0 which increases proportional to x,

$$<(\psi(x,t)-\psi(0,t))^2> = 2D_\psi x, \tag{4.9}$$

with a diffusion constant $D_\psi = \Lambda^2(C/4D)$. We will now discuss a number of systems which obey (4.4-4.7) and specify in each case the parameters Λ, D and C.

To find the time evolution of the fluctuations of the damped Sine-Gordon chain, we have studied the stochastic properties of the nucleation and recombination processes [40]. We derived a set of hydrodynamic equations for the time evolution of the local kink density $m(x,t)$ and the local antikink density $n(x,t)$,

$$\partial m/\partial t + \partial j_K/\partial x = j^{nuc} - j^{rec} + \phi, \tag{4.10}$$

$$\partial n/\partial t + \partial j_{AK}/\partial x = j^{nuc} - j^{rec} + \phi. \tag{4.11}$$

j_K and j_{AK} are the local kink and antikink currents, $j^{nuc} = j$ is the nucleation rate of new pairs given by (3.1) and j^{rec} is the recombination rate of pairs. If we subtract (4.11) from (4.10) we find a continuity equation for the difference between the kink and antikink density, $\rho = m - n$, and a current $j_K - j_{AK}$. The difference between the number of kinks M and the number of antikinks N is conserved and corresponds to the constant of motion P. (In a ring, $N = M$).

In the presence of a field F the kink flux is $j_K = -u(F)m$ and the antikink flux $j_{AK} = u(F)n$. The recombination current is given by $j^{rec} = 2umn$, as reasoned in connection with (3.1). The noise ϕ in (4.10, 4.11) describes shot noise due to the discrete and uncorrelated nature of the generation recombination events. The nucleation events occur at random in space and time. Ref. [40] showed that in the steady state the recombination events also occur randomly in space and time. In the steady state, we have $j^{nuc} = j^{rec}$. The generation and recombination events contribute, therefore, equally and independently to the noise. On the time and length scale of interest, here, the stochastic force is δ correlated in space and time, $<\phi(x,t)\phi(x',t')> = 2j^{nuc}\delta(x - x')\delta(t - t')$. The coarse grained displacement field θ can be found from the kink and antikink densities, n and m, by integration. $\partial\theta/\partial t$ is given by the imbalance of the kink and antikink fluxes, (see (3.2)), and the spatial variation is given by the imbalance of the kink and antikink densities, $\partial\theta/\partial x = 2\pi\rho$. With these specifications we find from (4.10) and (4.11) for the coarse grained displacement field [40],

$$\left(\frac{\partial^2\theta}{\partial t^2} - u^2\frac{\partial^2\theta}{\partial x^2}\right) + \frac{1}{2\pi}\left(\left(\frac{\partial\theta}{\partial t}\right)^2 - u^2\left(\frac{\partial\theta}{\partial x}\right)^2\right) = 4\pi u(j^{nuc} + \phi), \tag{4.12}$$

which upon linearization around the ensemble average motion $<\theta(t)> = 2\pi u\eta_0 t$ becomes,

$$\partial^2\psi/\partial t^2 + 2u\eta_0\partial\psi/\partial t - u^2\partial^2\psi/\partial x^2 = 4\pi u\phi. \tag{4.13}$$

Here, $\psi(x,t) = \theta(x,t) - <\theta(x,t)>$ is the deviation from the average motion and $\eta_0 = <n + m>$. Eq. (4.13) has the same form as the equation of motion of a linear, damped chain of harmonically coupled particles investigated by IMRY and GAVISH [44]. (This is the limiting case of the Sine-Gordon chain with $V_0 \to 0$). The system (4.13) has two types of long wavelength modes, one branch with frequency $\omega_1 = i4un_0$ given by the lifetime of the solitons and also a diffusive branch of modes with frequency $\omega_2 = -iD_\rho q^2$, where $D_\rho = u/2\eta_0$. The diffusive modes are found as solutions of

$$\partial\psi/\partial t - D_\rho\partial^2\psi/\partial x^2 = 2\pi\phi/\eta_0, \tag{4.14}$$

which is the long term approximation of (4.13). Eq. (4.14) corresponds to (4.7) with $D = D_\rho$, $\Lambda = 2\pi$, $\xi = \phi/\eta_0$ and $C = 2j^{nuc}/\eta_0^2 = u$. Therefore, we find that the mean square deviation from the average motion of a particle located at x behaves as

$$<(\psi(x,t) - \psi(x,0))^2> = 4\pi^2 u(t/\pi D_\rho)^{1/2} = 16\pi^{3/2} n_0 (D_\rho t)^{1/2}, \tag{4.15}$$

where $n_0 = \eta_0/2$ is the steady state density of kinks. From (4.14) we find, according to (4.9), the mean square amplitude of the difference in displacement of two particles located a distance x apart, $<(\psi(x,t) - \psi(0,t))^2> = <(\theta(x,t) - \theta(0,t))^2> = 8\pi^2 n_0 x$, i.e. (4.3).

At equilibrium $u = 0$ and the kinks can propagate only diffusively [43]. We have in this case $j_K = -D\partial m/\partial x + \xi_K$ and $j_{AK} = -D\partial n/\partial x + \xi_{AK}$, with $D = (\mu/2\pi)kT$, where μ is the mobility of the solitons and ζ_K, ζ_{AK} are fluctuations in the x directed motion of the solitons. In this case it is sufficient to consider the continuity equation (4.4), i.e. (4.10) minus (4.11). Together with (4.5), $\Lambda = 2\pi$ and $j_\rho = j_K - j_{AK}$, we find that the deviations, $\psi(x,t) = \theta(x,t) - <\theta(x,t)>$, obey Eq. (4.7) with $\xi = \xi_K - \xi_{AK}$ and $C = 4n_0 D$. Because $<\theta(x,t)>$ is independent of time, we have $\psi(x,t) - \psi(x,0) = \theta(x,t) - \theta(x,0)$ and hence,

$$<\theta(x,t) - \theta(x,0))^2> = 16\pi^{3/2} n_{eq}(Dt)^{1/2}, \tag{4.16}$$

where n_{eq} is the equilibrium density of kinks and $D = (\mu/2\pi)kT$. Eq. (4.16) differs from (4.15) only through the different diffusion constant, $D < D_\rho$, and the steady state density, $n_{eq} < n_0$.

The result (4.16) was also found by GUNTHER and IMRY [45] for a soliton gas with an infinite lifetime for the kinks and antikinks. For the case of finite lifetime they find that (4.16) grows as t. However, from our derivation (4.10), (4.11) it is seen that at equilibrium only the conservation law (4.4) is important. Recombination and annihilation drop out and cannot affect the power of t in (4.16).

SCHNEIDER and STOLL [46] followed the time evolution of a lightly damped Sine-Gordon chain on the computer and found that the mean square amplitude grows as $t^{4/3}$. In a later paper [47] they find, through scaling arguments, growth as $t^{1/2}$ for the heavily damped chain in accordance with (4.16) and growth as $t^{2/3}$ for an undamped chain. The $t^{4/3}$ is attributed to finite size effects. The finite chain was also considered by GUYER [48], who finds an asymptotic behavior in accordance with our results [43]. Guyer's prefactor is different from ours, as pointed out after (3.4).

Eqs. (4.4) - (4.6) describe the long term motion of a number of systems and define, therefore, a dynamic universality class. Consider N particles on a one dimensional lattice [49,50] with spacing a. The particles have a hopping rate $W = W_{i \to i+1} = W_{i \leftarrow i+1}$ and one lattice site can only be occupied by one particle; i.e. the particles are subject to a hard core interaction. The number of particles is conserved. ψ measures the deviations from a uniform reference configuration in which the particles are a distance n_0^{-1} apart, where n_0 is the equilibrium concentration of particles. Then (4.4)-(4.6) hold with $\Lambda = -1/n_0$, $\rho = n - n_0$ and a diffusion constant $D = a^2 W$. Interestingly one finds that the mean square amplitude $<(\psi(x,t)-\psi(x,0))^2>$ grows in the infinite system as $t^{1/2}$ for all densities $0<n_0<1/a$. That the low density limit $n_0 \to 0$ and the high density limit $n_0 \to 1/a$ behave in the same way is simply a consequence of the fact that the cases can be mapped onto one another by exchanging particles and holes. The hopping system can be regarded as a simple model of a kink gas; one identifies kinks with particles in the low density limit and with holes in the high density limit. The hard core interaction takes into account that kinks repel each other. Note that (4.3) and (4.9) must also hold for this system. With $C \propto D(1/a - n_0)n_0$ we find $D_\psi \propto (1/a - n_0)/n_0$. Furthermore, it has been shown that the reptation of polymers [51] is described by (4.4) - (4.6). This describes the motion of monomers caused by N defects, each storing a length Λ. The number of defects is conserved. In the absence of defects the monomers are located at $x = ma$, m integral, along a curvilinear abscissa. If $\psi(x,t)$ denotes the deviation from ma of the position of the mth monomer in presence of defects, then (4.4)-(4.6) hold again with $\rho = n$, the density of defects. The defect mobility gives rise to a diffusion constant $D = \mu kT$. With $C = 2n_0 D$, we find $D_\psi = \Lambda^2(n_0/2)$.

We would like to add two remarks. The results reported here for the Sine-Gordon soliton gas may be quite typical for a kink gas of a system with periodic potential (one local minimum and maximum per period); however, the soliton gas of a bistable potential (for instance, the θ^4 case) shows an entirely different behavior. In the case of a bistable potential a kink and an antikink must always follow each other in the x direction. This has the consequence that on a hydrodynamic scale $\rho = m - n = 0$ and, therefore, these systems do not exhibit long wavelength diffusion modes [52], in contrast to the results reported by Collins et al. [53]. We have, in this section, only considered the heavily damped case of the Sine-Gordon soliton gas. That the result (4.16) is also valid for smaller damping will be shown elsewhere [54]. The interesting question is whether the Hamiltonian system also belongs to this universality class.

Acknowledgement

The authors are particularly indebted to H. Thomas, who, through his collaboration with us in Refs. 26 and 40, and through subsequent interactions, helped to shape the work discussed here. We also thank P. M. Marcus, H. Thomas, L. W. Adams, Jr. and R. F. Voss for correcting our earlier results for the propagation velocity of driven kinks at large fields which, instead of the present Fig. 2, showed an asymptotic approach to the line $F = V_0$. P. Hänggi brought the work of P. M. Richards to our attention.

References

1. L. Prandtl, Z. Angew. Math. Mech. **8**, 85 (1928).
2. U. Dehlinger, Ann. Physik **2**, 749 (1929).
3. A. Seeger and P. Schiller, in *Physical Acoustics*, Vol. III, W. P. Mason, ed. (Academic Press, New York, 1966)p. 361.
4. H. Busch, Ann. Phys. **64**, 401 (1921).
5. K. E. Gray, J. Low Temp. Phys. **23**, 679 (1976).
6. R. Landauer, Phys. Rev. A **15**, 2117 (1977).
7. R. Landauer, Phys. Today **31**, 23 (1978).
8. R. A. Pasmanter, D. Bedeaux and P. Mazur, Physica **90A**, 151 (1978).
9. Y. D. Kalafati, I. A. Serbinov and L. A. Ryabova, JETP Lett. **29**, 583 (1979).
10. J. B. Gunn, IBM J. Res. Dev. **8**, 141 (1964) and IBM J. Res. Dev. **10**, 300 (1966).
11. B. W. Knight and G. A. Peterson, Phys. Rev. **155**, 393 (1967).
12. A. F. Volkov and S. M. Kogan, Sov. Phys. Usp. **11**, 881 (1969).
13. M. E. Levinshtein and M. S. Shur, Sov. Phys. Semicond. **5**, 1561 (1972); Sov. Phys. Semicond. **9**, 411 (1975).
14. E. Pytte and H. Thomas, Phys. Rev. **179**, 431 (1969).
15. M. Büttiker and H. Thomas, Z. Phys. B **33**, 275 (1979); Z. Phys. B **34**, 301 (1979).
16. R. Landauer, in *The Maximum Entropy Formalism*, R. D. Levine and M. Tribus, eds. (MIT Press, Cambridge, Mass., 1978), p.321.
17. R. Landauer, *Moebius Strip Coupling of Bistable Elements*, unpublished laboratory report of 1955, reprinted in 1976, available from the author, or as RC 6093 from Distribution Services at the author's laboratory.
18. G. E. Duvall and R. A. Graham, Rev. Mod Phys. **49**, 523 (1977).
19. T. J. Ahrens, Science **207**, 1035 (1980).
20. R. Landauer, *Phase Transition Waves: Solitons Versus Shock Waves*, J. Appl. Phys., to be published.
21. M. Büttiker and R. Landauer, Phys. Rev. Lett. **43**, 1453 (1979).
22. S. E. Trullinger, M. D. Miller, R. A. Guyer, A. R. Bishop, F. Palmer and J. A. Krumhansl, Phys. Rev. Lett. **40**, 206 (1978), 1603 (E) (1978).
23. R. A. Guyer and M. D. Miller, Phys. Rev. A **17**, 1774 (1978).
24. K. C. Lee and S. E. Trullinger, Phys. Rev. B **21**, 589 (1980).
25. M. Büttiker and R. Landauer, *Nucleation Theory of Overdamped Soliton Motion*, preprint.
26. M. Büttiker and H. Thomas, Phys. Lett. **77** A, 372 (1980).
27. P. M. Marcus and Y. Imry, Solid State Commun. **33**, 345 (1980).
28. L. W. Adams Jr., Phys. Rev. A **21**, 1648 (1980).
29. M. Urabe, J. Sci. Hiroshima Univ., Ser. A **18**, 379 (1955).
30. D. E. McCumber, J. Appl. Phys. **39**, 3113 (1968).
31. H. Risken and H. D. Vollmer, Z. Phys. B **35**, 177 (1979).
32. The $u(F)$ curves in Fig. 1 of Ref. 21 and Fig. 3 of Ref. 26 are erroneous. To obtain the correct u values from these figures they should be multiplied by $\sqrt{(1-(F/V_0)^2)}$. See Fig. 2 of this paper.
33. M. B. Fogel, S. E. Trullinger, A. R. Bishop and J. A. Krumhansl, Phys. Rev. Lett. **36**, 1411 (1976); Phys. Rev. B **15**, 1578 (1977).
34. J. Feder, K. C. Russell, J. Lothe and G. M. Pound, *Adv. Phys.* **15**, 111 (1966).
35. J. C. Burfoot and G. W. Taylor, *Polar Dielectrics and Their Applications* (University of Calif. Press, Berkeley, 1979).

36. J. S. Langer and V. Ambegaokar, Phys. Rev. **164**, 498 (1967).
37. D. E. McCumber and B. I. Halperin, Phys. Rev. B **1**, 1054 (1970).
38. R. Landauer and J. A. Swanson, Phys. Rev. **121**, 1668 (1961).
39. J. S. Langer in *Proceedings of the 1980 Sitges Symposium on Statistical Mechanics*, to be published.
40. C. H. Bennett, M. Büttiker, R. Landauer and H. Thomas, *Kinematics of the Forced and Overdamped Sine-Gordon Soliton Gas*, J. Stat. Phys., to be published.
41. J. F. Currie, J. A. Krumhansl, A. R. Bishop and S. E. Trullinger, *Statistical Mechanics of One-Dimensional Solitary-Wave-Bearing Scalar Fields: Exact Results and Ideal-Gas Phenomenology*, preprint.
42. T. Schneider and E. Stoll, "The Sine-Gordon Chain: Mass Diffusion", in *Solitons and Condensed Matter Physics*, ed. by A. R. Bishop and T. Schneider, Springer Series in Solid-State Sciences, Vol. 8 (Springer, Berlin, Heidelberg, New York 1978) p. 326
43. M. Büttiker and R. Landauer, J. Phys. C **13**, L325 (1980).
44. Y. Imry and B. Gavish, J. Chem. Phys. **61**, 1554 (1974).
45. L. Gunther and Y. Imry, Phys. Rev. Lett. **44**, 1225 (1980).
46. T. Schneider and E. Stoll, Phys. Rev. Lett. **41**, 1429 (1978).
47. T. Schneider and E. Stoll, *Long-time behavior and dynamic scaling properties of displacement fluctuations in the sine-Gordon chain*, Phys. Rev. B, to be published.
48. R. A. Guyer, Phys. Rev. B **21**, 4484 (1980).
49. P. M. Richards, Phys. Rev. B **16**, 1393 (1977).
50. S. Alexander and P. Pincus, Phys. Rev. B **18**, 2011 (1978).
51. P. G. de Gennes, J. Chem. Phys. **55**, 572 (1971).
52. J. A. Krumhansl and J. R. Schrieffer, Phy. Rev. B **11**, 3535 (1975).
53. M. A. Collins, A. Blumen, J. F. Currie and J. Ross, Phys. Rev. B **19**, 3630 (1979).
54. M. Büttiker and R. Landauer, *Fluctuations in the Sine-Gordon Chain at Thermal Equilibrium*, in preparation.

Inherent Effects of Discretization in an Interacting Kink-Phonon System

N. Theodorakopoulos and R. Klein

Fakultät für Physik, Universität Konstanz
D-7750 Konstanz, Fed. Rep. of Germany

1. Introduction

There is a considerable body of knowledge about classical 1-D systems capable of supporting nonlinear soliton-like excitations [1]. We know, for example, that the free energy of a discretized Sine-Gordon (SG) or ϕ^4 chain exhibits parts which may be directly related to independent "gases" of linear (phonon) and nonlinear (kink) modes respectively [2]. Linear stability analysis [3] tells us that kinks are transparent to small oscillations of the underlying continuum; interactions can be characterized, to first order, by a phase shift, which in turn reduces the density of states available to linear modes, to allow for the kink degrees of freedom. This energy sharing, when taken into account, improves the agreement between configurational phenomenology and exact (transfer integral) results [4]. We can even go one step further and examine second order effects, to conclude that solitons perform a diffusive motion under the influence of thermal phonons yet no energy is being exchanged between nonlinear and continuum linear modes [5].

In view of the fundamental nature of these results, in connection with both equilibrium statistical mechanics and microscopic nonlinear dynamics, it seems that there is an adequate motive to investigate and assess the role of discretization. As it turns out, new features appear [6] which are qualitatively different from those expected within the continuum framework. Here, we will concentrate on the change in the phonon density of states (Sect. 2) and the possible changes in the kink's position and state of motion (Sect. 3).

2. First order effects: the phonon density of states

We consider the equation of motion for the displacement of the n-th atom in a linear chain of spacing ℓ under the influence of a harmonic nearest-neighbor coupling and an on-site anharmonic potential $V(\phi)$

$$\ddot{\phi}_n = \left(\frac{d}{\ell}\right) (\phi_{n+1} - 2\phi_n + \phi_{n-1}) - V'(\phi_n) \quad . \tag{1}$$

The dimensionless constant d/ℓ measures the strength of the harmonic coupling relative to the on-site potential. We shall present our findings for the ϕ^4 case $V(\phi) = (\phi^2 - 1)^2/4$. Details of the calculation and results for the Sine-Gordon potential will be published elsewhere [7]. For values of the parameter $d/\ell >> 1$ the static kink solution of (1) can be well approximated by $\phi_K(y_n) = \tanh y_n$, where $y_n = n\ell/d\sqrt{2}$. The excited states of a kink can be obtained by the standard ansatz of linear stability analysis

$$\phi_n(t) = \phi_K(y_n) + f_n^j \exp(-i\Omega_j t) \quad . \tag{2}$$

It is then possible to obtain the translational and internal oscillation modes corresponding to $\Omega_j \cong 0$ and $3/2$, respectively, with any deviations vanishing as d/ℓ increases. In order to describe the scattering modes (phonons) without resorting to the continuum approximation we must solve the difference equations

$$f_{n+1}^Q + (W_n - 2)f_n^Q + f_{n-1}^Q = 2(\cos Q\ell - 1)f_n^Q \quad , \tag{3}$$

where $W_n = 3(\ell/d)^2 \mathrm{sech}^2 y_n$ and the r.h.s. of (3) can be obtained by observing that, away from the kink, the solution of the problem should reduce to the vacuum dispersion relation $\Omega_Q^2 = 2 + 2(d/\ell)^2(1 - \cos Q\ell)$ and

$$\lim_{n\to\pm\infty} f_n^Q = \begin{cases} e^{iQn\ell} + R(Q)\, e^{-iQn\ell} & \text{(4a)} \\ T(Q)\, e^{iQn\ell} \quad . & \text{(4b)} \end{cases}$$

We may now distinguish three cases according to the value of the wave vector. If $|Q\ell| << 1$ (Region I) the continuum approximation is valid and results are well known. Region II is defined by the inequality $(d/\ell)\sin|Q\ell| >> 1$ and covers, in the limit $d >> \ell$, the bulk of the Brillouin zone. Here it can be shown that both the ϕ^4 and the SG kink retain their reflectionless property ($R=0$) towards phonons satisfying the condition II. Phase shifts are however quantitatively different from continuum predictions [6]. Region III covers the immediate vicinity of the Brillouin zone edge $|Q\ell-\pi| << 1$. Here it is possible to reduce (3) to a differential equation by assuming $f_n^Q = (-1)^n \tilde{f}_n^Q$, where $\tilde{f}$ varies smoothly. This procedure allows us to identify

$$\tilde{f}_n^Q = C_\nu(q)\, P_\nu^{iq}(\tanh y_n) \tag{5}$$

where P_ν^{iq} is the associated Legendre function [8], $\nu = -1/2 + i\kappa$, $\kappa^2 = 23/4$ (ϕ^4) or $7/4$ (SG), $q = (Q\ell-\pi)\sqrt{2}\, d/\ell$ and C is a normalization factor which can be derived by matching (5) to the asymptotic conditions (4). This matching also demonstrates the possibility of reflection for $|q| \lesssim \kappa$ since it yields

$$|R|^{-2} - 1 = (\sinh \pi q / \cosh \pi\kappa)^2 \quad . \tag{6}$$

The phase shift $\delta(Q) = \arg T(Q)$ and the reflection coefficient $R(Q)$ determine now the asymptotic properties of the kink-phonon interaction. Accordingly the change in the phonon density of states due to the presence of a static kink is now given (Fig. 1) by

$$\Delta\rho(Q) = \frac{1}{2\pi} \frac{d}{dQ} \left[\delta(Q) + \cos^{-1}|T(Q)|\,\right] \quad . \tag{7}$$

3. Second order effects: the kink's state of motion

The lack of Lorentz invariance forces us to abandon the continuum approach based on Noether's theorem [5] and perform second-order perturbation theory [9]. This approach yields an equation of motion for the amplitude B_T of the translational mode (proportional to the kink displacement)

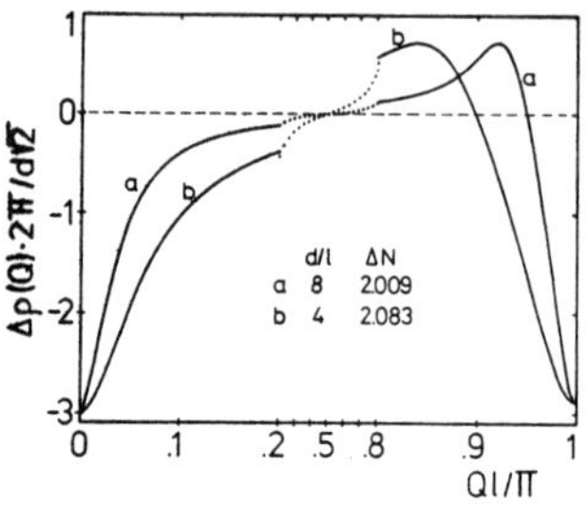

Fig.1 The change in the phonon density of states is plotted as a function of Q. The predictions of continuum theory correspond to the left portion of the graph (region I). Discrete lattice effects appear in region II (notice the different Q-scale) and are most distinct in region III, where a major rearrangement of available states occur. The number of "removed" phonon states $\Delta N = -\int_{-\pi/\ell}^{\pi/\ell} dQ\, \Delta\rho(Q)$ is given for two values of d/ℓ.

$$\ddot{B}_T(t) = -\frac{3\sqrt{3}}{2}\,\ell \sum_n \phi_K(y_n)\,\mathrm{sech}^2 y_n \cdot \eta^2(y_n) \quad , \tag{8}$$

where

$$\eta(y_n,t) = \frac{\alpha}{2}\int d\bar{Q}\; A(\bar{Q}-Q)\; f_n^{\bar{Q}}\; \exp(-i\Omega_{\bar{Q}}t) + \mathrm{c.c.} \tag{9}$$

represents a wave packet strongly peaked around Q which approaches the initially static kink from the left. The sum in (8) can be approximated by an integral since, as it turns out, in all three regions defined above, $\eta^2(y_n)$ does not vary appreciably over a lattice constant. This allows us to evaluate the final kink displacement $\Delta y = -(\sqrt{3}/2)B_T(\infty)/d\sqrt{2}$ for any phonon wave vector. Our results are summarized below.

Region II:

$$\Delta y = -\left[E_{ph}(Q)/E_K\right]\,(\ell/d)^2 3\csc^2 Q\ell \tag{10}$$

where $E_K = 2/3$ is the rest energy of the kink and $E_{ph}(Q)$ the energy carried by the phonon wave-packet (9).

Region III: The limiting form of $B_T(t)$ contains a term growing linearly with t; the kink displacement is then time dependent, $\Delta y(t) = \Delta y + vt$, where Δy and v can be identified, respectively, as a static displacement (Fig. 2) and a final kink velocity (Fig. 3). The corresponding analytical expressions are

$$\Delta y = -\left[E_{ph}(Q)/E_K\right]\,(2|R|^2 - 1)\; Y(q;\kappa) \tag{11a}$$

$$v = -\left[E_{ph}(Q)/E_K\right]\,2|R|^2\, u_Q \tag{11b}$$

where $u_Q = (\partial\Omega_Q/\partial Q)/d\sqrt{2}$ is the group velocity of the phonon wave packet, $Y(q;\kappa) = \mathrm{Re}\{\psi(1/2+iq-i\kappa) + \psi(1/2+iq+i\kappa) - 2\psi(iq)\}$, and ψ is the digamma function.

The results (11) possess, apart from their particular analytical structure, a qualitative significance. It becomes evident that the breakdown of the reflectionless property of a static kink towards zone edge phonons is accompanied by a change in the state of motion of an initially static kink - as a result of its interaction with such phonons. This manifestation of inherent discrete lattice properties provides us with a microscopic mechanism, a channel, via which solitons may eventually reach a thermal distribution of their velocities - in the framework of a Hamiltonian system.

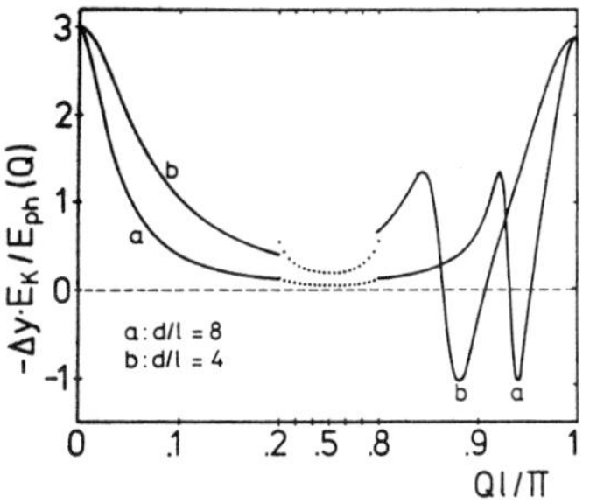

Fig.2 The static displacement of a kink that results from its collision with a phonon of wave-number Q and energy $E_{ph}(Q)$ is plotted as a function of Q. Discretization manifests itself most clearly near the zone boundary (region III), whereas the continuum model would allow only for the extension of the curves of region I ($Q\ell \ll 1$), which fall off monotonically.

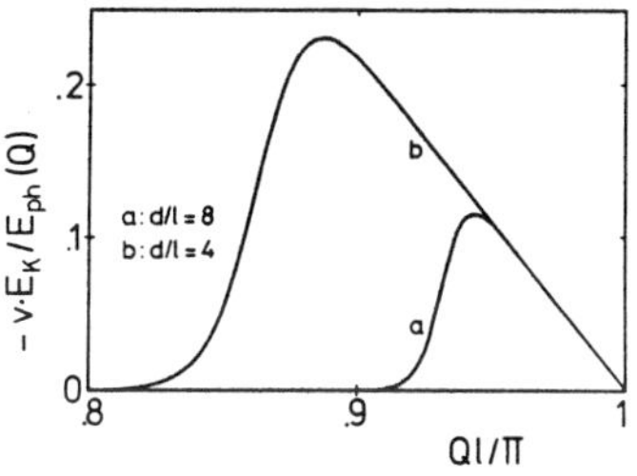

Fig.3 An initially static kink (of rest energy E_K) acquires a final velocity v after colliding with a phonon which lies near the zone boundary. This is a strictly discrete lattice effect (see also Fig.2) occurring near the zone boundary; here -v is plotted against Q.

References

1. A.R. Bishop and T. Schneider (Eds.), Solitons and Condensed Matter Physics, Springer Series in Solid-State Sciences, Vol. 8 (Springer, Berlin, Heidelberg, New York 1978)
2. J.A. Krumhansl and J.R. Schrieffer, Phys. Rev. B11,3535(1975)
3. R. Rajaraman, Phys. Reports 21,229(1975)
4. A.R. Bishop in Ref. 1
5. N. Theodorakopoulos, Z. Phys. B33,385(1979)
6. N. Theodorakopoulos, W. Wünderlich, R. Klein, Sol. State Comm. 33,213(1980)
7. N. Theodorakopoulos, to be published
8. A. Erdélyi et al., Higher Transcendental Functions, Vol.1, McGraw-Hill (1953)
9. W. Hasenfratz and R. Klein, Physica 89A,191(1977).

Ising Models, Solitons, and the Devil's Staircase

Per Bak

Fysisk Laboratorium I, H.C. Ørsted Institute, Universitetsparken 5
DK-2100 Copenhagen Ø, Denmark

1. Introduction

Periodically modulated structures are very common in solid state physics. Helical and sinusoidal magnetic structures, charge-density waves in layered compounds and quasi-one-dimensional conductors, and rare gas monolayers adsorbed on graphite are among the best known examples [1].

To understand the nature of modulated "incommensurate" structures it is important to perform calculations on relatively simple model systems. It turns out that there exists a simple Ising model which serves this purpose: a quasi-one-dimensional Ising model with competing interactions in one particular direction, first introduced by Elliott [2], displays a modulated phase.

Figure 1 shows the two-and three-dimensional versions of the model. The spins, $S=\pm1$, interact through a ferromagnetic nearest neighbor interaction, J_1 . In one direction, however (the z direction for the 3d model and the x direction for the 2d model) there is an antiferromagnetic interaction $J_2<0$. The competition between J_1 and J_2 stabilizes the modulated phase. In this article recent work on this model will be reviewed, and new results in two dimensions will be presented. Section 2 deals with the 3d model and section 3 is devoted to its 2d cousin.

The Ising model with nearest neighbor interactions only has been intensively studied and in two dimensions it has even been solved. Its phase diagram includes a ferromagnetic phase and a paramagnetic phase. The simple generalization of the model defined by the addition of the antiferromagnetic bond gives rise to a variety of interesting physical features, connected with the

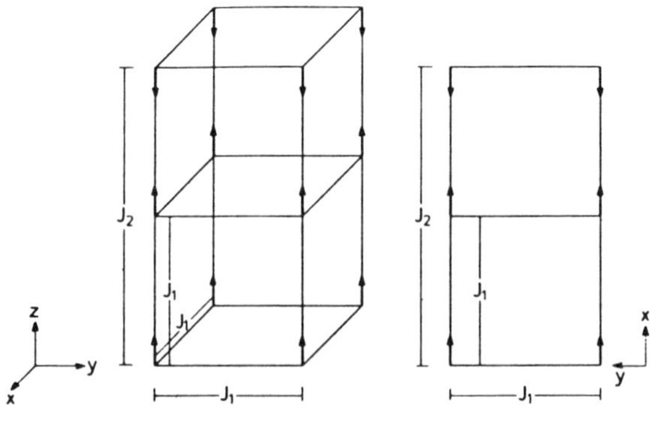

Fig.1 Three- and two-dimensional versions of the Ising model with competing interactions

appearance of the modulated phase. The 3d version exhibits an infinity of phases where the periodicity is commensurate with the lattice [3-8]. We shall use the very expressive phrase "the Devil's Staircase (DS)" for this behaviour [9,10]. The rare-earth magnet CeSb exhibits a large number of commensurate phases and could be a realization of the DS [3,11,12]. The 2d version has very different properties: At non-zero temperature the devil's staircase gives way to a floating modulated phase. This result is derived through an interesting analogy with the quantum sine-Gordon equation. The transition from the modulated phase to the paramagnetic phase is of the Kosterlitz-Thouless type [13]. We find no support for the assertion that the model displays a 2d Lifshitz point [14,15].

2. 3d Ising Model with Competing Interactions

The Hamiltonian is

$$H = - \sum_{<ij>} J_{ij} S_i S_j \tag{2.1}$$

with the interactions defined in Fig. 1. The mean-field phase diagram(Fig. 2) consists of three areas: (i) a ferromagnetic phase, (ii) a disordered paramagnetic phase, and (iii) a modulated phase characterized by a wave vector $\vec{q} = 2\pi/a(0,0,q)$ (or $2\pi/a(q,0)$). These phases are separated by a Lifshitz point P [15,16].

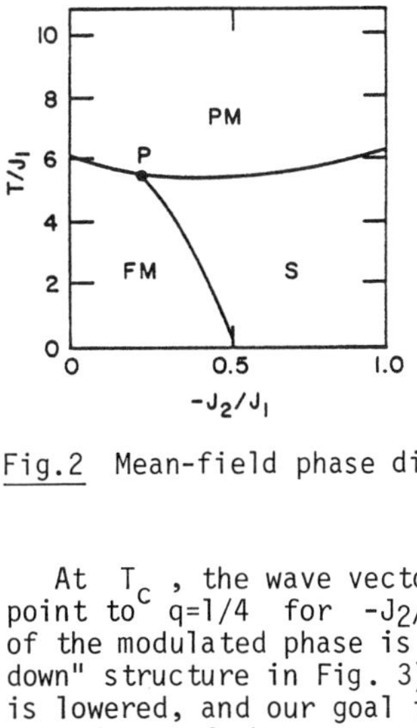

Fig.2 Mean-field phase diagram

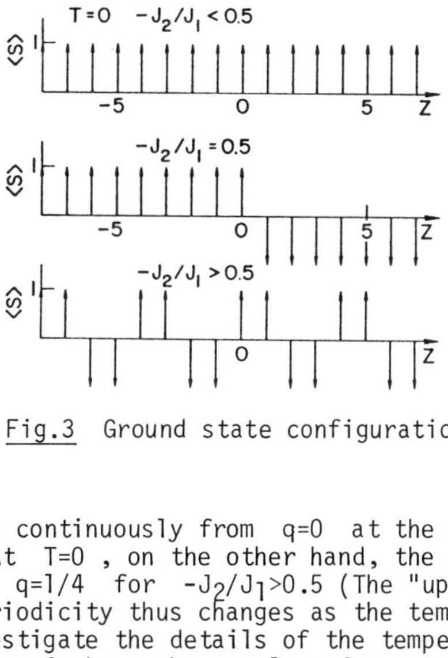

Fig.3 Ground state configurations

At T_c , the wave vector changes continuously from $q=0$ at the Lifshitz point to $q=1/4$ for $-J_2/J_1 \to \infty$. At $T=0$, on the other hand, the periodicity of the modulated phase is given by $q=1/4$ for $-J_2/J_1>0.5$ (The "up-up-down-down" structure in Fig. 3). The periodicity thus changes as the temperature is lowered, and our goal is to investigate the details of the temperature dependence of the wave vector. Figure 4 shows the results of a numerical mean field calculation for selected values of J_2/J_1 . For details see Refs. [3,4]. Several unusual features are evident. Firstly, the wave vector has a stepwise temperature dependence. The $q=1/4$ phase is stable up to a rather high temperature, then q deviates rapidly from this value. Also, the curve is not always monotonic. For $J_2/J_1 = -0.6$ the wave vector first decreases as the temperature is lowered, then it increases. More and more commensurate phases emerge as the calculation is refined, indicating the existence of a DS. Some rational values of q do not occur. For example, the periodicity goes from $q=2/11$ to $q=1/6$ without assuming the intermediate value $q=3/17$. Near T_c the numeric calculation breaks down, but we shall see that in this limit the MF theory can be solved analytically [4,17], and the DS is stable in this limit.

Most features of the numerical calculations can be explained using a soliton theory. Roughly, the calculation proceeds as follows [4]. Starting from the mean-field equations the free energy is expanded in terms of order parameters $S_{\pm q}$ describing "spin density waves" $S_{\pm q}\exp(\pm i\,\vec{q}\cdot\vec{r})$, $\vec{q} = 2\pi/a(0,0,q)$. To investigate the stability of the $q=1/4$ phase we calculate the free energy associated with a spin structure

$$S(\vec{r}) = \frac{1}{\sqrt{2}} S_{1/4}(\vec{r})\exp(2\pi i(0,0,1/4)\cdot\vec{r}) + cc \tag{2.2}$$

where $S_{1/4}(\vec{r})$ is a slowly varying function of $\vec{r}$. The resulting effective Landau-Ginzburg free energy functional becomes

$$F = \int d\vec{r}\, F(\vec{r}) ,$$

$$F(\vec{r}) = \frac{1}{2}r\,|S_{1/4}|^2 + \frac{1}{8}T|S_{1/4}|^4 + \frac{1}{2}c|(\nabla_z - \frac{ia}{2c})S_{1/4}|^2$$

$$+ \frac{1}{2}c'|\nabla_\perp S_{1/4}|^2 + \frac{1}{48}T(S_{1/4}^4 + S_{-1/4}^4) \tag{2.3}$$

where $r = T-4J_1 + 2J_2$, $a = 2J_1$, $c = -4J_2$, $c' = J_1$.

The last term is an umklapp term and gives rise to the locking at $q=1/4$. Following [18,19] we now assume that near the $q=1/4$ commensurate phase the amplitude of the order parameter is constant whereas the phase is allowed to be space dependent:

$$S_{\pm 1/4}(\vec{r}) = A\exp(\pm i\,\phi(z)) . \tag{2.4}$$

The free energy per unit area perpendicular to z becomes

$$F = \int dz\, cA^2\Big[\frac{1}{2}(\frac{d\phi}{dz} - \delta)^2 + v_4(1 + \cos M\phi)\Big] \tag{2.5}$$

where $M=4$, $\delta=J_1/4J_2$, $v_4 = -TA^2/96\,J_2$, $A^2=3(4J_1-2J_2-T)/T$.

The amplitude A minimizes the free energy (2.3) in the commensurate phase. The first term favors the incommensurate phase with wave vector $q=1/4+\delta/2\pi$ ($\phi=\delta z$). The second term favors the commensurate phase, $\phi=\pi/4$. A transition between the two phases take place because of the competition between the two terms. The phase function which minimizes (2.5) satisfies the sine-Gordon equation

$$\phi''(z) + 4v\sin 4\phi = 0 \tag{2.6}$$

which has the soliton solution

$$\phi(z) = \frac{\pi}{4} - \tanh^{-1}\exp(4\sqrt{v}z) \tag{2.7}$$

describing a domain wall separating two almost commensurate regions, as can be seen from inserting (2.7) into (2.4) and (2.2). In general, the solutions are regularly spaced solitons, with density $\bar{q}$. The free energy can be expressed as a function of $\bar{q}$ which is also the deviation of the wave vector from the commensurate one:

$$F = cA^2 \left[\left(4\frac{\sqrt{V}}{\pi} - \delta\right) + \frac{16\sqrt{V}}{\pi} \exp\left(\frac{-2\pi\sqrt{V}}{\bar{q}}\right) \bar{q} \right] . \quad (2.8)$$

The first term is the soliton energy, and the second term is a repulsion between solitons. When the soliton energy is positive the commensurate phase is stable ($\vec{q}=0$). When the energy is negative the incommensurate phase is

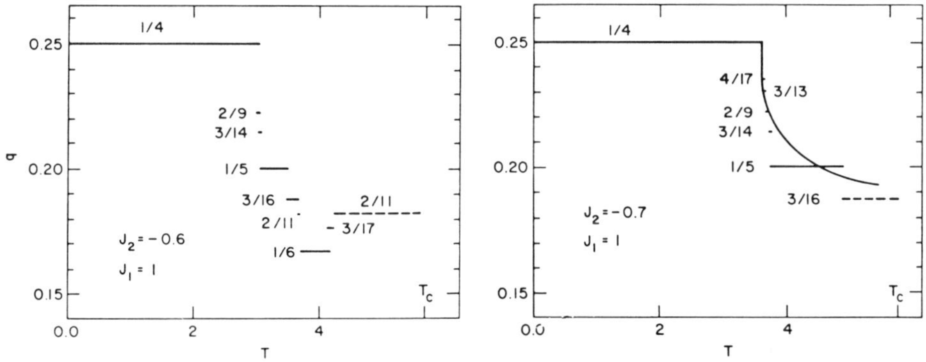

Fig.4 Wave vector vs. temperature as calculated numerically. The full curve is the result of the analytic "soliton" calculation

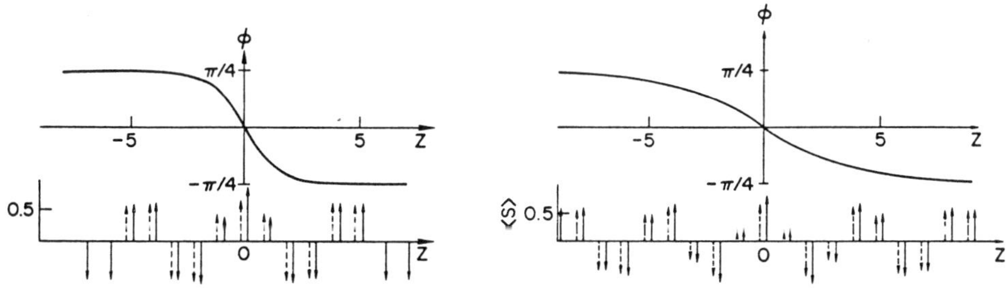

Fig.5 Soliton solutions to (2.6). The broken arrows show the numerical MF solutions ($J_2 = -0.7J_1$, $J_2 = -2J_1$)

stable. The soliton spin configurations (2.7) agree with the numerical results (Fig.5). This serves to justify the fundamental ansatz (2.4). The temperature dependence of the wave vector is shown in Fig.4 for $J_2/J_1=-0.7$. The soliton has a characteristic width

$$\Delta = -4J_2/J_1 \quad (2.9)$$

and the transition temperature is

$$T_{CI} = 4J_1 - 2J_2 + \pi^2 J_1^2/8J_2 \quad . \quad (2.10)$$

The transition from the $q=1/6$ phase (three-up-three-down-layers) can also be discussed in terms of solitons. The free energy is again given by (2.5) but with

$$M=6 \;,\; \delta = \frac{\sqrt{3}(J_1+2J_2)}{-J_1+4J_2} \qquad (2.11)$$

$$A^2 = 2(5J_1-J_2-T)/T$$

$$v_6 = A^4T/60(J_1-4J_2)$$

$$c = J_1/2 - 2J_2 \quad .$$

The transition temperature is

$$T_{CI} = 5J_1 - J_2 + \frac{3\pi}{4}\left\{(J_1+2J_2)\,\frac{5(5J_1-J_2)}{(J_1-4J_2)}\right\}^{\frac{1}{2}} \quad . \qquad (2.12)$$

The stability of higher order commensurate phases, $q = N/M$, is governed by higher order umklapp terms of order M for M even and $2M$ for M odd. The widths of the commensurate phases (vs. J_2) is proportional to $\sqrt{v_M} \sim A^{(M-2)/2} \sim (T_c-T)^{(M-2)/4}$. The widths thus decay exponentially with M. It can be shown that this is more than sufficient to ensure that the various commensurate phases will not overlap [20] for small enough (T_c-T). Thus, near the critical temperature all commensurate phases (with $0 \leq q < 1/4$) will appear in the phase diagram, generating the devil's staircase. Between the infinity of commensurate phases an infinity of incommensurate phases will be stable! Furthermore, it can be argued that in a relatively narrow regime between a commensurate phase and the incommensurate phase the solitons will be pinned to the lattice, forming the so-called stochastic phases [9,21]. The arguments leading to the stability of the DS near T_c presented here are all mean-field. Renormalization group calculations in $4-\varepsilon$ dimensions indicate that the staircase will survive critical fluctuations near T_c [22]. The exponents governing the widths of the C phases will change.

By combining the numerical calculations at low temperatures with the soliton theory at higher temperatures the phase diagram in Fig.6 has been constructed. The dark areas indicate higher order commensurate phases and incommensurate phases. At low temperatures the phase diagram includes first order transitions lines between different commensurate phases, and from the ferromagnetic to the modulated phase. The staircase thus becomes more and more harmless as the temperature is lowered, but it is not known precisely how the various C phases disappear.

Villain and Gordon [6] and Fisher and Selke [7] have investigated the phase diagram near the point where an infinity of commensurate phases becomes degenerate at $T=0$. Villain explains the first order transitions in terms of attractive or oscillating interactions between solitons. Fisher and Selke show explicitly - using a low temperature expansion - that only phases given by $q = \ell/2(2\ell+1)$, ℓ integer, survive asymptotically close to the point $T=0$, $J_2 = -J_1/2$.

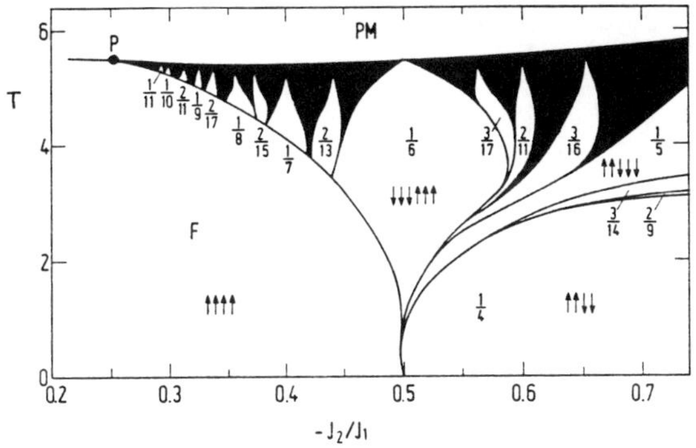

Fig.6 Phase diagram for 3d Ising model with competing interactions

The critical behavior near T_c is that of the 3d-xy model [4,23]. The two components of the order parameter correspond to the amplitude and the phase of the spin-density-wave (2.2). In two dimensions we expect similarly the critical behavior to be that of the 2d-xy model which makes it particularly interesting in view of recent theories on this class of models [13,24-26].

3. The 2d Ising Model with Competing Interactions

Our starting point is again a free energy expansion similar to (2.3). To discuss the transition from the paramagnetic to the modulated phase we expand around the wave vector which minimizes the second order term, and the linear gradient term disappears:

$$F(\vec{r}) = \tfrac{1}{2}r|S_q|^2 + \tfrac{1}{2}c|\frac{d}{dx}S_q|^2 + \tfrac{1}{2}c'|\frac{d}{dy}S_q|^2 + \dots \qquad (3.1)$$

where $r = T - 2J_1 + 2J_2$

$$c = -4J_2 + \frac{J_1^2}{4J_2}$$

$$c' = J_1 \quad .$$

Since fluctuations are important in 2d, even far from T_c, we shall now consider (3.1) as a Hamiltonian. The trace of the exponential of the Hamiltonian should be taken to calculate the free energy. An alternative way of deriving an effective Hamiltonian is by means of the Hubbard transformation [27] leading to exactly the same result. As is usual in two dimensions amplitude fluctuations will be ignored, i.e. we shall include only the configurations

$$S_q = A \exp i\phi(\vec{r}) \qquad (3.2)$$

in the calculations. Since we expect the transition temperature to be far below the mean-field transition temperature we fix A (quite arbitrarily) at the value 1, and the Hamiltonian takes the form:

$$H = \int dxdy\left[(-2J_2 + \frac{J_1^2}{8J_2})(\frac{d\phi}{dx})^2 + \frac{J_1}{2}(\frac{d\phi}{dy})^2\right] . \quad (3.3)$$

This is indeed the Hamiltonian used by Kosterlitz and Thouless to discuss the xy transition [13]. The ordered phase has no true long range order, but the order-parameter correlation functions decay algebraically in space. A phase transition to a disordered phase takes place as a consequence of vortex separation at a critical temperature T_c . If the vortex energy renormalization is ignored, the critical temperature becomes

$$T_c \simeq \pi\sqrt{(-2J_2 + \frac{J_1^2}{8J_2})(\frac{J_1}{2})} . \quad (3.4)$$

The critical curve has been plotted in Fig.7. In the extreme anisotropic limit, $J_2 = -0.25J_1$ (where mean-field theory predicts a Lifshitz point) T_c goes to zero. This may reflect the general belief that the lower critical dimensionality, below which the Lifshitz point is not stable, is d=2.5. In any case, the situation in this limit is not clear. Generally, the phase transition temperature is higher than the one found by Selke and Fisher using Monte Carlo techniques [14]. This may well be due to our rather crude approximations.

To investigate the stability of the q = 1/4 phase we expand the Hamiltonian around q = 1/4

$$H = \int dxdy\left[-2J_2(\frac{d\phi}{dx} - \delta)^2 + \frac{J_1}{2}(\frac{d\phi}{dy})^2 + \frac{T}{24}(1 - \cos 4\phi)\right] , \ \delta = J_1/4J_2 \quad (3.5)$$

where A=1 has again been inserted. The umklapp term takes the form of a $\cos M\phi$ term in the Hamiltonian. The effect of such terms has been studied extensively by José et al. [24]. Here, we shall use a transfer matrix technique which reveals some interesting connections between seemingly very different models. Employing this method in the y-direction, the calculation of the free energy associated with (3.5) amounts to the calculation of the ground state energy of the quantum Hamiltonian

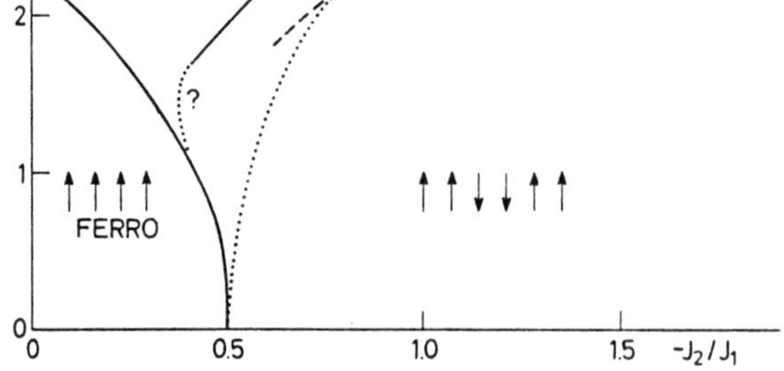

Fig.7 Phase diagram for 2d Ising model with competing interactions

$$\frac{\hat{H}}{T} = L_y \int dx \left[C\pi^2 + A\left(\frac{d\phi}{dx} - \delta\right)^2 + B(1 - \cos M\phi) \right] \qquad (3.6)$$

with $C = \frac{T}{2J_1}$, $A = \frac{-2J_2}{T}$, $B = \frac{1}{24}$, $M = 4$, and L_y is the length of the system in the y-direction. Here, π is the momentum conjugate to ϕ . This is the one-dimensional quantum sine-Gordon equation. For $\delta=0$, the nature of the ground state changes when the renormalized quantum soliton energy becomes zero. In the 2d picture this describes the situation where the free energy associated with domain wall formation becomes zero. The quantum soliton energy is [28]

$$m_S = \sqrt{8AB}\,\left(1 - \frac{M^2}{8\pi}\sqrt{\frac{C}{A}}\right) \quad . \qquad (3.7)$$

The last term is a quantum correction to the 1d classical sine-Gordon domain wall energy. For M=4 , the transition given by (3.7) coalesce with the Kosterlitz-Thouless transition given by (3.4) in the absence of the $\cos 4\phi$ term. This is in perfect agreement with the results derived by José et al. [24]. For $\delta>0$ the transition takes place for $m_S = \delta A$. In the 2d picture, δ is a domain wall chemical potential, and the transition takes place when the domain wall free energy balances the chemical potential:

$$A\delta = \sqrt{8AB}\,\left(1 - \frac{M^2}{8\pi}\sqrt{\frac{C}{A}}\right) \quad . \qquad (3.8)a$$

By inserting appropriate values we find

$$T_{CI} = 2J_1 - 2J_2 - \frac{\pi^2 J_1^2}{8J_2} \Big/ \left(1 - \frac{T_{CI}}{\pi\sqrt{J_1|J_2|}}\right) \quad . \qquad (3.8)b$$

The coefficient in the brackets gives a correction to the mean field transition temperature. The commensurate-incommensurate transition temperature given by (3.8) has also been plotted in Fig.7. Because of the finite δ , there is a (narrow) temperature range where the incommensurate phase is stable. The phase boundary of the ferromagnetic phase was calculated using the theory of Müller-Hartmann and Zittartz [29].

The stability of higher order commensurate phases can be discussed in terms of obvious generalizations of (3.6). The general result is that these phases cannot be stable near T_c . The condition $m_S = 0$ gives an upper critical temperature T_M for the stability of higher order commensurate phases:

$$T_M \sim \frac{16T_c}{M^2} \quad . \qquad (3.9)$$

The incommensurate phase thus generally remains stable vs. $\cos M\phi$ perturbations. The I phase is thus an unpinned floating phase in the terminology of Halperin and Nelson [25]. It can be argued that the floating phase is in fact stable all the way to $T = 0$ [30]. The Hamiltonian (3.6) has been used by Bak and Fukuyama to discuss the commensurability effects on a 1d quantum model of a modulated system [31]. It was found that the high order commensurate phases do not survive quantum fluctuations, just as we have shown here that the high order commensurate phases (and the DS) do not survive temperature induced fluctuations in a 2d classical system!

References

1. See P. Bak, "Solitons in Incommensurate Systems", in Solitons and Condensed Matter Physics, ed. by A.R. Bishop and T. Schneider, Springer Series in Solid-State Sciences, Vol. 8 (Springer, Berlin, Heidelberg, New York 1978) p.216, and references therein.
2. R.J. Elliott, Phys.Rev. 124, 346 (1961).
3. P. Bak and J. von Boehm, Phys.Rev.Lett. 42, 122 (1979); J.Appl.Phys.50, 1970 (1979).
4. P.Bak and J. von Boehm, Phys.Rev. B21, 5297 (1980).
5. P. Bak and J. von Boehm, Bull.Am.Phys.Soc. 24, 429 (1978).
6. J. Villain and M.Gordon, J.Phys.C 13,3117 (1980).
7. M.E. Fisher and W. Selke, Phys.Rev.Lett. 44, 1502 (1980).
8. E.B. Rasmussen, Thesis, University of Aarhus, Denmark, and to be published.
9. S. Aubry, "The New Concept of Transitions by Breaking of Analyticity in a Crystallographic Model", in Solitons and Condensed Matter Physics, ed. by A.R. Bishop and T. Schneider, Springer Series in Solid-State Sciences, Vol..8 (Springer, Berlin, Heidelberg, New York 1978) p.264.
10. B. Mandelbrot, Fractals: Form, change, and Dimension (Freeman, San Francisco, 1970).
11. P. Fisher, B. Lebech, G. Meier, B.D. Rainford, and O. Vogt, J.Phys.C11, 345 (1978).
12. J. Rossat-Mignod, P. Burlet, H. Bartholin, J. Villain, T.S. Wang, and O. Vogt, Phys.Rev. B16, 446 (1977).
13. J.M. Kosterlitz and D.J. Thouless, J.Phys.C6, 1181 (1973); J.M. Kosterlitz, J.Phys.C7, 1046 (1974).
14. R.M. Hornreich, R. Liebmann, H.G. Schuster, and W. Selke, Z.Phys.B35, 91 (1979); W. Selke and M.E. Fisher, Z.Phys.B, to be published.
15. W. Selke, Z.Phys.B29, 1133 (1978).
16. W. Selke and M.E. Fisher, Phys.Rev.B20, 257 (1979).
17. P.Bak in "Proceedings of the 1980 Conference of the Condensed Matter Division of the European Physical Society, Antwerp April 9-11 (1980)."
18. W.L. McMillan, Phys.Rev.B14, 1496 (1976).
19. P.Bak and V.J. Emery, Phys.Rev.Lett. 36, 978 (1976).
20. V.L. Pokrovsky, Solid State Commun. 26, 77 (1978).
21. V.L. Pokrovsky, private communications.
22. A. Aharory and P. Bak, to be published.
23. D. Mukamel and S. Krinsky, Phys.Rev.B13, 5065 (1976); P. Bak, Phys.Rev. B.13 5086 (1976).
24. J. José, L.M. Kadanoff, S. Kirkpatrick, and D. Nelson, Phys.Rev.B17, 1477 (1978).
25. B.I. Halperin and D. Nelson, Phys.Rev.Lett. 41, 121 (1978).
26. V.L. Pokrovsky and A.L. Talapov, ZhETF (1980).
27. J. Hubbard, Phys.Rev.Lett. 3, 77 (1959).
28. P.F. Dashen, B. Hasslacher, and A. Neveu, Phys.Rev.D10, 4114, 4130 (1974).
29. E. Müller-Hartmann and J. Zittartz, Z.Phys. B27, 261 (1977).
30. P. Bak and J. Villain, to be published.
31. P. Bak and H. Fukuyama , Phys.Rev. B21, 3287 (1979).

Part III

Magnetic Chains

Quantum Spin Chains

Jill C. Bonner, Physics Department, University of Rhode Island
Kingston, RI 02881, USA

H.W.J. Blöte, Laboratorium voor Technische Natuurkunde
Technische Hogeschool Delft, Delft, Netherlands

Hans Beck, Institut de Physique de l'Université de Neuchâtel
CH-2000 Neuchâtel, Switzerland

Gerhard Müller, Institut für Physik der Universität Basel
CH-4056 Basel, Switzerland

Introduction

Serious scientific interest in one-dimensional (1-D) physics arose in the early 1960's. This interest was stimulated by exact as well as accurate numerical solutions to a variety of quantum spin chain problems [1]. The potential relevance of such solutions to real experimental systems was first demonstrated by Griffiths [2] in conjunction with workers at the Kamerlingh Onnes Laboratorium, Leiden. Theory and experiment were shown to be in excellent agreement for a naturally quasi-1-D Heisenberg spin 1/2 antiferromagnet, copper tetrammine sulphate [$Cu(NH_3)_4SO_4 \cdot H_2O$]. Further stimulus to the new field of quasi-1-D magnetism was provided by an annotated collection of reprinted papers on a variety of 1-D model systems, including lattice gases, dynamical disordered crystal lattices, many-fermion gases (electron gases) as well as magnets. The collection appeared in book form, and remains today an important introduction to 1-D theory [3].

Interest in quantum spin chains has developed in surges. Initial enthusiasts were primarily experimental physicists who attempted to identify naturally occurring quasi-1-D magnetic insulators or who performed "magnetic engineering" by substituting large non-magnetic organic "spacer" molecular complexes like pyridines or pyrazines in place of, say, water molecules. The chains of magnetic ions were thus forced further apart and the 1-D characteristics of the substance much enhanced. The goal was to provide experimental realizations of a variety of spin chain models. As the zoo of experimental animals neared completion, interest among physicists started to wane, but compensation was provided by growing interest among coordination chemists. Since the theory of superexchange remains on a far from secure footing, chemists are tending to investigate empirically the variation of exchange constant with bond length and bond angle (and possibly other factors like ligand orientation). They make use of homologous families of 1-D magnets, relying on the complete and accurate statistical mechanical knowledge not available in higher-D systems. One important source of uncertainty in the processing of the data is thus removed.

In the early 1970's a new class of naturally anisotropic, quasi-1-D systems came into vogue. These are organic conductors, semi-conductors, and even insulators [4]. Materials like the famous TTF-TCNQ are called organic metals because of their high conductivity at low temperature with its negative temperature coefficient. Considerable excitement and also some controversy was generated by a prediction by Little [5] that organic conductors might show "high temperature" superconductivity at, say, 50 K. Despite several years of intense effort, the Little prediction has not come close to being realised. However, research is continuing with renewed stimulation by the recent discovery of a new member of the class, which goes superconducting at 12 kbar pressure with a transition at 0.9 K [6]. What interests us here is that even the conducting members of this large class of quasi-1-D organics can be described by 1-D models directly related to quantum spin chains.

It appears, however, that currently the greatest interest in quantum spin chains arises from their intimate relation to non-trivial lattice gauge field theories having one space and one time dimension. The relevant field theory solutions or excitations commonly have soliton character and are of high interest to particle physicists.

Several reviews of the state of the art in 1-D physics have appeared rather recently. Early analytical work beginning in the 1930's has been reviewed by THOMPSON [7] and more recent theoretical developments by BONNER [1], and theory with emphasis on spin dynamics by STEINER, VILLAIN and WINDSOR [8] and BIRGENEAU and SHIRANE [9]. Other reviews with strong emphasis on dynamics include articles by RICHARDS [10] and HONE and RICHARDS [11]. A series of reviews by DE JONGH features quasi-1-D theory and experiment [12]. Quasi-1-D experimental systems from the viewpoint of a chemist are reviewed by CARLIN [13]. Finally, a recent article by KOGUT [14] gives a clear and detailed discussion of the relation of quantum spin chains and related models to lattice field theories.

The focus of this paper will be several very recent developments in the theory of spin chains, including both static and dynamical properties. Random 1-D systems are not considered. The striking feature of recent work of non-random systems has been the importance of quantum effects, and classical models will be discussed only in relation to differences in behavior between them and their quantum counterparts. Specifically, we shall concentrate on (1) the current analytic status of spin chains, (2) 1-D models exactly solvable by the famous Bethe Ansatz and also mappings between models which give several solutions for the price of one, (3) the current status of numerical studies on spin chains, (4) alternating (dimerized) chains, relevant to organic conductors and insulators, (5) a new quantum renormalization group method, and (6) a novel approach to quantum spin dynamics.

Spin Chains: Current Analytic Status

The effective spin Hamiltonian for the general quantum spin chain may be written:

$$\mathcal{H} = -2J \Sigma_{i=1}^{N} \{a s_i^z s_{i+1}^z + b s_i^x s_{i+1}^x + c s_i^y s_{i+1}^y\}. \qquad (1)$$

We have bilinear spin coupling, assume nearest-neighbor spin interactions only, and the effective exchange constant is J, modified for the spin couplings in different directions by factors a, b and c. The Ising model is obtained by putting b = c = 0 (in general, by letting any two of a,b,c equal zero). The XY model is obtained by putting a = 0 (in general by letting any one of a,b,c equal zero). If a = b = c, we have complete rotational symmetry in spin space and obtain the Heisenberg model which may, of course, be written in vector form

$$\mathcal{H} = -2J \Sigma_{i=1}^{N} \vec{s}_i \cdot \vec{s}_{i+1}. \qquad (2)$$

If J>0, the spins may lower their energy by aligning parallel and we have a Heisenberg ferromagnet. If J<0, the antiparallel spin arrangement is favored and we have an antiferromagnet. If a, b and c are not equal we have spin anisotropy. If $a > b \simeq c$, we have uniaxial (easy axis) anisotropy. On the other hand, if $a < b \simeq c$, easy plane anisotropy results. The effects of an applied magnetic field can be represented by the addition of a general Zeeman term $-g\beta\vec{H}\cdot\Sigma_{i=1}^{N}\vec{s}_i$ to Hamiltonian (1). Finally, the spin value may range over $1/2 \leq s \leq \infty$.

The static properties commonly measured by experimentalists are the initial (zero field) susceptibility χ, the specific heat at constant field C_H, and the magnetization isotherms as a function of applied field, M_T vs H. In dynamics the Fourier transforms of the spin-spin correlation functions are important. The 1-D spin-1/2 Ising model was first solved in 1925 (ISING [15]). A complete description including correlation function properties may be found in THOMPSON [7]. The 1-D spin-1/2 XY model was solved analytically in the early 1960's, independently by LIEB, SCHULTZ and MATTIS [16] with emphasis on correlation functions (see also later work by PFEUTY[17]), and by KATSURA [18] with emphasis on the thermal and magnetic properties. KATSURA [18] and PFEUTY [17] also obtained exact results for the transverse Ising model. This interesting model results from taking $b = c = 0$ in (1) and retaining, say, only the X component of the Zeeman term. Since spin-spin interaction and applied field are mutually perpendicular, this Ising model has a quantum character and is often described as the simplest quantum mechanical model. The case of general a, b and c in (1) is usually called the Baxter XYZ model after BAXTER [19] who pioneered the analytic attack. These various solutions illustrate the special character of 1-D vs 3-D systems. Major features are the presence of characteristic rounded maxima in the specific heat and the perpendicular or antiferromagnetic parallel susceptibility. Such effects are attributed to short-range order since these systems show critical singularities only at $T = 0$. It should be noted, however, that if the interaction range is sufficiently increased, even 1-D model systems have critical points such that $T_c > 0$ [20].

Let us now review the analytic status of the general, spin-1/2 Ising-Heisenberg-XY linear chain with symmetry ($b = c$, say), often called the XXZ model. The mathematical analysis required for this general model is more complicated than that required to solve the 2-D Ising model (ONSAGER [21]). Since the 2-D Ising model solution has been so fruitful in the development of the modern theory of critical phenomena, it would seem profitable to pursue the 1-D quantum chain solution in equivalent detail. A breakthrough by YANG and YANG about 1966 [22] resulted in a detailed solution as a function of anisotropy and magnetic field for $T = 0$. However, this was sufficient to determine the character of the interesting critical singularities. The finite temperature properties presented a greater problem. Progress has been made by JOHNSON and McCOY [23] and JOHNSON [24], and independently by TAKAHASHI [25], on the basis of a formulation due to GAUDIN [26]. Nevertheless, information is still lacking, most notably for the Heisenberg ferro- and anti-ferromagnetic limits, and the correlation functions in general.

A parameter range of the XXZ model which has just yielded to analytic attack is the Ising-Heisenberg ferromagnet in zero and very small magnetic field. This corresponds to $a > b = c$ and $J > 0$ in (1). For analytic work, (1) is more conveniently written in the form

$$\mathcal{H} = -\Sigma_{i=1}^{N}\{s_i^x s_{i+1}^x + s_i^y s_{i+1}^y + \Delta(s_i^z s_{i+1}^z - 1/4)\} - H\Sigma_{i=1}^{N} s_i^z. \qquad (3)$$

Interesting and complex results have been obtained for the elementary excitations and low-temperature thermodynamics [27]. These results are based on the Bethe Ansatz-Gaudin formalism, and reveal complex cross-over behavior (a) as a function of spin anisotropy Δ and (b) as a function of magnetic field H. Fig. 1 shows the $H = 0$ excitations as a function of the anisotropy γ (the inverse of Δ). Near the Ising limit the thermal properties, e.g. specific heat,

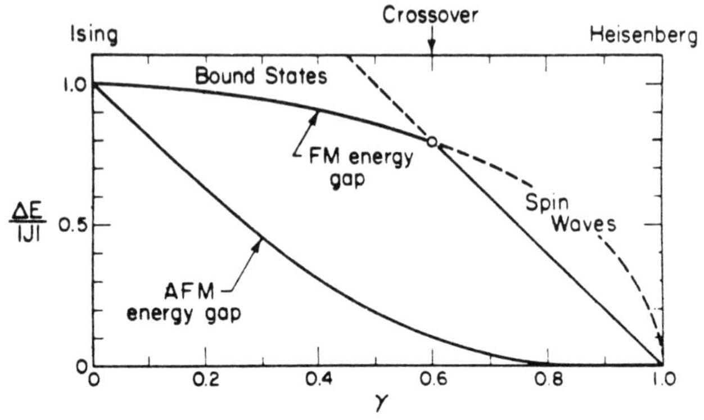

Fig.1 Cross-over behavior in the thermal excitations of the Ising-Heisenberg ferromagnet.

are dominated by the bound state excitation curve given by $E_B=\frac{1}{2}(\Delta^2-1)^{\frac{1}{2}}$ [i.e. $E_B/J=(1-\gamma^2)^{\frac{1}{2}}$]. As the Heisenberg limit is approached, a cross-over occurs at $\gamma=0.6$ ($\Delta=5/3$) and thereafter the specific heat is dominated by a different class of excitations, unbound or spin wave in type, given by $E_{SW}=(\Delta-1)$ [i.e. $E_{SW}/J=2(1-\gamma)$]. Interestingly, in the case of the magnetic excitations, i.e. susceptibility, domination by the bound states occurs over the whole anisotropy range from Ising to Heisenberg limits.

The situation is even more complicated when a field is applied. For sufficently large fields, both χ and C_H are spin-wave dominated. For smaller fields (χ), and smaller fields with $\Delta>5/3$ (C_H), there are two bound state dominated regions, between which cross-over occurs. These results are potentially interesting in connection with the quantum soliton problem, and to field theorists. Unfortunately, the important Heisenberg limit seems a more formidable problem, not tractable by this method. Pioneering work is underway on two very different approaches which may be powerful enough to treat the Heisenberg limit (and find correlation functions). They are (a) the quantum inverse scattering method [28] and (b) a new and very abstruse approach pioneered by the Japanese [29].

Finally, mention should be made of a new approach which may be regarded either as an approximate treatment of a spin-1/2 1-D lattice or an exact treatment of a spin-1/2 1-D continuum model [30]. It is based on techniques for solving the field theoretic Tomonaga-Luttinger models [31]. The approach is now termed "bosonization," and though quantitative accuracy is hard to assess, the general validity of the method has been established [32].

Bethe Ansatz Models and Mappings

The Bethe Ansatz was formulated in 1931 by BETHE [33] as a basic ingredient of the eigenfunctions of the Heisenberg spin 1/2 linear chain. The analytic results above for the general anisotropic and XYZ chains are all based on the Bethe Ansatz approach. Rather recently it was realized that the (generalized) Ansatz was the key to the analytic solution of a wider class of models than the quantum spin chain. Exact solutions came thick and fast for systems such as Fermi and Bose gas delta function models [34], the linear Hubbard model [35], 1-D plasma which crystallizes as a Wigner solid [36], and the Lai-Sutherland model [37] which includes the Hubbard model and a dilute magnetic model as special cases. Very recent work has mapped the Kondo problem into a 1-D Bethe Ansatz model whose solution is closely related to the problem of the Heisenberg spin chain [38].

A striking feature of current research is the large number of mappings or mathematical isomorphisms which have been discovered between the spin-1/2 quantum chain, including its various limiting cases, and a wide assortment of 1-D and higher-D models. Specifically, the mapping between 1-D quantum and 2-D classical models proceeds via the fact that the Hamiltonian matrix of the one class commutes with the transfer matrix of the other class (see XYZ proof by SUTHERLAND [38]). The various mappings have already been systematically reviewed in the context of soliton theory [40] and will simply be listed here. The general magnetic model of Hamiltonian (1) can be mapped into the following 2-D classical models: classes of ferro- and antiferroelectric models; classical Coulomb gas; 2-D Ising model; surface roughening model; the dimer problem [41]; lattice theories of polymer melting [41]; the percolation problem [42]; the coloring problem [42]; and the Potts model [43], under suitable special parameter choices and relevant parameter ranges. Hamiltonian (1) may also be mapped into the following 1-D models: the 1-D spinless fermion gas with or without interactions; the 1-D quantum many-fermion Luttinger and Tomonaga models; generalised Hubbard-type models for organic conductors with hopping and on-site repulsive or attractive terms [44] [45], including the Hubbard dimer gas [46]. Note the Kondo problem solution represents essentially the first 1-D to 3-D mapping.

Finally we note that the mapping between 2-D classical vertex models and 1-D quantum models which can take place at zero or non-zero [47,48] temperature offers a radically new approach to the quantum chain problem. Unlike quantum systems, classical systems can be directly simulated on a computer and then the results can be transcribed to the quantum equivalents [49].

Quantum Spin Chains: Numerical Studies

The Bethe Ansatz approach to uniform spin chain problems applies only to quantum models with completely integrable Hamiltonians. This includes the spin-1/2 Heisenberg and XY models. It does not appear to be applicable to models with $S > 1/2$. Even for spin 1/2, the Bethe Ansatz approach cannot be applied to the Heisenberg alternating chain, a nearest-neighbor exchange model where there are two unequal exchange constants which alternate along the chain. Since the alternating chain is important to the theory of spin-Peierls transitions in quasi-1-D organics, and there is experimental interest in effective spin-1 systems (containing the Ni^{2+} ion), spin-3/2 systems (Cr^{3+} ion) and spin-5/2 systems (Mn^{2+} ion), reliable theoretical calculations are needed. Since an analytic approach is not feasible at this time, we must have recourse to a numerical BONNER-FISHER approach [50]. This involves calculating the properties exactly of a sequence of small finite spin chains, with both periodic and free-end boundary conditions, and steadily increasing system size N. Extrapolations are then made to the thermodynamic limit. The largest system in the sequence and hence number of systems available for extrapolation, is spin dependent, since the calculation involves symmetry reduction and subsequent diagonalization of Hamiltonian matrices of dimension $(2s+1)^N x (2s+1)^N$. However, BLÖTE has been able to perform extrapolations for spin values s = 1/2, 1, 3/2, 2, up to 5/2, for ground state energies [51], specific heat [52], susceptibility, and T = 0 magnetization isotherms vs field [53]. The variation of specific heat (see [52]) and susceptibility (shown in Fig.2) with spin is not very striking, except in one important respect. The $s=\infty$ curves all show non-physical features at low temperatures. For example, the classical Heisenberg specific heat [54]

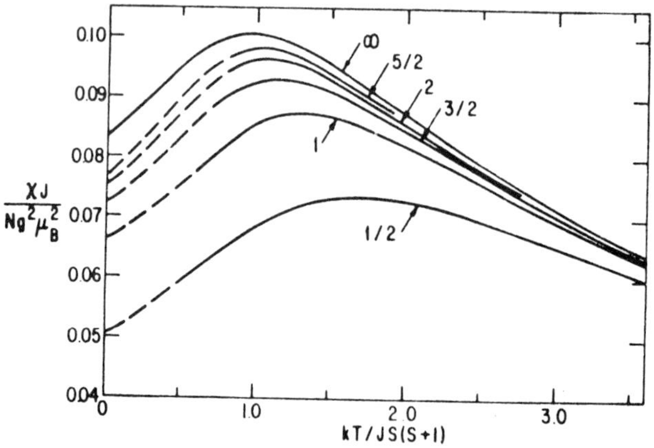

Fig.2 Extrapolated finite chain susceptibilities for Heisenberg antiferromagnets showing dependence on spin.

rises monotonically to a non-zero value as $T \to 0$, instead of vanishing in accordance with the third law of thermodynamics, as do the curves for $s < \infty$.

The accuracy of this extrapolation method, for the optimal case of spin 1/2, as estimated by comparison with subsequent analytically based calculations, is discussed in [1]. Ref. [1] also includes some detailed comparisons of the various quantum spin models.

Alternating Chains

Alternating spin chains are something of a novelty to physicists. Originally they were studied from a chemical viewpoint, to explain the properties of certain organic free radicals [55]. From such studies evolved spin-exciton theory [56]. Some recent applications have been to copper nitrate [57,58] and copper bromide imidazole [59]. Perhaps the most exciting application has been to members of a family of organometallic complexes, insulating relatives of the organic conductor TTF-TCNQ. These are the TTF bisdithiolenes, denoted TTF·BDT(M), where M = Cu or Au, the first experimental systems to display a spin-Peierls transition [60]. This is a magnetoelastic rather than purely magnetic phase transition. In zero field an assembly of Heisenberg, spin-1/2, uniform antiferromagnetic chains undergoes a transition as the temperature is lowered, as a result of lattice distortion. Below the transition the system is described as an assembly of alternating chains, where the degree of alternation is a function of temperature. The Hamiltonian for an alternating Heisenberg chain may be written:

$$\mathcal{H} = -2J\sum_{i=1}^{N/2}\{\vec{S}_{2i-1}\cdot\vec{S}_{2i} + \alpha\vec{S}_{2i}\cdot\vec{S}_{2i+1}\}, \qquad (4)$$

where α is a convenient alternation parameter, i.e. the ratio of interaction strengths of the two exchange constants. When $\alpha=0$, the alternating chain reduces to a non-interacting assembly of spin-pairs or dimers (dimer limit). When $\alpha=1$, we recover the familiar uniform Heisenberg chain of (2).

In the absence of some generalized Ansatz serving as a basis for analytical attack, this model has been treated theoretically by a variety of approximate techniques. Two of the more successful methods are the BULAEVSKII Hartree-Fock approach [61] subsequently incorporated into spin-Peierls theory by PYTTE [62] and BRAY ET AL. [63]; and an alternating variant of the LUTHER-PESCHEL-

LUTTINGER continuum model [30], employed in the spin-Peierls work of CROSS and FISHER [64]. There are differences between them, and the Luttinger approach, which is not a mean-field approach, is claimed to have greater accuracy. However, there is at present no available experimental means of distinguishing between the two models. Hence an extensive numerical finite chain study was undertaken on periodic alternating chains of up to 12 spins [65] in an attempt to extend earlier studies by DUFFY and BARR [55]. Attention was paid both to spectral excitations and thermodynamic properties. A comparative study of extrapolated susceptibilities over the entire α range vs calculations based on the Bulaevskii paper [66] has already appeared. The behavior of the T=0 magnetization isotherms as a function of α is shown in Fig.3. In general, we remark that in comparison with the more exact calculations the mean-field Bulaevskii theory is surprisingly successful. No spurious T>0 phase transition appears for antiferromagnetic systems. Thermodynamic predictions are qualitatively correct but quantitatively in error by roughly 15% over the whole alternation range. No similar direct comparisons can be made with the CROSS-FISHER theory [64], since at present a Luttinger-type approach cannot predict amplitudes, which are cut-off dependent and therefore arbitrary.

The numerical calculations strongly suggest that the presence of alternation results in the appearance of an energy gap between the non-degenerate singlet ground state and the lowest excited states (triplet excitons), i.e. the gap vanishes only in the uniform limit, α=1. Fig. 4 shows finite N curves

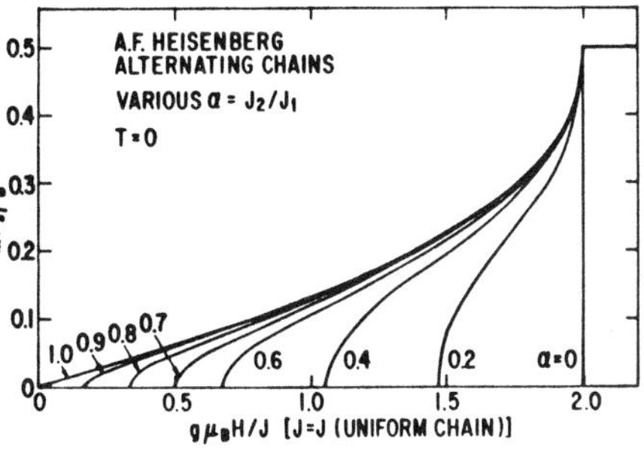

Fig.3 Zero-temperature magnetization isotherms for alternating Heisenberg linear antiferromagnet.

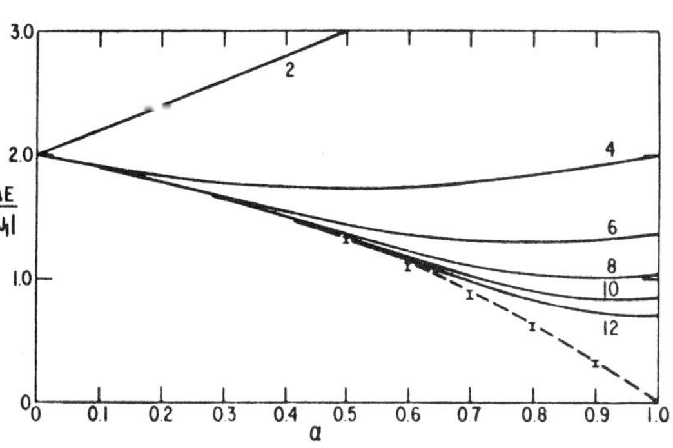

Fig.4 Finite chain extrapolation for energy gap of alternating linear Heisenberg antiferromagnet.

for alternating chains of 2 through 12 spins, together with the extrapolated result, and results of BULAEVSKII [61] for comparison. The existence of such a gap is crucial to spin-Peierls theory and also spin exciton theory in physical chemistry. However, since the alternating linear chain still possesses full isotropic spin symmetry, doubts have been raised on the basis of intuition derived essentially from classical (spin-∞) 1-D systems, about the existence of such a gap [67]. It is interesting that the XY alternating chain, which can be solved exactly for both s=1/2 and s=∞, shows a gap in the quantum case and no gap in the classical case, for all non-zero alternation. Numerical evidence is very strong that the same situation occurs for the Heisenberg antiferromagnetic alternating chain.

Quantum Renormalization Group Method

The real space renormalization group (RG) method pioneered by NIEMEIJER and VAN LEEUWEN [68] has been very successfully applied to 1-D [69] and 2-D [70] Ising models. However, a successful real space RG method for quantum systems has been slow in coming, essentially because of the familiar problem of non-commuting operators. A new method devised recently by field theorists [71] and shortly thereafter applied to a type of Kondo problem [72], offers hope. The class of quantum spin models provides obvious test cases for the new method, which will be reviewed in this light. We note again that the lack of exact solutions for uniform chains with $1/2 < s < \infty$ and alternating chains with $s < \infty$ makes reliable approximate methods valuable.

The spirit of the method is as follows. The lattice is sub-divided into blocks of N_S sites such that the eigenvalues and eigenvectors of each block may be calculated exactly. The basis of each block (2^{N_s} levels, s = 1/2) is truncated to some number N_L of levels, and the coupling between the blocks is written within the truncated basis. If the temperature region of interest is T = 0, where the critical singularities are located for 1-D, short-range, quantum models, then the truncated basis need only contain the ground state and dominant set of first excited states. If, therefore, both N_S and N_L are small, analytic RG recursion relations can be formulated and solved. In general, however, use of a computer is required. This is true if information at higher temperatures is needed. High temperature calculations may use directly all the block levels instead of just the lowest [73], or may use an approximate technique which preserves the important information contained in the higher levels at each recursion step [74].

The new method has been exhaustively tested on the simplest quantum model, the transverse Ising model, with encouraging results [71,75]. A more stringent test is posed by the more complex quantum models like the 1-D alternating Heisenberg antiferromagnet and the 1-D Ising-Heisenberg spin-anisotropic antiferromagnet (Eq.(1) with $a > b = c$ and $J < 0$). Comparison can be made with finite chain calculations and, in the case of the Ising-Heisenberg chain, with exact analytic results [77,76,23]. In Fig. 5a, we see again the extrapolated finite chain energy gap ΔE for Heisenberg antiferromagnetic alternating chains together with the exact result for N = 12. In comparison we see the energy gap predictions for two types of RG calculation. One curve corresponds to N_S=9 (odd) with N_L=2, giving an unstable fixed point at $\alpha^*=1$ (uniform limit) and a stable fixed point at $\alpha^*=0$ (dimer limit), consistent with an energy gap for all $\alpha < 1$, in agreement with numerical predictions. However, the value of the gap is quantitatively unreliable near the dimer limit (where finite chain convergence is excellent). This is attributable to the 2-level truncation which does not utilize the singlet-triplet character of the low-lying levels. In comparison we show an N_S=4 (even), 4-level truncation RG which does preserve the singlet-triplet symmetry. This RG is in very good agreement with the extrapolated curve out to $\alpha \simeq 0.6$, but predicts a gap vanishing at $\alpha = 0.962$ rather than unity. This must be interpreted as resulting from automatic failure of an important symmetry ($\alpha \Leftrightarrow \alpha^{-1}$) whenever N_S is even. In Fig. 5b, comparable results for the Ising-Heisenberg antiferromagnet with spin anisotropy parameter γ=b/a are shown. Crosses show the energy gap obtained by extrapol-

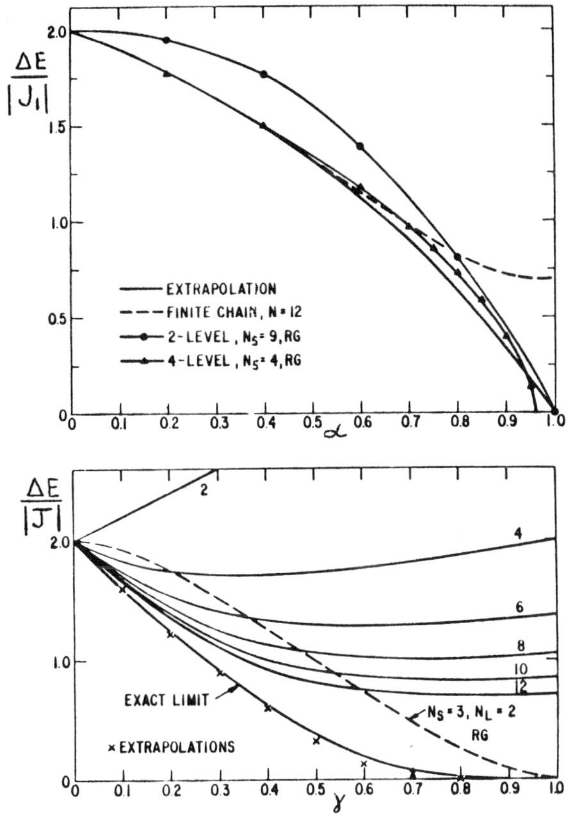

(a)

(b)

Fig.5 (a) N_S odd and even RG calculations in comparison with extrapolated result for alternation energy gap. (b) Finite chain extrapolations and RG calculation for Ising-Heisenberg anisotropy energy gap compared with exact result.

ation of the finite chain sequence N= 4, 6, 8, 10, 12. The extrapolations predicted that the gap vanished only at $\gamma=1$ [50], a prediction confirmed by subsequent analytic calculations [76,23] which gave a gap (solid curve) vanishing exponentially as $\gamma\to 0$. An N_S = 3, N_L = 2 RG result (dashed) is characterized by fixed points $\gamma^*=1$ (unstable) and $\gamma^*=0$ (stable), and hence vanishes at $\gamma=1$. However the curve is clearly in poor agreement with the exact result. Finite temperature properties using this method have been obtained by JOSÉ and HIRSCH [74] in the context of a study of random anisotropic antiferromagnet chains. The accuracy of the method was tested for the limiting case of the uniform Heisenberg chain by comparison with numerical extrapolations [50]. For both specific heat and susceptibility, agreement was fair over most of the temperature range (i.e. to within 15-20%). There is a more serious problem at very low temperatures. It appears that the fundamental structure of these RG's may lead to non-physical behavior. Specifically, the susceptibility vanishes instead of going to a finite limiting value known exactly [78] and the specific heat vanishes with zero slope instead of linearly [23,50].

Method	Ground State Energy	Energy Gap
Hartree-Fock[69]	$\xi_0 \sim \delta^2 \ell n^2 \delta$	$\Delta E \sim \delta \ell n \delta$
Cross-Fisher[72]	$\xi_0 \sim \delta^{4/3}$	$\Delta E \sim \delta^{2/3}$
N_{odd}RG [77]	$\xi_0 \sim \delta^{1.5}$	$\Delta E \sim \delta^{.76}$
N_{even}RG [77]	$\xi_0 \sim \delta^{1.6}$	$\Delta E \sim \delta^{.64}$
Finite Chains[80]	$\xi_0 \sim \delta^{3/2}$	$\Delta E \sim \delta^{3/4}$

Table 1 Comparison of spin-Peierls exponents

Quantities which can be regarded as critical singularities for the spin-Peierls problem are (a) the value of deviation of the ground state energy per spin from its exactly known [79] uniform value with the onset of dimerization (call this ξ_0) and (b) quantitatively how the energy gap between ground and first excited states (ΔE) behaves near the uniform limit as a function of alternation. Predictions of the various approximate methods are collected in Table 1. In order to obtain analytic results, the finite chain extrapolations have been augmented by RG based finite size scaling techniques [80]. A point of interest is the difference between the analytic finite chain exponents and the analytic Cross-Fisher exponents which are claimed to be exact [64].

Quantum Spin Dynamics

Spin dynamical calculations on 1-D systems have relied heavily on classical ($s = \infty$) theory [9,10,11] despite abundant evidence that quantum effects can be extremely important at low temperatures. A new approach to the spin dynamics of the 1-D isotropic S=1/2 Heisenberg antiferromagnet [81-83] does not involve the many-body techniques usually employed, but instead is based on analytic calculations of excitation energies and densities of states combined with finite chain calculations of matrix elements. Specifically the results are obtained on the basis of four main techniques: (1) analytic calculations of a special class of Bethe-Ansatz states; (2) exact finite chain calculations for chains of 4 through 10 spins; (3) derivation and application of two different kinds of selection rules; and (4) the use of various kinds of sum rules. Major emphasis has been put on zero-temperature studies in zero and non-zero field. Finite chain calculations have been used to extend the dynamical study to finite temperatures.

In zero field an approximate analytic expression is available for the dynamical correlation function which fits excellently into the fragmentary picture of the few available exact results. This expression has then been exploited to yield a variety of further results of theoretical and experimental interest. The T=0 expression is governed by a two-parameter continuum of spin-wave-type triplet (s=1) excitations. Explicitly,

$$S_{\mu\mu}(q,\omega) = \frac{1}{[\omega^2-(\frac{1}{2}\pi J)^2 \sin^2 q]} \cdot (\omega - \tfrac{1}{2}\pi J \sin q) \cdot (\pi J \sin(\tfrac{1}{2}q) - \omega). \tag{5}$$

This expression "switches on" at the lower continuum boundary given by the formula $E_1(q) = \frac{1}{2}\pi J|\sin q|$ (first derived by DES CLOIZEAUX and PEARSON [84]), where there is a divergence. A tail of decreasing spectral weight extends to the upper continuum boundary, $E_2(q) = \pi J|\sin(q/2)|$, where there is a cut-off, dependent on the value of the constant A. This expression is not rigorous, but exact sum rules are violated by only a small amount. Expression (5) predicts increasing asymmetry in the spectral weight distribution as $q \to \pi$, a purely quantum effect, which has recently been verified by neutron scattering data [85]. The result for the integrated intensity derived from (5) is in much better agreement with neutron scattering data on the linear chain antiferromagnet cobalt chloride dipyridine (CPC) [86] than the corresponding semiclassical result.

In Fig.6 we show an exact calculation of $S_{zz}(q,\omega)$ for a cyclic chain of N = 10 spins. For each q, $S_{zz}(q,\omega)$ is a collection of δ functions. The

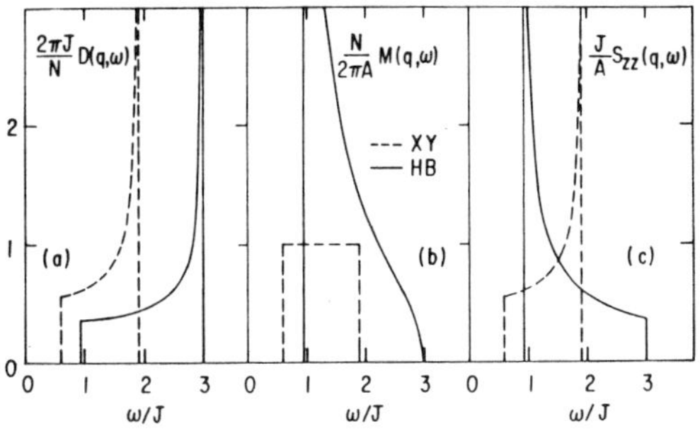

Fig.6 Exact calculation of $S_{zz}(q,\omega)$ for a chain of N=10 spins in zero field.

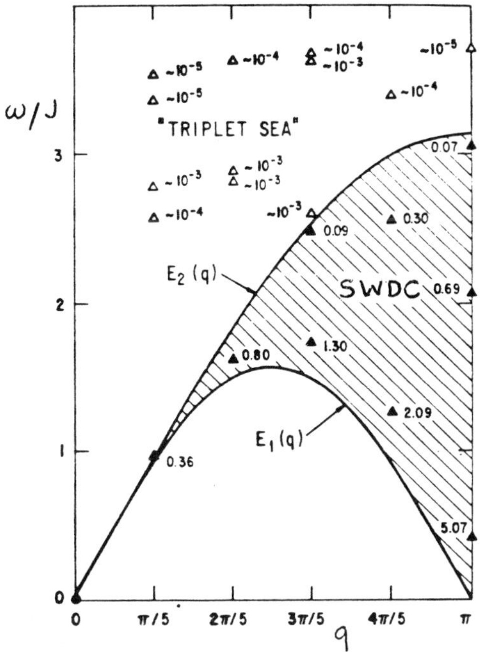

Fig.7 Comparison of T=0 density of state functions, matrix element functions, and dynamical correlation functions for Heisenberg vs XY models.

triangles denote energy and wave number of the triplet excitations, and the numbers represent the corresponding spectral weight. No type of excitation other than triplet carries any spectral weight. We may note the following: (a) almost all the spectral weight is concentrated within the boundaries of the spin-wave double continuum (SWDC); (b) the matrix elements increase in magnitude as the energy decreases to the lower boundary $E_1(q)$ (and also increase in magnitude as $q \to \pi$); and (c) there is some very small but finite weight above $E_2(q)$. Exact sum rule arguments of HOHENBERG and BRINKMAN [87] and finite chain calculations [88] indicate that this weight should persist in the thermodynamic limit. Some interesting information is available on states of this type near the critical field, which calls for further investigation of their nature for all fields. The value of the constant A may be approximately determined from sum rules. Three simple sum rules link $S(q,\omega)$ to the static susceptibility, the ground state energy, and the integrated intensity. The resulting values for A differ among themselves by about 20% with a mean of $A \sim 1.18$, reflecting failure to take account of states of the "triplet sea."

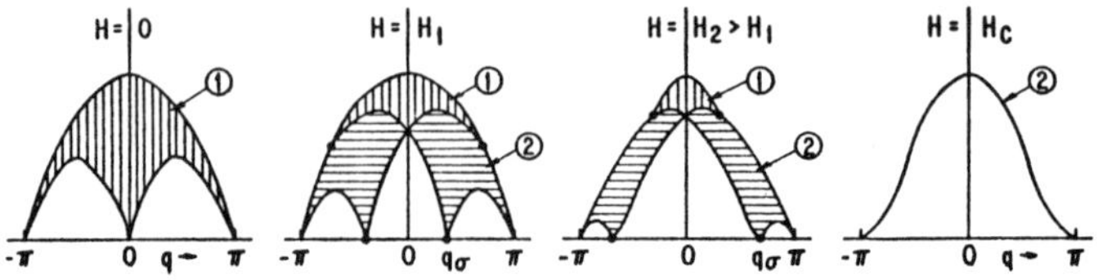

Fig.8 Behavior of the two continua dominating $S_{zz}(q,\omega)$ as a function of magnetic field.

Fig. 7 shows an interesting comparison between Heisenberg and XY models at T=0 for density of states $D(q,\omega)$; matrix element function $M(q,\omega)$; and dynamical correlation function $S_{zz}(q,\omega)$. For both models $D(q,\omega)$ diverges at the upper SWDC boundary. The XY matrix elements are constant over the SWDC, but diverge strongly at the lower boundary for the Heisenberg model. Hence we have the interesting difference between the models that $S(q,\omega)$ diverges at the lower boundary for the Heisenberg model but at the upper boundary for the XY model.

For non-zero field there are striking differences between the longitudinal and the transverse correlations, both involving different continua of excitations. With increasing field, the longitudinal fluctuations are more and more suppressed, whereas the transverse fluctuations become more important. Exact selection rules show that for H=0 and finite N, six different SWDC contribute to the dynamics, two for the longitudinal and four for the transverse correlations, giving rise to considerable complexity. In Fig. 8 the behavior of the two continua for $S_{zz}(q,\omega)$ is shown as a function of H. As H increases, the importance of continuum 1 decreases. It is most interesting that in the thermodynamic limit it may be proved that additional selection rules come into play, reducing the number of contributing continua from 6 to 3. In Fig. 8, continuum 1 loses all weight for $H > 0$ in the limit leaving only continuum 2. In zero field, however, continuum 1 continues to dominate exclusively!

The theory for $H > 0$ predicts multiple structures in the scattering intensity, various features of $S_{zz}(q,\omega)$ and $S_{xx}(q,\omega)$ which are simply related to the magnetic field, e.g. soft mode locations and energy relations between scattering peaks, and quite different behavior for the corresponding integrated intensities, $I_{zz}(q)$ and $I_{xx}(q)$. All these features are susceptible to future experimental observation.

References

1. J.C. Bonner, J. Appl. Phys. 49, 1299 (1978)
2. R.B. Griffiths, Phys. Rev. 135, A659 (1964)
3. E.H. Lieb and D.C. Mattis, Mathematical Physics in One Dimension (Academic Press, N.Y.) 1966
4. See Low Dimensional Cooperative Phenomena, Ed. H.J. Keller, NATO A.S.I. (Plenum Press, N.Y.) 1975
5. W.A. Little, Phys. Rev. 134, A1416 (1964)
6. D. Jérome, A. Mazaud, M. Ribault and K. Bechgaard, J. Phys. (Paris) Lett. 41, L95 (1980)
7. C.J. Thompson, Phase Transitions and Critical Phenomena, Vol. 1, Eds. C. Domb and M.S. Green (Academic Press, London) 1972, Chap. 5
8. M. Steiner, J. Villain and C.G. Windsor, Adv. Phys. 25, 87 (1976)
9. R.J. Birgeneau and G. Shirane, Physics Today 31, No. 12, 32 (1978)
10. P.M. Richards, Proc. Int. School of Physics Enrico Fermi, Ed. K.A. Muller (North Holland, Amsterdam) p. 539 (1973)
11. D.W. Hone and P.M. Richards, Ann. Rev. Mat. Science 4, 337 (1974)

12. L.J. de Jongh and A.R. Miedema, Adv. Phys. 23, 1 (1974). L.J. de Jongh, J. Appl. Phys. 49, 1305 (1978). L.J. de Jongh, Proc. European Physical Society Meeting, Antwerp, 1980 (Plenum Press, to be published)
13. R.L. Carlin and A.J. Van Duynevelt, Magnetic Properties of Transition Metal Compounds, Inorganic Chemistry Concepts, Vol.2, (Springer, Berlin, Heidelberg, New York 1977) Sect. VIB
14. J. Kogut, Rev. Mod. Phys. 51, 659 (1979)
15. E. Ising, Z. Physik 31, 253 (1925)
16. E. Lieb, T. Schultz and D. Mattis, Ann. Phys. (NY) 16, 407 (1961)
17. P. Pfeuty, Ann. Phys. 57, 79 (1970)
18. S. Katsura, Phys. Rev. 127, 1508 (1962)
19. R.J. Baxter, Phys. Rev. Lett. 26, 834 (1971); Ann Phys. (NY) 70, 323 (1972)
20. J.F. Nagle and J.C. Bonner, J. Phys. C3, 352 (1970). G.S. Joyce, Phys. Rev. 146, 349 (1966). M.E. Fisher, S. Ma and B. Nickel, Phys. Rev. Lett. 29, 917 (1972).
21. L. Onsager, Phys. Rev. 65, 117 (1944)
22. C.N. Yang and C.P. Yang, Phys. Rev. 147, 303 (1966); ibid. 150, 321,327 (1966); ibid. 151, 258 (1966).
23. J.D. Johnson and B.M. McCoy, Phys. Rev. A6, 1613 (1972)
24. J.D. Johnson, Phys. Rev. A9, 1743 (1974)
25. M. Takahashi, Prog. Theor. Phys. (Jpn) 50, 1519 (1973)
26. M. Gaudin, Phys. Rev. Lett. 26, 1301 (1971)
27. J.D. Johnson and J.C. Bonner, Phys. Rev. Lett. 44, 616 (1980)
28. L.D. Fadeev, Leningrad preprint, to be published. M. Fowler, to be published. H. Bergknoff and H.B. Thacker, Phys. Rev. Lett. 42, 135 (1979)
29. L.P. Kadanoff and M. Kohmoto, to be published. M. Sato, T. Miwa and M. Jimbo, Proc. Jpn. Acad. 53A, 147, 153, 183 (1977)
30. A. Luther and I. Peschel, Phys. Rev. B 12, 3908 (1975)
31. See commentary and paper in Ref. 3
32. M. Fowler, J. Phys. C 13, 1459 (1980)
33. H. Bethe, Z. Physik 71, 205 (1931)
34. M. Takahashi, Prog. Theor. Phys. (Jpn) 46, 1388 (1971); C.K. Lai, Phys. Rev. A 8, 2567 (1973)
35. M. Takahashi, Prog. Theor. Phys. 43, 1619 (1970); and ibid. 50, 1519 (1973)
36. B. Sutherland, Phys. Rev. Lett. 34, 1083 (1975); and ibid 35, 185 (1975)
37. B. Sutherland, Phys. Rev. B 12, 3795 (1975)
38. N. Andrei, Phys. Rev. Lett. 45, 379 (1980)
39. B. Sutherland, J. Math. Phys. 11, 3183 (1970)
40. A. Bishop, Physica Scripta, 20, 409 (1979)
41. J.F. Nagle and J.C. Bonner, Ann. Rev. Phys. Chem., 27, 291 (1976)
42. H.N.V. Temperley and E.H. Lieb, Proc. Roy. Soc. 80, 322 (1971)
43. M.J. Stephen and L. Mittag, Phys. Lett. 41A, 357 (1972)
44. V.J. Emery, Phys. Rev. B 14, 2989 (1976)
45. M. Fowler, Organic Conductors and Semiconductors, ed. by L. Pāl, G. Grüner, A. Janossy, and J. Sólyom, Lecture Notes in Physics, Vol. 65 (Springer, Berlin, Heidelberg, New York 1977)
46. M. Fowler and M.W. Puga, Phys. Rev. B 18, 421 (1978)
47. M. Suzuki, Progr. Theor. Phys. (Jpn) 56, 1454 (1976)
48. B.M. McCoy and T.T. Wu, Il Nuovo Cimento LVI B, 311 (1968)
49. M. Barma and B.S. Shastry, Phys. Rev. B 18, 3351 (1978); D.P. Landau, private communication
50. J.C. Bonner and M.E. Fisher, Phys. Rev. 135, A640 (1964)
51. H.W.J. Blöte, Physica 93B, 93 (1978)
52. H.W.J. Blöte, Physica 79B, 427 (1975). See Fig. 4.
53. J.J. Smit, L.J. de Jongh, D. Klerk and H.W.J. Blöte, Colloques Internes CNRS 242, Physique Sous Champs Magnetiques Intenses, 253 (1974)
54. M.E. Fisher, Amer. J. Phys. 32, 343 (1964); T. Nakamura, J. Phys. Soc. Japan 7, 264 (1952)

55. W. Duffy, Jr., and K.P. Barr, Phys. Rev. 165, 647 (1968)
56. Z.G. Soos, Ann. Rev. Phys. Chem. 25, 121 (1974); J. Chem. Phys. 46, 4284 (1967)
57. K.M. Diederix et al., Phys. Rev. B 19, 240 (1979)
58. J.C. Bonner, S.A. Friedberg, H. Kobayashi, and B.E. Myers, Proc. 12th Int. Conf. on Low Temp. Phys., Kyoto, Japan, p. 691 (1970)
59. J.J. Smit, L.J. de Jongh, J.A.C. van Ooijen, J. Reedijk and J.C. Bonner, Physica 97B, 229 (1979)
60. I.S. Jacobs et al., Phys. Rev. B 14, 3036 (1976)
61. L.N. Bulaevskii, Sov. Phys. JETP 17, 684 (1963)
62. E. Pytte, Phys. Rev. B 10, 4637 (1974)
63. J.W. Bray et al., Phys. Rev. Lett. 35, 744 (1975)
64. M.C. Cross and D.S. Fisher, Phys. Rev. B 19, 402 (1979)
65. J.C. Bonner and H.W.J. Blöte, Phys. Rev. B, to be published
66. J.C. Bonner, H.W.J. Blöte, J.W. Bray and I.S. Jacobs, J. Appl. Phys. 50, 1810 (1979)
67. M. Drawid and J.W. Halley, A.I.P. Conf. Proc. 34, 208 (1976)
68. Th. Niemeijer and J.M.J. van Leeuwen, Phase Transitions and Critical Phenomena, Eds. C. Domb and M.S. Green (Academic Press, London) Vol. 6, 1976
69. D.R. Nelson and M.E. Fisher, A.I.P. Conf. Proc. 18, 888 (1973)
70. H.J. Hilhorst, M. Schick and J.M.J. van Leeuwen, Phys. Rev. Lett. 40, 1605 (1978)
71. S.D. Drell, M. Weinstein and S. Yankielowicz, Phys. Rev. D 14, 487 (1976) and subsequent work
72. R. Jullien, J.N. Fields and S. Doniach, Phys. Rev. B 6, 4889 (1977).
73. S-T. Chui and J.W. Bray, Phys. Rev. B 18, 2426 (1978)
74. J.E. Hirsch and J.V. José, J. Phys. C 13, L53 (1980); J.E. Hirsch, Phys. Rev. B, to be published
75. R. Jullien, P. Pfeuty, J.N. Fields and S. Doniach, Phys. Rev. B 18, 3568 (1978)
76. J. des Cloizeaux and M. Gaudin, J. Math. Phys. 7, 1384 (1966)
77. H.W.J. Blöte, J.C. Bonner and J.N. Fields, J. Mag. Mag. Mat. 15-18, 405 (1980); J.N. Fields, H.W.J. Blöte and J.C. Bonner, J. Appl. Phys. 50, 1808 (1979)
78. R.B. Griffiths, Phys. Rev. 133, A768 (1964); C.N.Yang and C.P. Yang, Phys. Rev. 151, 258 (1966)
79. L. Hulthén, Arkiv. Mat. Astron. Physik 26A, 1 (1938)
80. M.P. Nightingale and H.W.J. Blöte, to be published; H.W.J. Blöte, private communication
81. G. Müller, H. Beck and J.C. Bonner, Phys. Rev. Lett. 43, 75 (1979)
82. G. Müller, H. Thomas, H. Beck and J.C. Bonner, Phys. Rev.B, to be published
83. J.P. Groen, T.O. Klaassen, N.J. Poulis, G. Müller, H. Thomas, and H. Beck, Physica, to be published
84. J. des Cloizeaux and J.J. Pearson, Phys. Rev. 128, 2131 (1962)
85. I.U. Heilman et al, Solid State Comm. 31, 607 (1979)
86. Y. Endoh et al., Phys. Rev. Lett. 32, 170 (1974)
87. P.C. Hohenberg and W.F. Brinkman, Phys. Rev. B 10, 128 (1974)
88. G. Müller and H. Beck, J. Phys. C 11, 483 (1978)

Support for this work has been provided by the U.S. National Science Foundation, grant number DMR77-24136, NATO, the Dutch National Science Foundation (F.O.M.), and the Swiss National Science Foundation. The hospitality of the Aspen Center for Theoretical Physics, where this paper was prepared, is gratefully acknowledged.

Dynamic Correlations in Classical Heisenberg Chains

Stephen W. Lovesey

Rutherford and Appleton Laboratories
Chilton, Oxfordshire, UK

1. Prologue

Up until a few years ago the study of one dimensional (1d) magnetic models was largely of academic interest. This situation has been completely changed with the discovery of many highly anisotropic magnetic insulators which display predominantly 1d magnetic features above a three dimensional ordering temperature [1,2]. This temperature, which is determined by the strengths of interchain couplings, can be as low as 1K. Here we survey the theory of dynamic correlations in some classical spin models.

In the absence of an external field, a purely 1d magnet does not support long-range order at a finite temperature. Zero temperature can be regarded as the critical temperature. The characteristic feature of 1d magnets is the strong short-range correlations that develop for temperatures which are small compared with the exchange coupling ($T \ll J$). An alternative viewpoint is that, the phase precursive to the ordering transition in three dimensional magnets is extended in 1d magnets over a temperature interval $\sim J$. There are special difficulties in describing spontaneous spin fluctuations in highly correlated disordered magnets. Conventional treatments of the equations-of-motion for the spin variables are inapplicable to such systems since they take the mean exchange field as the leading-order term, and this is zero in the absence of long-range order.

The problem of adequately describing dynamic correlations in classical 1d magnets is particularly tantalizing because all static correlation functions can be calculated exactly. Moreover, when the Hamiltonian contains just a simple isotropic exchange coupling between nearest neighbour spins, and no anisotropy terms, the static correlation functions can be expressed in simple closed form. If we know the crystal structure and force constants of a solid then the spectrum of harmonic lattice vibrations, or phonon spectrum, is easily calculated. Anharmonic effects can be evaluated systematically using conventional many-body perturbation techniques. An analogous situation exists in a 1d magnet only when the temperature is small compared with the external field and a spin condensate is formed. Otherwise the thermally induced disorder is an essential feature in the dynamic response.

Even so, physical intuition leads us to expect, and experiments and computer simulations confirm, that in the limit of low temperatures and zero field a 1d magnet will support collective modes similar to the spin waves in the ordered state. The thermal disorder inherent in a 1d magnet will result in a finite lifetime of the collective modes.

Before we attempt to make the intuitive ideas more quantitative we pause to define some basic notation. The exchange Hamiltonian used here is

$$H = -\tfrac{1}{2} J \sum_{j\delta} S_j . S_{j+\delta} \tag{1.1}$$

where S_j is a unit vector representing the spin at the j'th lattice site, and δ extends to the two nearest neighbours. Models with $J > 0$ and $J < 0$ are referred to as having ferromagnetic (F) and antiferromagnetic (AF) couplings, respectively.

In subsequent discussions it is convenient to use a set of energy and frequency units in which $|J|$ is unity. We shall also measure wavevectors and inverse correlation lengths in units of the inverse lattice spacing.

2. Purely Heisenberg Exchange Coupled Models

The spin relaxation function is denoted by $F_m(k,\omega)$ where k and ω are the wavevector and frequency variables. The isothermal spin susceptibility $\chi(k)$ is the integral of $F_m(k,\omega)$ over all ω for a fixed k,

$$\chi(k) = \frac{1}{\pi} \int_{-\infty}^{\infty} d\omega \; F_m(k,\omega) \quad . \tag{2.1}$$

The exact result for $\chi(k)$ is given in the table. The divergence of $\chi(0)$ and $\chi(\pi)$ for $J > 0$ and $J < 0$, respectively, in the limit $T \to 0$ characterises $F(J > 0)$ and $AF(J < 0)$.

If the magnet supports a collective mode with a long lifetime then its frequency is, to a good approximation, ω_o where the normalised second moment

$$\omega_o^2 = \int_{-\infty}^{\infty} d\omega \; \omega^2 \; F_m(k,\omega) \Bigg/ \int_{-\infty}^{\infty} d\omega \; F_m(k,\omega) \quad . \tag{2.2}$$

From the results given in the table(1), and $T << 1$,

$$\begin{aligned} \omega_o &= 2\{1 - \cos(k)\} \; (1 - T) \qquad & (F) \\ &= 2 \sin(k) \; (1 - T) \qquad & (AF) \end{aligned} \quad . \tag{2.3}$$

In the limit $T \to 0$, these dispersion relations are the same as those obtained for ordered 1d magnets. We can easily estimate the width of the collective mode contribution to $F_m(k,\omega)$, which is proportional to the inverse lifetime, from the mean-square fluctuation about ω_o. Denoting the width by Γ, we have for long lifetimes the excellent approximation,

$$\begin{aligned} \Gamma^2 &= \int_{-\infty}^{\infty} d\omega \; (\omega^2 - \omega_o^2)^2 \; F_m(k,\omega) \Big/ \pi \, \omega_o^2 \, \chi(k) \\ &= \omega_1^2 - \omega_o^2 \quad . \end{aligned} \tag{2.4}$$

The second equality defines the normalised fourth moment of $F_m(k,\omega)$. Using the results for ω_1^2 given in the table, we find for $T << 1$,

---

(1) The low T expansion of the first ten moments are given in [3].

Table - Integrals of the spin and energy relaxation functions for exchange coupled 1d magnets

	Spin
$\chi(k)$, $TC_m(k)$	$(1 - u^2)/3T(1 + u^2 - 2u\gamma)$
ω_0^2	$4uT(1 - \gamma)(1 + u^2 - 2u\gamma)/(1 - u^2)$
ω_1^2	$\frac{2}{3}\{5 - 3\gamma - v(1 - 3\gamma + 3/u) - u[6\gamma - (2 + v)(2\gamma^2 - 1)]\}$
	Energy
$\chi(k)$, $TC_m(h)$	$(1 + \gamma)(1 - 3u^2 + 2v)/6T$
ω_0^2	$12T^2u^2(1 - \gamma)/(1 - 3u^2 + 2v)$
ω_1^2	$2(5u^2 + u^2v + 3v^2 - u^2(4 + 5v)\gamma)/3u^2$

The Hamiltonian is defined in (1.1), and the results are for $J > 0$, $\gamma = \cos(k)$. The integrals are defined in (2.1), (2.10), (2.2) and (2.4). Temperature T is measured in units of $|J|$, and $u = \coth(1/T) - T$ and $v = 1 - 3|u|T$. For AF, $u \to -u$. Note that the integrals of $F_e(k, \omega)$ are independent of the sign of J. The results for ω_0^2 and ω_1^2 for the energy are from Ref. [7], and the remaining results are from Refs. [9,20].

$$\Gamma^2 = 6T\{1 - \cos(k)\} \quad (F)$$
$$= 2T\{3 + \cos(k)\} \quad (AF) \quad . \tag{2.5}$$

From these results we conclude that the lifetime of the collective mode in both F and AF models tends to infinity as $1/T^{\frac{1}{2}}$ as $T \to 0$ ($k \neq 0$ for F).

The result $\Gamma \to 0$ as $T \to 0$ is the key to a calculation of the dynamic response which is correct to leading order in T[4]. For F the collective mode frequency has the same dispersion as ω_0 in (2.3) but the temperature renormalisation factor is $(1 - T/2)$ instead of $(1 - T)$, ie the thermally induced softening is less than a naive estimate based on the second moment. The half-width also differs from the previous estimate, and it is

$$\Delta = T \sin(k), \quad k < \pi \quad (F) \tag{2.6a}$$

except near the zone boundary where it approaches the value

$$\Delta = T^{3/2}, \quad k = \pi \qquad (F) \quad . \tag{2.6b}$$

Although the temperature renormalisation factor in the collective mode frequency is the same for F and AF, the width in the latter is quite different from the result (2.6) for F and it is independent of k, namely,

$$\Delta = T \qquad (AF) \quad . \tag{2.7}$$

More subtle differences occur between the line shapes for the two types of exchange coupling. For AF and $k << T$ the ordinary spin diffusion result for $F_m(k,\omega)$ is recovered and the spin diffusion constant

$$D = 2/T \qquad (AF) \quad . \tag{2.8}$$

The borderline between diffusive and resonance behaviour, where $F_m(k,\omega)$ is dominated by the collective mode, lies at a wavevector $\sim$ T. Returning to F, a diffusive form is found for $k \to 0$ and $\omega << 2T$, and

$$D = \pi/\{2\ \ln(1/T)\} \qquad (F) \quad . \tag{2.9}$$

The derivation of this result for the diffusion constant is not completely rigorous but the logarithmic temperature behaviour is consistent with computer simulation data[4]. The dynamic scaling hypothesis for $F_m(k,\omega)$ in the limit $T \to 0$ is apparently violated for both F and AF. The violation in the latter case arises from a small feature of the lineshape, whereas for F there is a very strong violation since, for example, scaling requires the diffusion constant to tend to a finite value as $T \to 0$ in contrast to the result (2.9).

It is well known that two spin wave bound states occur in ordered magnets. Their excitation is not a first-order process for F, but they manifest themselves as resonances in the self-energy which make the spin wave lineshape asymmetric [5]. The situation is quite different for AF since there are two degenerate spin waves which precess in opposite directions, and the excitation of a bound state formed by one of each type of spin wave is a first-order process[6].

Thus far we have discussed only the spectrum of spontaneous spin fluctuations. A second fundamental variable is, of course, the exchange energy density whose mean-square fluctuation is proportional to the wavevector dependent specific heat at constant magnetisation, $C_m(k)$. Denoting the energy relaxation function by $F_e(k,\omega)$, the analogue of (2.1) is

$$TC_m(k) = \frac{1}{\pi} \int_{-\infty}^{\infty} d\omega\ F_e(k,\omega) \quad . \tag{2.10}$$

In the limit $T \to 0$, we obtain from the table,

$$C_m(k) = \tfrac{1}{2}\{1 + \cos(k)\} \tag{2.11}$$

whereas

$$\chi(k) = \frac{1}{3}\{1 \mp \cos(k)\}^{-1} \tag{2.12}$$

where the upper and lower signs are for F and AF, respectively. The divergence of $\chi(k)$ at the ordering wavevector is not reflected in $C_m(k)$; indeed, $C_m(k)$ is the same for both F and AF at all temperatures.

From the results given in the table, we see that the normalised second and fourth moments of $F_e(k,\omega)$ are both independent of T in the limit $T \to 0$, and their difference, Γ^2, is finite[7]. In consequence, $F_e(k,\omega)$ is not a pair of delta functions for $T \to 0$, unlike $F_m(k,\omega)$, although it is singular. For both F and AF[4],

$$F_e(k,\omega) = \tfrac{1}{2} T\{1 + \cos(k)\}\{\theta_k^2 - \omega^2\}^{-\frac{1}{2}}, \quad \omega^2 < \theta_k^2$$
$$= 0, \quad \omega^2 > \theta_k^2 \tag{2.13}$$

where $\theta_k = 4 \sin(k/2)$, and $T << 1$. The square root singularity in $F_e(k,\omega)$ leads to pronounced oscillations in its temporal Fourier transform. A similar singularity is obtained in the spectrum of vibrations of a spatially disordered harmonic chain[8].

A knowledge of the exact static correlation functions can be used to construct expressions for the relaxation functions that are useful over a range of temperatures, and Mori's theory of a generalised Langevin equation provides an appropriate formal framework[11]. Viewed in one way, Mori's theory is a convenient method of constructing the collisional self-energy for a given set of dynamic variables. A judicious choice of variables renders the self-energy, or memory function, benign and therefore adequately represented by a simple function of ω. In most cases, one can only expect to achieve success for a range of parameters, eg an expression for $F_m(k,\omega)$ that is reliable for F, $\omega^2 > Tk^2$ and modest temperatures.

In its simplest guise the Mori theory provides a moment expansion for the Laplace transform of the relaxation function in the form of a continued fraction. Consider first how such an expansion can be applied to $F_m(k,\omega)$. The low-order moments appear in the combination ω_o^2 and Γ^2, while higher moments occur in combinations that are independent of T in the limit $T \to 0$. In addition, the higher order contributions are relatively insensitive functions of the wavevector. Hence, a theory which correctly incorporates ω_o^2 and Γ^2 will include the salient features of the dynamics[9]. Such a theory is, by construction, reliable only for small times, or large frequencies, and since the collective mode has a positive dispersion the wavevector should not be small. The theory is most likely to fall short in estimating the amplitude of the collective mode contribution to $F_m(k,\omega)$, while the frequency and, to a lesser extent, the width of the contribution will be quite accurately described. The precise form of the termination of the continued fraction will mainly effect the collective mode amplitude[10]. However, the only reliable way of improving the approximation for the relaxation function is to include more exact moments, although it is not clear that the continued fraction is in general uniformly convergent, ie a termination at the fifth order might be no better, or even worse, than terminating at the fourth order[3].

If the first few terms of an expansion procedure fail to give an adequate representation of the experimental or computer simulation data then it might be argued that the expansion is made about the wrong starting point. Hence, if the objective is to account for hydrodynamic phenomena it is probably expedient to abandon a straightforward continued fraction expansion in terms of the magnetisation in favour of an expansion starting with a more appropriate combination of variables [4]. Even so, high-order continued fraction expansions are useful for a quite separate numerical problem. Computer simulation data is usually in the form of the time development of correlation functions for times typically not much longer than 10/J [12,13,14], and continued fraction expansions are a reliable method for constructing the relaxation functions from the data [13].

Computer simulation studies have confirmed the low temperature results (2.7) - (2.9) [4,13]. At higher temperatures ($T \sim 0.3$) LOVELUCK and WINDSOR [13] conclude that all published theories are inadequate.

3. Single Site Anisotropy

Most theoretical and computer simulation studies of purely exchange coupled magnets focus on results for AF because of various experimental data available for the almost ideal 1d AF magnet TMMC [1,2]. Another well-characterised system, $CsNiF_3$, is described by a ferromagnetic exchange and a single site anisotropy [2,15].

$$B \sum_j (S_j^z)^2 \qquad (3.1)$$

with $b = B/J = 0.21^{(2)}$. For $T \ll 1$, the ferromagnetically aligned spins lie in the x-y plane to minimise the anisotropy energy. On the other hand, if B is negative the spins align parallel to the z-axis, and the static properties display Ising-like behaviour as functions of temperature [16,30].

Within the molecular field approximation, the anisotropy (3.1) with $B > 0$ reduces the critical temperature for z - z fluctuations. Since the critical temperature for $B = 0$ corresponds to $T = 0$ it is therefore not surprising that the inverse correlation length κ_z for $B \neq 0$ is finite at $T = 0$, and [16,30]

$$\kappa_z = (2b)^{\frac{1}{2}} \quad \text{with} \quad \chi_z(0) = 1/2\,B \qquad (F) \,. \qquad (3.2)$$

The ordering occurs in the x - y plane so $\kappa_x \to 0$, and $\chi_x(0)$ diverges in the limit $T \to 0$.

The salient features of z - z spin fluctuations (or the out-of-plane fluctuations) can be understood by using the argument described at the beginning of section 2. The normalised second moment of the relaxation function in the limit $T \to 0$ is [18]

$$\omega_o^2 = 4\{1 - \cos(k)\}\{1 + b - \cos(k)\} \qquad (F) \qquad (3.3)$$

and the dispersion of the collective mode in $CsNiF_3$ is indeed described very accurately by ω_o. The quantity Γ, defined as in (2.4), is much smaller (for a given low T and k) for finite B than for $B = 0$. Hence, the linewidth of out-of-plane fluctuations is predicted to be small compared with that for a purely exchange coupled system, and this is borne out by experiment [2,15].

---

(2) Ni ions have a spin = 1 and therefore quantum corrections are more important for $CsNiF_3$ than for TMMC in which the Mn ions have a spin = 5/2. The corrections effect the value asigned to the parameter B. The value quoted leads to good agreement between the measured collective mode frequency and ω_o given in (3.3) [15].

There are also some interesting differences predicted between in-plane and out-of-plane fluctuations [17].

For AF the ordering wavevector is at $k = \pi$, and the wavevector dependent susceptibility relative to this wavevector is $\chi(\pi - k^*)$. In place of (3.3)

$$\omega_o^2 = 4\{1 - \cos(k)\}\{1 + b + \cos(k)\} \quad \text{(AF)} \tag{3.4}$$

in the limit $T \to 0$, and there is an **energy gap** at $k^* = 0$ which is believed to supress the peak at $\omega = 0$ in the spin relaxation function [18].

4. External Magnetic Field

An external magnetic field parallel to the z-axis contributes an energy

$$-h|J|\sum_j S_j^z \quad , \tag{4.1}$$

With $h \neq 0$ and $T \to 0$ there is long-range order, and for a ferromagnetic coupling the average value of S^z is [19]

$$\langle S^z \rangle = 1 - T/K \qquad \text{(F)} \tag{4.2}$$

where $K^2 = h(h + 4)$. The corresponding longitudinal and transverse susceptibilities are [20]

$$\chi_z(k) = (T/K)(h + 2)/[K^2 + 2\{1 - \cos(k)\}] \text{(F)} \tag{4.3}$$

and

$$\chi_x(k) = [h + 2\{1 - \cos(k)\}]^{-1} \qquad \text{(F)} \quad . \tag{4.4}$$

These results show that K and $h^{\frac{1}{2}}$ are the longitudinal and transverse inverse correlation lengths, respectively, in the limit of $h \neq 0$ and $T \to 0$.

An antiferromagnetically coupled chain displays a particularly rich structure as a function of field [20]. For $T << 1$ and modest fields $\chi_z(k^*)$ displays a peak at $k^* = 0$, and therefore the ordering wavevector is $k = \pi$. The inverse correlation length κ_z^* increases rapidly with h, and it diverges at $h = 2\surd 2$ for $T << 1$. A study of $\chi_z(k^*)$ as a function of k^* for increasing values of h, and a given small T, shows that it has a very weak wavevector dependence for $h = 2\surd 2$, and for larger fields it peaks at $k^* = \pi$, ie a nuclear reciprocal lattice vector. This behaviour is interpreted as a low T transition from antiferromagnetic to ferromagnetic order, at a critical field $= 2\surd 2$. The transverse inverse correlation length remains essentially constant, at its zero field value, up to $h = 2\surd 2$, beyond which it increases approximately linearly with h [33].

In the presence of a magnetic field it is necessary to distinguish between the specific heats at constant magnetisation and field which are related, respectively, to fluctuations in the local temperature and entropy [20]. The deviation of the ratio of the wavevector dependent specific heats at $k = 0$ from unity is a measure of the field induced coupling between the longitudinal magnetisation and the energy density. For F, and $h \neq 0$ and $T \to 0$ [21]

$$C_m(k) = p^2\{1 + \cos(k)\}/(1 + p^2) \qquad \text{(F)} \tag{4.5}$$

and

$$C_h(k) - C_m(k) = \left[2ph\{h + 3 + \cos(k)\}\right]^2/K \quad (F) \tag{4.6}$$

where $p = (h + 2 - K)/2 \leq 1$. The wavevector dependence of $C_m(k)$ is the same as in zero field, and $C_m(0)$ decreases rapidly from unity with h. While $C_h(k)$ decreases with k it is finite at the zone boundary in contrast with $C_m(k)$. Moreover, $C_h(0) = 1$, and therefore $C_h(0)/C_m(0)$ increases rapidly with h, demonstrating that the field induced coupling between the energy and longitudinal magnetisation densities increases rapidly with field in a ferromagnet. The ratio of the specific heats is relatively insensitive to temperature[22]. For AF it is found that[22] $C_m(k)$ has essentially the same wavevector dependence as in zero field, and it is zero at $k = \pi$, whereas $C_h(k)$ displays a pronounced peak at $k = \pi$, in common with $\chi(k)$. Not surprisingly, $C_h(0)/C_m(0) \simeq 1$ for a wide range of fields. The value of h at which the ratio begins to depart from unity depends quite sensitively on T since increasing the temperature softens the antiferromagnetic order and promotes the ferromagnetic response.

In view of these last remarks on the static properties of AF magnets in a magnetic field we do not expect that a modest field will produce significant changes in $F_m(k,\omega)$ for small T. The collective mode contribution remains sharp, and its width and dispersion are similar to the zero field case except near $k^* = 0$ where the field creates an energy gap in the dispersion[22,23]. For $h \gg 1$ a sizable ferromagnetic response is induced, and $F_m(k,\omega)$ is then akin to the relaxation function for F.

For F, longitudinal spin fluctuations, $h \neq 0$ and $T \to 0$[22]

$$\omega_o^2 = 4p^2\{1 - \cos(k)\}\left[K^2 + 2\{1 - \cos(k)\}\right]/(1 + p^2) \quad (F) \tag{4.7}$$

and

$$\Gamma^2 = 2K\{1 - \cos(k)\}\{1 + 3p^2 + 2p^2\cos(k)\}/(h + 2) \quad (F) \ . \tag{4.8}$$

For $h = 0$, $p = 1$ and $K = 0$, and in this limit (4.7) reproduces the linear spin wave dispersion. The fact that Γ is finite for $h \neq 0$ and $T \to 0$ is at odds with physical intuition. We would expect the lifetime of both the longitudinal and transverse collective modes to be infinite as $T \to 0$. The flaw in the argument that leads to the prediction of a finite lifetime for longitudinal fluctuations as $T \to 0$ is that it completely neglects the static coupling between the longitudinal magnetisation, M_k, and the exchange energy, E_k. A continued fraction expansion for $F_m(k,\omega)$ in terms of M_k, carried to a high order will, of course, yield correct results. However, a straightforward low-order expansion in M_k is totally unreliable, for the reasons just explained. The essential physics is simply that, the field induced coupling of E_k and M_k implies that for $h \neq 0$ a fluctuation in one variable creates a fluctuation in the other. Of course, E_k and M_k are dynamically coupled even in zero field. For example, the time derivatives of E_k and M_k have a finite correlation for $h = 0$, whereas E_k and M_k are uncorrelated in the zero field limit.

For $h \neq 0$ it is clearly expedient to define a pair of new variables that are uncorrelated. If we choose one of the variables to be M_k then the second is proportional to the local temperature whose fluctuations determine $C_m(k)$ [20].

We have noted already that $F_e(k,\omega)$ is singular for $h = 0$ and $T \to 0$, and that its temporal Fourier transform oscillates strongly. These oscillations persist at modest temperatures[23], and for $h \neq 0$ it has been demonstrated that they manifest themselves in the longitudinal spin relaxation function as an additional low frequency peak [14,21,24]. A similar phenomenon is unlikely to be observed with AF because the field induced coupling of E_k and M_k is very small [21,24].

The normal mode frequencies and amplitudes for $h \neq 0$ can be derived by constructing simultaneous equation-of-motion for M_k and the local temperature [21,24]. In the limit of $h \neq 0$ and $T \to 0$, the normal modes have an infinite lifetime, as required. Denoting the normal mode frequencies by ω_1 and ω_2 then, to a good approximation,

$$\omega_1^2 = 2\{1 - \cos(k)\}\{3 + p^4 - 4p^2 \cos(k)\}/(1 + p^2) \qquad (F) \qquad (4.9)$$

and

$$\omega_2^2 = 2\{1 - \cos(k)\}(1 + p^2) \qquad (F) . \qquad (4.10)$$

For $h = 0$, $\omega_1 = 4 \sin^2(k/2)$ and $\omega_2 = 2\sqrt{2} \sin(k/2)$, and we therefore interpret the mode with frequency ω_1 as predominantly a spin wave. The amplitude of the energy, or temperature, mode in the spin relaxation function is a strong function of k, and it is a significant feature over only a relatively small fraction of the zone [14].

5. Diluted Chains

The magnetic correlations within a chain are obviously disturbed by impurities, and we might expect more dramatic impurity effects in 1d than in a higher dimension. For example, a single non-magnetic impurity in a chain divides the chain in two unconnected parts and it therefore changes the correlation length. For a binary system there is in general a total of three different exchange constants. If the temperature is large compared with the magnitudes of the exchange constants then the impurities are unlikely to have a very pronounced effect on the dynamic correlations. In the opposite limit of small T the impurity induced modes will feature in the dynamic response.

Static correlation functions for a binary mixed chain, and isotropic exchange couplings, can be evaluated in closed form. The result for the susceptibility can be used to interpret quasi-elastic neutron scattering data which yields the correlation length [25], and the low-order frequency moments used to calculate the relaxation function in a continued fraction expansion [26]. Calculations for AF and up to 15% of magnetic impurities show that the spin wave contribution is sharp at modest temperatures, and this is in accord with experimental findings [25]. The theory also predicts a very strong quasi-elastic component in the relaxation function.

A low-order continued fraction expansion is inappropriate at small temperatures since the impurity induced modes are a potentially important feature in the dynamic response in this limit. The problem of the dynamic response of an ordered magnet containing a single magnetic impurity can be solved exactly (in the linear spin wave approximation) even though the defect is spatially extended [27]. The frequency dependence of the impurity modes is complicated since it is determined by the perturbation parameters and the response, or dynamic susceptibility, of the host magnet. The modes are damped if they interact with the host spin waves, but they can also lie outside the frequency response of the host in which case they have an infinite lifetime. The most important perturbation parameter is the difference between the exchange couplings for the host spins, and the host and impurity spin.

The exact solution of the single impurity problem is the starting point of a treatment of a model with many impurities where the problem is to perform the appropriate configurational averaging of the relaxation function [28]. For a low concentration, c, a reliable estimate of the relaxation function can be obtained with an average t-matrix theory. In this approximation, the spin wave is found to dominate the dynamic response [29]. If the defect is due soley to a difference in the host-impurity and host-exchange couplings, and the ratio of the two coupling constants is (1 + f), then the impurity induced spin wave width for F is (in units of the host-exchange coupling)

$$\Delta = 4cf^2(1 - \gamma_k)\sin(k)\{f^2(3 - \gamma_k^2) + 2f(1 + 2\gamma_k - \gamma_k^2) + 1 + \gamma_k^2\}/\{(1 + \gamma_k)(1 + f)^4 + 4f^2(1 - \gamma_k)(\gamma_k + f)^2\} \quad (F) \qquad (5.1)$$

where $\gamma_k = \cos(k)$. The corresponding expression for AF is

$$\Delta = cf^2 \sin(k/2)\sin(k)\{(1 + f)^2\cos^2(k/2) + 2\sin^2(k/2)\}/\{(1 + f)^4\cos^2(k/2) + 4f^2\sin^4(k/2)\} \quad (AF) \qquad (5.2)$$

Both these expressions are linear in the impurity concentration because of the nature of the approximation, and vanish at the extremes of the Brillouin zone, k = 0 and π. Thermal fluctuations will add to the widths [31], and also create a strong central peak in the relaxation function for small k. In addition to damping the impurities will also induce a shift in the spin wave peak away from the unperturbed frequencies, $2\{1 - \cos(k)\}$ for F and $2\sin(k/2)$ for AF.

In addition to the spin wave response, the relaxation function can contain structure due to the impurity modes associated with a single defect. This structure can be in the form of a strong asymmetry in the spin wave response or an additional peak depending on the perturbation parameters; in the present case a single parameter f that ranges from -1 to infinity. Impurity modes which lie outside the spin wave band of the host ($0 < \omega < 4$ for F, and $0 < \omega < 2$ for AF) have a completely negligible amplitude for small c. The damping of the modes within the band is proportional to the spin wave density of states, and for 1d magnets significant structure in the relaxation function is predicted when the single defect impurity mode frequencies occur around the middle of the band. Two types of mode occur for F and their frequencies are,

$$\omega = -2(1 + f)/f, \quad \text{for } f < -1/3$$

and (F) (5.3)

$$\omega = 2(1 + f)/f, \quad \text{for } f > 1 \quad .$$

For AF there is only one mode in the band and its frequency is

$$\omega = -2(1 + f)/f, \quad \text{for } f < -1/2 \qquad (AF) \quad . \qquad (5.4)$$

In the event that an impurity mode yields a second peak in the relaxation function its position will be shifted slightly from the values (5.3) and (5.4) by interactions with the host spin waves.

The usefulness of the approximation scheme just described is put in some question by exact results for two impurities in an F chain, since the calculations show that impurity states in the spin wave band interfere even for an infinite separation of impurities [32]. At a finite separation there is significant interference between modes both within and above the spin wave band of states.

Acknowledgements

I am grateful to Drs B and H De Raedt and Professor A Sjölander for sending copies of their work. Drs E Balcar and R B Stinchcombe kindly commented on the first draft of this paper.

References

1. D. Hone and P.M. Richards, Ann.Rev.Mat.Sci. 4, 337 (1974).
2. M. Steiner, J. Villain and C.G. Windsor, Adv.Phys. 25, 87 (1976).
3. H. De Raedt, Phys.Rev. B19, 2585 (1979).
4. G. Reiter and A. Sjölander, J.Phys. C13, 3027 (1980).
5. R. Silberglitt and A.B. Harris, Phys.Rev. 174, 640 (1968).
6. S.W. Lovesey and E. Balcar, Phys.Lett. 30A, 84 (1969).
7. B. De Raedt and H. De Raedt, Phys.Rev. B21, 4108 (1980).
8. K. Kim and M. Nelkin, Phys.Rev. B7, 2762 (1973).
9. S.W. Lovesey, J.Phys. C7, 2008 (1974).
10. K. Tomita and H. Mashiyama, Prog.Theor.Phys. 48, 1133 (1972); ibid 51, 1312 (1974).
11. For a review of Mori's theory see, for example, S.W. Lovesey, Condensed Matter Physics: Dynamic Correlations, Vol 49 Frontiers in Physics (Benjamin/Cummings, Massachusetts, 1980).
12. P. Heller and M. Blume, Phys.Rev.Lett. 39, 962 (1977).
13. J.M. Loveluck and C.G. Windsor, J.Phys. C11, 2999 (1978).
14. J.M. Loveluck and E. Balcar, Phys.Rev.Lett. 42, 1562 (1979).
15. M. Steiner, J.Mag. and Mag.Mat. 14, 142 (1979).
16. J.M. Loveluck and S.W. Lovesey and S. Aubry, J.Phys. C8, 3841 (1975).
17. B. De Raedt and H. De Raedt, Phys.Rev. B17, 4344 (1978).
18. J.M. Loveluck and S.W. Lovesey, J.Phys. C8, 3857 (1975).
19. M. Blume, P. Heller and N.A. Lurie, Phys.Rev. B11, 4483 (1975).
20. S.W. Lovesey and J.M. Loveluck, J.Phys. C9, 3639 (1976).
21. S.W. Lovesey, Z.Physik B37, 307 (1980).
22. S.W. Lovesey and J.M. Loveluck, J.Phys. C9, 3659 (1976).
23. H. De Raedt and B. De Raedt, Phys.Rev. B21, 304 (1980).
24. S.W. Lovesey and J.M. Loveluck, J.Phys. C12, 4015 (1979).
25. J.P. Boucher, W.J. Fitzgerald, K. Knorr, C. Dupas and J.P. Renard, J.Phys. Lett. 39, L-86 (1978).
26. T. Tonegawa, J.Phys,Soc.Jap. 47, 1471 (1979).
27. T. Wolfram and J. Callaway, Phys.Rev. 130, 2207 (1963).
28. R.J. Elliott, J.A. Krumhansl and P.L. Leath, Rev.Mod.Phys. 46, 465 (1974).
29. S.W. Lovesey, Rutherford Laboratory Report RL-80-035 (1980).
30. K. Nakamura and T. Sasada, J.Phys. C11, 331 (1978).
31. E. Patzak, J.Mag. and Mag. Mat. 15-18, 1039 (1980).
32. T. Tonegawa, J.Phys.Soc. Japan 33, 348 (1972).
33. R.B. Stinchcombe, J.Phys. C13, L159 (1980).

Dynamics of One-Dimensional Magnets: Neutron Scattering Studies

M. Steiner

Hahn-Meitner-Institut für Kernforschung , Glienicker Str. 100
D-1000 Berlin 39, Fed. Rep. of Germany

Abstract

Spin dynamics can be studied most directly by means of inelastic neutron scattering. Since spin dynamics is the essential phenomenon in 1-D magnets, where only short range order exists, results of neutron scattering experiments are of fundamental importance for the understanding of such systems. This paper reviews such results for spin wave excitations in different systems with respect to the existence condition and dispersion of the spin waves, and the influence of anisotropies and quantum effects on them. In addition the present status of neutron scattering experiments on nonlinear excitations in different 1-D magnets will be discussed for different types of nonlinear systems.

1. Introduction

The fundamental difference between 1-D and 3-D magnets, the lack of long range magnetic order, LRO, in 1-D systems implies that the short range order, SRO built up in 1-D systems is varying in space and time for any temperature. Thus all ordering phenomena are of a dynamic type and understanding of the physical properties of 1-D-systems cannot be achieved without the understanding of the underlying dynamical properties of the system. It appeared from detailed studies on quasi-1-D magnets that these systems provide a nearly ideal testing field for theoretical approaches to these problems [1,2,3]. This is mainly due to the fact that real systems have been found which resemble very closely idealized theoretical models. On the other hand these idealized models are to a much greater extent tractable theoretically than 3-D systems under the same conditions, e.g. in the critical region, what in turn stimulated new and more detailed experiments. This very fruitful interplay between theory and experiment has made this field one of the most fascinating fields in 1-D physics, especially in view of the most recent development: it has been shown theoretically that under certain conditions the linear spin wave approach is not sufficient and a complete nonlinear theory is necessary. The new nonlinear excitations which follow from this theory have been observed in different systems, thus opening a new field for the study of nonlinear phenomena.

One of the most powerful experimental tools to study 1-D magnets is inelastic neutron scattering. This is due to the fact that the cross section for thermal neutrons is proportional to the space and time Fourier transform of the spinpair correlation function G(r,t) [4]

$$\frac{d^2\sigma}{d\omega d\Omega} \sim S(q,\omega) = \int dr e^{-iqr} \int dt e^{-i\omega t} \langle \vec{S}_\perp(o\rho) \vec{S}_\perp(r,t) \rangle \tag{1}$$

with

$$G(r,t) = \langle \vec{S}(0,0)\vec{S}(r,t)\rangle = f(T) \tag{2}$$

and $\vec{S}_\perp$ being the spin component perpendicular to the momentum transfer $\vec{Q}$. Since 1-D magnets do not show LRO, $\langle M \rangle \equiv 0$, G(r,t) is the simplest nonzero quantity in such a system which contains all the information about the dynamics. On the other hand G(r,t) can be calculated theoretically for different model Hamiltonians

$$\mathcal{H} = -2J\sum_i \vec{S}_i \vec{S}_{i+1} + A\sum_i (S_i^z)^2 \tag{3}$$

where J and A are the exchange and anisotropy energy respectively. If S>1/2 classical approaches are usually used, whereas for S = 1/2 quantum calculations have to be used. Thus by means of inelastic neutron scattering the theoretical predictions for the different models can be checked directly. In contrast to other experimental techniques inelastic neutron scattering allows for the measurement of S(q,ω) over the whole Brillouin zone.

The results presented in the following sections have been chosen in view of demonstrating the power of the experimental method and of the discussion of a consistent physical picture of the dynamics in 1-D magnets. In section 2 realization of 1-D magnets in 3-D crystals will be discussed, in section 3 and 4 experimental results for spin wave and nonlinear excitations will be reported, respectively and in section 5 a synopsis of the dynamics in 1-D magnets will be given together with an outlook on that field as seen from a neutron scatterer's point of view.

2. Realization of 1-D Magnets in 3-D Crystals

Since truly one-dimensional systems are clearly unattainable experiments can only be performed on more or less ideal quasi-1-D systems. In the magnetic case very good quasi-1-D magnets can be produced by making use of the properties of the interaction mechanism between magnetic ions in insulating materials, which will be considered here only In Fig.1 two typical crystal structures of 1-D magnets are shown. Figure 1(a) shows the structure of TMMC and $CsNiF_3$ (hexagonal P63/mmc) and 1(b) that of CPC (for details, see [1]). Crystal field effects in the crystal are mostly responsible for the anisotropy term A in (3). In different chemical compounds it is possible to find

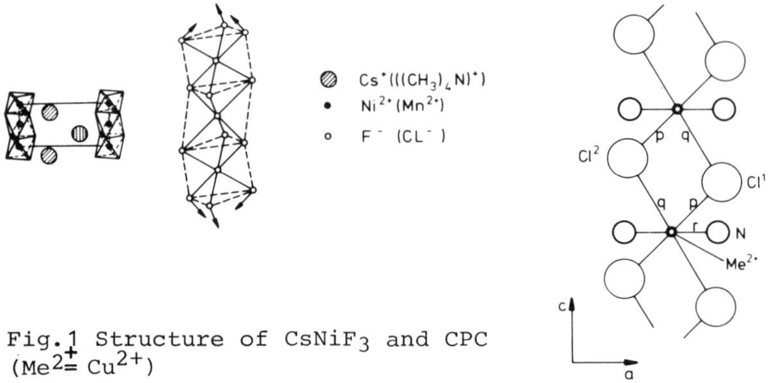

Fig.1 Structure of $CsNiF_3$ and CPC ($Me^{2+} = Cu^{2+}$)

very good realizations of 1-D magnets which often very closely approach the simple Hamiltonian (3), [1]. This allows very close comparison between theory and experiments.

3. Spin Wave Excitations

Spinwaves are defined as the lowest lying excited states of a magnetically ordered system. Thus one of the principal requirements for the existence of such excitations is LRO in the system. Since 1-D magnets do not show LRO the usual picture of, and the usual theories for spin waves cannot be simply transformed from 3-D to 1-D magnets. The rather simple and physically appealing picture of considering the spin waves as the only elementary excitations in the system can be preserved under certain restrictions, however. The SRO in the 1-D system is characterized by the correlation length ξ, defined as the distance between two spins, whose probability to be fully aligned is 1/e. Within this characteristic length ξ the system can be considered "ordered" and consequently spin waves obeying

$$q_{s\omega} > 1/\xi = \kappa$$

should be similar to the usual spin waves. For $q_s \leq \kappa$ the excitations become overdamped, finally loose there propagating character, and the dynamics become diffusive. Since κ is strongly temperature dependent [1] the range in q available for the spin waves depends on T as well. At any temperature there is a critical wavevector q_{crit} which separates the regimes for propagating and diffusive dynamics. In addition, the temperature dependence of κ is reflected in the width of the spin waves as observed by inelastic n-scattering [5],[6].

3.1 Experimental Results

3.1.1 Spinwaves in the 1-D Planar Ferromagnet $CsNiF_3$

A very impressive way to demonstrate the one dimensionality of a system is to measure the dispersion of the spin waves in dif-

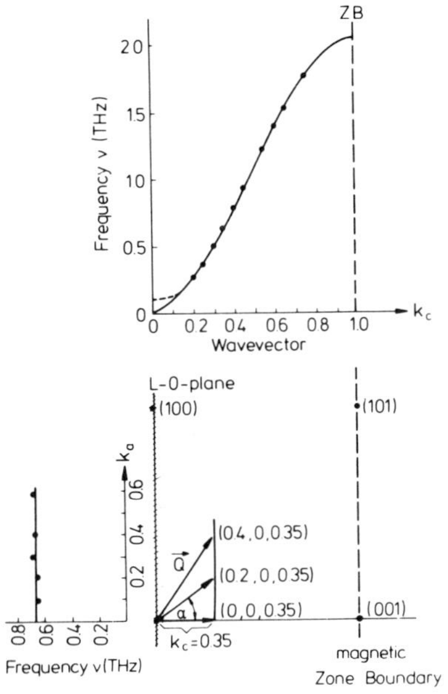

Fig.2 Spin wave dispersion in $CsNiF_3$ at H = O (——) and at 4.5K.

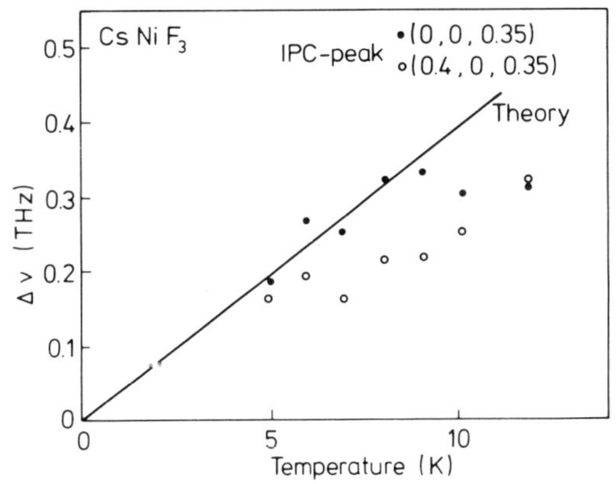

Fig.3 The width Δ of the spin waves versus T; theory is given by the full line

Fig.4 Variation of the lineshape of the spin waves for different q_a with q_c= 0.03r.l.u.; theory is the solid line ▷

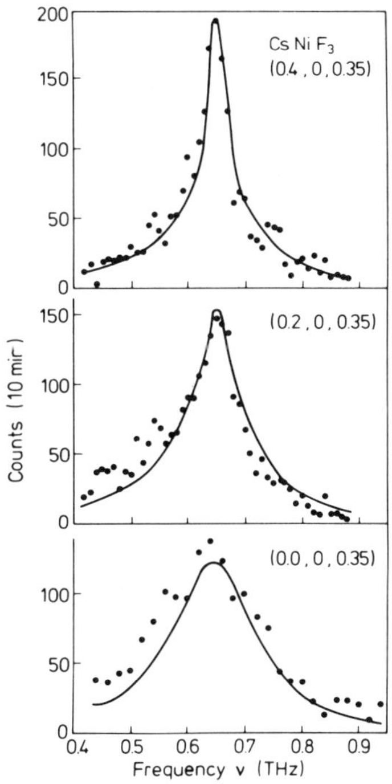

ferent crystal directions. This is shown in Fig.2, where the lack of dispersion for $q \parallel a^*$ is obvious, whereas for $q \parallel c^*$ the results follow the simple dispersion law [6]

$$\hbar\omega = E(q) = 2S\ [(2J - 2J\cos q_c)(2J - 2J\cos q_c + A)]^{1/2} \qquad (4)$$

from what the following parameters for (3) can be derived

$J/k = 11.8$ K; $A/k = 4.5$ K; $S = 1$
(Quantum correction for $S = 1$ yields $A/k = 9$ K)

this being the most direct way to derive the microscopic parameters in the Hamiltonian (3). The influence of the finite correlation length as discussed above can be seen in the increase of the measured width Δ of the spin wave peaks with temperature in Fig.3. The finding $\Delta \sim T$ [5] implies $\Delta \sim \kappa$ as expected from the theory [6].

Since neutrons see the spin component perpendicular to the momentum transfer $\vec{Q}$ only, it is possible to study the influence of the anisotropy A on the correlation function. Figure 4 shows how the lineshape changes when $\vec{Q}$ is tilted away from $\vec{Q} \parallel c^*$: With increasing angle between $\vec{Q}$ and c^* a new very sharp peak appears with increasing intensity in addition to the broad peak, characterized by κ, which stems from S^x and S^y only. The sharp, in fact resolution limited peak is due to the S^z component and should reflect the lifetime of the spin wave in this nonordered system [5],[6].

3.1.2 Spin Waves in the 1-D-X-Y AFM TMMC

TMMC ($((CH_3)_4N)MnCl_3$) is probably the best studied 1-D magnet [1,2,3]. The microscopic parameters have been determined from the spin wave dispersion [7],[8] as being ($S = 5/2$)

$J/k = 13$ K and $D = 0.016$

for the following Hamiltonian

$$\mathcal{H} = J \sum_i (S_i^x S_{i+1}^x + S_i^y S_{i+1}^y + (1 - D) S_i^z S_{i+1}^z) . \qquad (5)$$

In this compound the prediction of a critical wavevector $q_{critical}$ defining the stability region for spin waves as

$$q_s > q_{critical}(T)$$

has been tested in detail. Figure 5 shows a typical result obtained at $q_s = 0.030$ r.l.u. at different temperatures as shown in the figure as well [2]. It was deduced from such measurements that $q_{critical} = 1.5\,\kappa$ what is in good agreement with linear spin wave theory.

Very interesting predictions were made from linear spinwave theory for the case of applying a strong magnetic field in the x-y-plane: Two magnon modes are expected around twice the spinwave energy due to the divergent density of states at $q_c = o$. The identification as two magnon states is especially simple in such a 1-D system. Experimental results in high magnetic fields could indeed be interpreted within this two magnon picture [9].

In Fig.6 experimental results are shown, where the different peaks in the spectrum are identified according to the linear theory for one and two magnon modes.

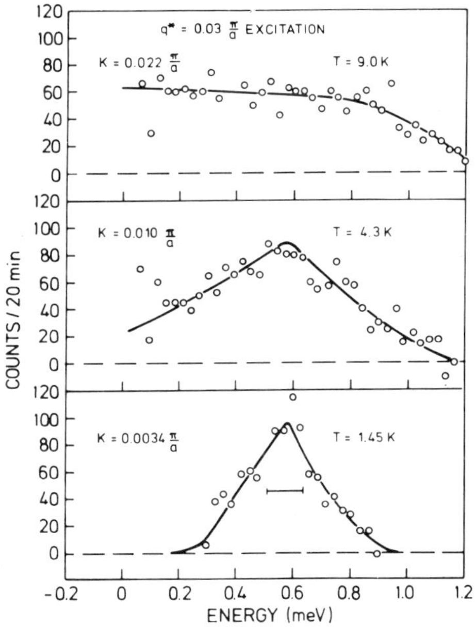

Fig.5 Damping of the spin wave at q_c=0.03 in TMMC at different temperatures

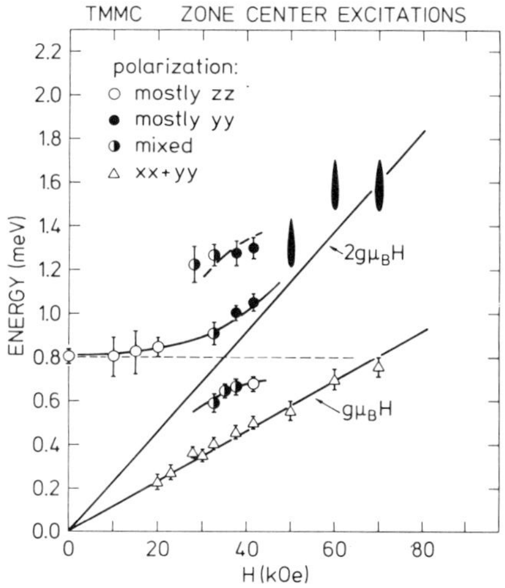

Fig.6 One and two magnon excitations in TMMC at 1.8 K: O, Δ one magnon excitation, all other: two magnon or one magnon-two magnon mixed states

3.1.3 Quantum effects in 1-D spinwaves

It has been found, that fundamental deviations from classical spinwave theory, i.e. more than just scaling of some parameters, should only be expected for S = 1/2-systems. In the S = 1-system $CsNiF_3$, for example, the only quantum effect is a scaling of the anisotropy constant A: $A_{classical} = 1/2\ A_{quantum}$.

First experiments on a quantum system, the S = 1/2 Heisenberg AFM CPC ($CuCl_2 \cdot 2NC_5H_5$), have shown that the spinwave peaks have strongly asymmetric lineshapes displaying a strong high energy tail [10], as shown in Fig.7.

This has been subsequently explained by different theoretical approaches [11],[12]. Common to both approaches is a physical picture, very different from the picture used in classical linear spinwave theory, where the single, one-spinwave excitation is the dominant feature: In the quantum chain there exist different excited states which contribute at a certain wavevector q with different weights. That means, that in quantum systems one considers an excitation continuum rather than a single well defined excitation. A similar situation is found in Ising systems with effective spin 1/2. $CsCoCl_3$ is a very good realization of such a quantum system [13],[14],[15]. Again theory pre-

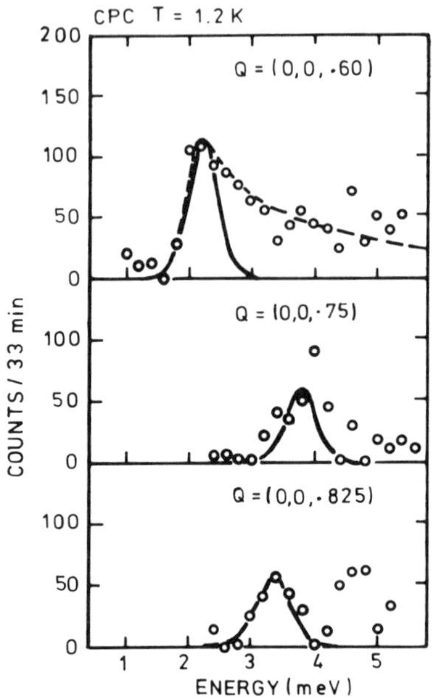

Fig.7 Lineshape of spinwaves in CPC at 1.2K. Solid lines indicate lineshape for classical linear spinwave theory

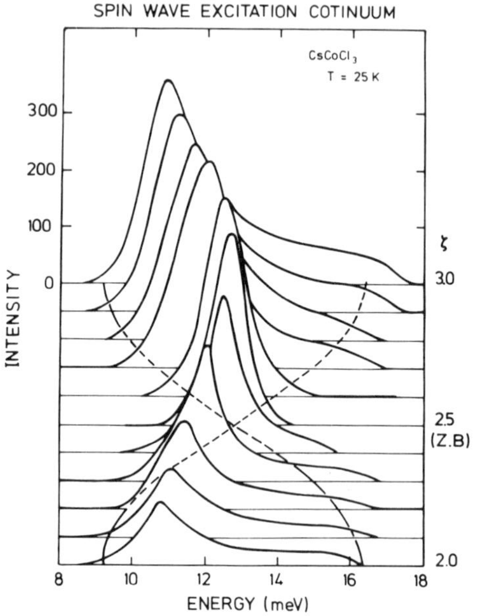

Fig.8 Lineshapes in the Ising AFM $CsCoCl_3$ at 25 K

dicts a continuum of excited states [16]. This band is broadest at the zone center and has zero width at the zone boundary. This is in distinct contrast to the often used single excitation result [17], where only one excitation is used in analogy to the classical spinwave result in non-Ising systems. Recent experimental studies on $CsCoCl_3$ clearly show the predicted asymmetry in the measured profiles as shown in Fig.8 [18]. The physics leading to the band of excitations is very different in the Heisenberg and Ising case, however. In the Heisenberg case the excitations can be viewed as the usual spinwaves and bound states of spinwaves [12], whereas in the Ising case the dominant excitations are spin flips, which form through a small non Ising term in the Hamiltonian propagating domain walls [19].

In concluding this section on spinwave excitations the following points appear to be the most important. The classical spinwave picture can be used in 1-D magnets very successfully as long as the system is of the X-Y or Heisenberg type and the spin value not 1/2. Under these circumstances the dominating excitations are quasi spinwaves with the restrictions $q_s > q_{critical}$, which follow the predicted dispersion and show the expected temperature dependence of the width. In the frame of this classical linear theory, the next possible excited states are the two magnon states, which have a well defined structure due to the divergence of density of states at the zone center and the zone boundary. In both the quantum system with S = 1/2 and the Ising case with small transverse interaction there appears a band of excitations due to the fact that not only the lowest energy excitations but a whole series of excitations contributes to the scattering. Important for the study of elementary excitations and their properties is, that their properties have to be investigated all over the Brillouin zone in order to understand their physical background and to test theoretical predictions. Inelastic neutron scattering experiments, covering this whole range, support the theoretical picture.

4. Nonlinear Excitations

Linear spinwave theory is by definition valid only if, in classical terms, the amplitude is small. Such theories have been used nearly exclusively in solid state physics with great success especially at temperatures much below a characteristic temperature of the system. The thermodynamics of such linear elementary excitations are well known even at the lowest temperatures where quantum statistics have to be used. It is known, however, that linear theory is not adequate for the description of systems with strong fluctuations, as around a phase transition. In order to overcome this problem KRUMMHANSL and SCHRIEFFER 20 have proposed a new approach to such systems: they suggest to use the full nonlinear equation of motion for the description of such systems. Though this in general is a too difficult mathematical task to be useful, there may exist systems for which the nonlinear Hamiltonian can be solved under reasonable conditions and thus predictions could be made, which can be tested experimentally [21] , [22] . 1-D magnets are especially promising candidates for the search of the elementary excitations solely due to the nonlinearity of the system, since no LRO exists which could suppress the fluctuations,which drive the system into nonlinearity. The thermodynamics for such systems is by far not complete as for the linear case but is being developed very rapidly [3] , [22] , [23] , [24] . Theories predict that the nonlinear modes either yield a separate contribution to the n-cross section [25] , [26] , [27] or should introduce characteristic changes into quantities measurable by neutron scattering [28] , [29] . Consequently n-scattering enables the most direct test of theories for nonlinear excitations in 1-D magnets. These excitations, solitons and breathers, are a continuing challenge for both experimentalists and theorists. The results available for three different types of nonlinear excitations are presented in the following.

4.1 2π-Solitons in the 1-D Planar Ferromagnet $CsNiF_3$

A 1-D-planar ferromagnet like $CsNiF_3$ in an external magnetic field can be described by the sine-Gordon equation [25] :

$$\frac{\partial^2 \phi}{\partial z^2} - \frac{1}{c^2} \frac{\partial^2 \phi}{\partial t^2} = m^2 \sin \phi \qquad (6)$$

with $c^2 = J \cdot A \cdot a\, S^2$, $m^2 = g\mu_B H^x/2JaS$, ϕ being the angle between the spin S and H^x. The large amplitude solutions in this case are 2 π-solitons and breathers. The solitons give rise to a central peak in $S(q,\omega)$ behaving as a gas of thermally activated noninteracting quasiparticles with finite size [25], [26], [27]: this cross section has the following characteristics:

i) the width is proportional to q at fixed (T, H)
T at fixed (q, H)
$H^{-1/2}$ at fixed (q, T)

ii) the intensity is exponential in T and $H^{-1/2}$.

These predictions were tested experimentally and found to agree with experiment [30], [31]. The exponential T-dependence of the intensity and the linear q-dependence of the width of the central peak are shown in Fig.9 together with theoretical results (full lines). In Fig.10 the q-dependence of the intensity is shown which reflects the shape of the soliton. The solid line is the best fit and the broken lines are theoretical results. The only free parameter is the rest energy E_0 of the soliton goven as

$$E_o = 8m\cdot a\cdot 2J = (g\mu_B H^x \cdot 2J)^{1/2}\cdot 8 \ .$$

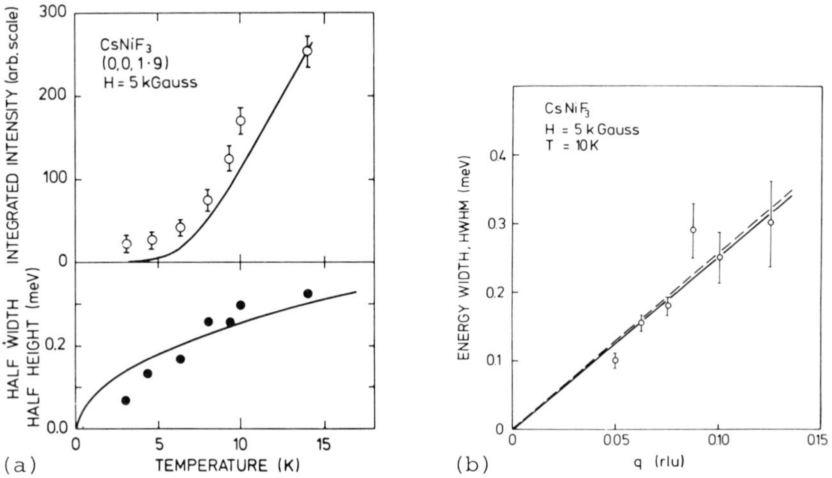

Fig.9 Intensity and width versus temperature (a) and width versus q_c (b) for the central peak in $CsNiF_3$ at 5 kG (solid lines are fits of the theory)

From the known microscopic parameters of $CsNiF_3$ one finds

$$E_o/k = 34 \text{ K} \ .$$

The best fit to the experimental result yields

$$E_o/k = 28 \text{ K}$$

besides for $J(q)$, where $E_o/k = 55$ K was found. The found results can be interpreted as being due to n-scattering from a gas of thermally activated soliton quasiparticles whose size was found to be smaller than expected. This might be due to relativistic effects [26], [27]. The reduced E_o found is probably due to different corrections to be applied to the classical continuum model used [23], [32], [33].

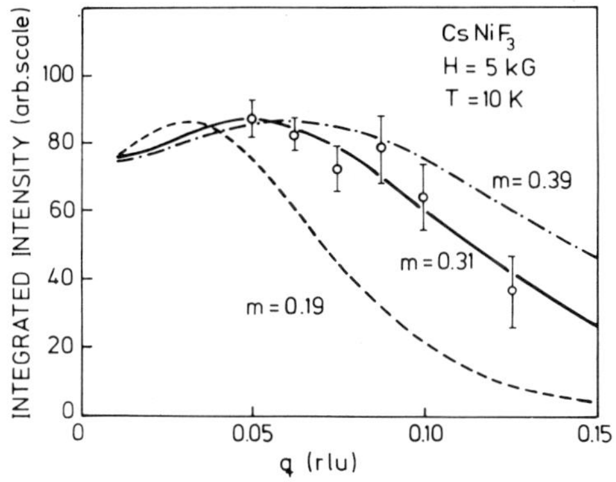

Fig.10 Integrated intensity versus q for the central component in $CsNiF_3$ at 10 K, 5 kG. Solid line is best fit (m=0.31) broken lines calculated for m=0.19 (theoretical value) and m=0.39

4.2 π-Solitons in the 1-D-Planar Antiferromagnet TMMC

Similarly as in the ferromagnetic case the planar antiferromagnetic chain in an external field can be mapped on to a sine-Gordon model [3], [28] with the only difference that now the solitons are π-solitons instead of 2π-solitons in the ferromagnet. These π-solitons transform one of the degenerate "Neèl-states" into the other therefore introducing disorder into the system. The most direct way to observe solitary behaviour in 1-D antiferromagnets is therefore to investigate the correlation length ξ and the energy width of the planes in reciprocal space. Such experiments have been done recently on TMMC showing the predicted exponential temperature and field dependence [34]. In Fig.11 this exponential dependence is demonstrated for the energy width Γ_D and the width in reciprocal space Γ_S. The soliton rest energy found from these results is about 20 % lower than the expected value $E_{s,o}$ = 3.4 K at 10 kG.

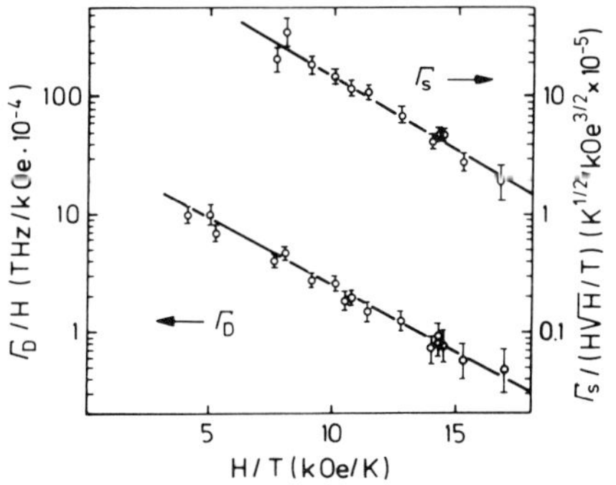

Fig.11 (H/T)-dependence of Γ_D and Γ_S in TMMC

4.3 Solitons in the 1-D-Ising Antiferromagnet $CsCoCl_3$

As for the planar anisotropy there exist two degenerate antiferromagnetic domains in the Ising chain separated by domain walls. As shown by Villain [35] these domains propagate via a small transverse term in the Hamiltonian giving rise to a central peak in $S(q,\omega)$. This can be regarded as a similar phenomenon as the moving solitons described above. Recently this problem has been dealt with again theoretically [16] as well as experimentally [18] in studies on $CsCoCl_3$. These results show the predicted central peak, yet with much less structure as predicted by Villain. In Fig. 12 this central peak is shown as obtained from experiment (a) and from theory (b) showing the agreement between theory and experiment.

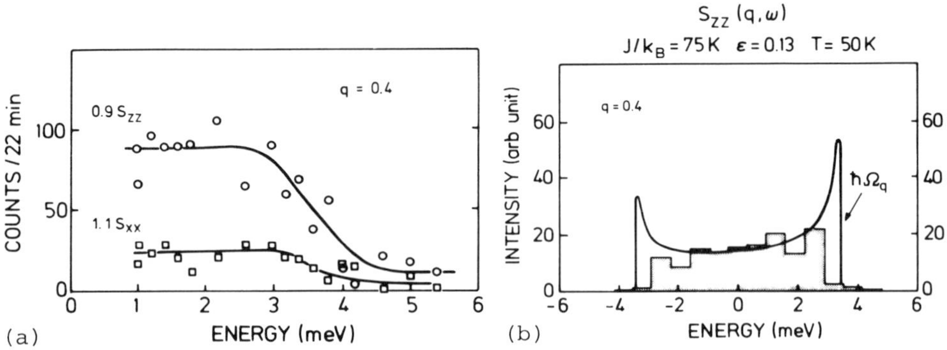

Fig.12 The central peak in $CsCoCl_3$ at 50 K: a) experimental b) theoretical

In concluding this section on n-studies on solitons one should note generally that the theoretical and the experimental situation in the study of nonlinear excitations is much less advanced as for the linear excitation. It is clear, however, that the nonlinear excitations are an essential part of the dynamics of 1-D magnets and cannot be neglected. There are, however, a large number of principal questions open in both theory and experiment: What is the influence of the breathers? How does the finite anisotropy influence the nonlinear modes? What is the relativistic effect? What happens when the density of the solitons becomes high, etc.? To answer these questions big efforts have to be made in theory and experiment.

5. Synopsis

Neutron scattering experiments have contributed significantly to the understanding of the dynamics of linear spin systems. Whereas in the beginning these studies were concentrated on spinwave-like excitations, the efforts are now concentrated on nonlinear excitations. Since the conditions leading to strong nonlinearity are found in many other solid state systems, e.g. around phase transitions, it seems necessary to develop nonlinear physics. In this frame linear and nonlinear modes should not be considered separately, but would emerge consistently from the same nonlinear theory. As test systems magnetic chains appear unique since the variety of J, A, S in real systems is large and since the external magnetic field allows to tune the nonlinearity in a way to use the experimental possibilities most efficiently. This will hopefully enable an experimental investigation of the open questions concerning nonlinear modes. Although the solitons have attracted most of the interest recently, one should not forget that there are still unsolved problems in the "linear" regime including e.g.: spin diffusion or relaxation at finite q for the different models, multimagnon states, S = 1/2 Heisenberg systems, mixed systems [36]. Spin dynamics in 1-D magnets will remain a challenge for n-scatterers, who have to design experiments, which can unambiguously destinguish between linear and nonlinear modes.

Acknowledgement

I am grateful for very stimulating discussions to H.J. Mikeska, J.K. Kjems, K. Hirakawa, G. Shirane, T. Schneider, and H. Yoshizawa. Thanks are due to all colleagues for sending their results prior to publication.

References

1. Steiner, M., Villain, J., Windsor, C.G. Adv. Phys. 25, 87, 1976.
2. Shirane, G., Birgeneau, R.J., B. Physica 86-88 B, 639, 1977.
3. Mikeska, H.J. J. Magn. Magn. Mat. 12, 35, 1979.
4. Marshall, W., Lovesey, S.W. Theory of Thermal Neutron Scattering Oxford Univ. Press, 1971.
5. Steiner, M., Dorner, B., Villain, J. J. Phys. C, 8, 165, 1975.
6. Villain, J. J. Phys. (Paris), 35, 27, 1974.
7. Hutchings, M.T., Shirane, G., Birgeneau, R.J., Holt, S.L. Phys. Rev. B5, 1999, 1972.
8. Heilmann, I.U., Birgeneau, R.J., Endoh, Y., Reiter, G., Shirane, G., Holt, S.L. Solid State Commun. 31, 607, 1979.
9. Heilmann, I.U., Shirane, G., Birgeneau, R.J., Holt, S.L., Kjems, J.K. private comm.
10. Heilmann, I.U., Shirane, G., Endoh, Y., Birgeneau, R.J., Holt, S.L. Phys. Rev. B18, 3530, 1978.
11. Mikeska, H.J. Phys. Rev. B12, 2794, 1975
Mikeska, H.J., Pesch, U. J. Phys. C12, L37, 1979.
12. Müller, G., Beck, H., Bonner, J.C. J. appl. Phys. 50, 7405, 1979.
13. Tellenbach, U. J. Phys. C,11, 2281, 1978
ibid. 11, 2287, 1978.
14. Hirakawa, K., Yoshizawa, H. J. Phys. Soc. Jap. 46, 455, 1979.
15. Buyers, W.J.L., Yamanaka, J., Nagler, S.E., Armstrong, R.L. Sol. State Comm. 33, 857, 1980.
16. Ishimura, N., Shiba, H. Progr. Theoret. Phys. 63, 743, 1980.
17. de Cloiseaux, J., Gaudin, M. J. Math. Phys. 7, 1384, 1966.
18. Yoshizawa, H., Hirakawa, K., Satija, S.K., Shirane, G. to be published.
19. Villain, J. Physica, 79B, 1, 1979.
20. Krummhansl, J.A., Schrieffer, J.K. Phys. Rev. B11, 3535, 1975.

21. see A.R. Bishop, T. Schneider, eds., *Solitons and Condensed Matter Physics*, Springer Series in Solid-State Sciences, Vol.8 (Springer, Berlin, Heidelberg, New York 1978)
22. Bishop, A.R., Krummhansl, J.A., Trullinger, S.E. Physica D. 1, I, 1980.
23. Maki, K., Takayama, H. Phys. Rev. B20, 3223, 1979
ibid. B20, 5002, 1979
Takayama, H., Maki, K. Phys. Rev. B20, 5009, 1979.
24. Schneider, T., Stoll, E. Phys. Rev. Lett. 41, 1429, 1978.
25. Mikeska, H.J. J. Phys. C11, L29, 1978.
26. Leung, K.M., Huber, D.L. Sol. State Comm. 32, 127, 1979.
27. Steiner, M. J. appl. Phys. 50, 7395, 1979.
28. Maki, K. preprint.
29. José, V.J., Sahni, P.S. Phys. Rev. Lett. 43, 78, 1979.
30. Kjems, J.K., Steiner, M. Phys. Rev. Lett. 41, 1137, 1978.
31. Kakurai, K., Kjems, J.K., Steiner, M. in: Ordering in Strongly Fluctuating Condensed Matter Systems (ed. T. Riste) Plenum Press, New York, p. 141, 1980.
32. Fesser, K. submitted to Z. Phys. B.
33. Theodorakopoulos, N., Wunderlich, W., Klein, R. Sol. State Comm. 33, 213, 1980.
34. Boucher, J.P., Regnault, L.P., Rossat Mignod, J., Renard, J.P., Bauillot, J., Stirling, W.G. Sol. State Comm. 33, 171, 1980.
35. Villain, J. Physica 79B, 1, 1975.
36. Endoh, Y., Shirane, G., Birgeneau, R.J., Ajiro, Y. Phys. Rev. B19, 1476, 1979.

Solitons in One-Dimensional Magnets with Various Symmetries

H.J. Mikeska

Institut für Theoretische Physik, Universität Hannover
D-3000 Hannover, Fed. Rep. of Germany

In the last few years one-dimensional (1d) magnets have turned out to be suitable systems to investigate nonlinear modes theoretically and to relate the results to experiments and to computer simulations [1-3]. Such an investigation consists of obtaining analytic solutions to the full nonlinear equations of motion and of considering these modes in thermal equilibrium in order to find their contribution to observable quantities, e.g. correlation functions.

Nonlinear modes in 1d magnets are of two types: kink modes (among them solitons in the strict sense) require a finite amount of energy and are topologically stable; pulse modes do not exhibit an energy gap and may be continuously deformed to localised small amplitude solutions. In the following we will treat in a continuum description classical magnetic chains with an isotropic exchange interaction

$$H_0 = - J \sum_n \vec{S}_n \vec{S}_{n+1} \tag{1}$$

and, in addition, various symmetry reducing interactions. We will present several examples for both types of modes and, in a special case, discuss the crossover from kink to pulse mode as the interaction constants vary. We will finally discuss the thermodynamics of kinks as modified by linearised modes (magnons) and indicate a consistent low temperature treatment of pulse mode contributions to correlation functions.

1. Kink Modes in xy Ferromagnets

Consider a 1d ferromagnet with the Hamiltonian $H = H_0 + H_{xy} + H_{sb}$. H_{xy} serves to establish the xy plane as an easy plane and H_{sb} breaks the symmetry in the xy plane by energetically favouring the x direction. Below we will give three examples with this symmetry, where the equations of motion, for sufficiently strong easy plane anisotropy, can be reduced to the Sine-Gordon (SG) equation

$$\phi_{zz} - c^{-2} \phi_{tt} = m^2 \sin\phi \tag{2}$$

with $\phi(z,t)$ being related to $\Phi(z,t)$, the angle measuring the direction of the spin in the xy plane. This is true for the following examples:

(i) $H_{xy} = A \sum_n (S_n^z)^2$, $H_{sb} = - C \sum_n (S_n^x)^2$ (3)

$m^2 = 2C/J$, $c^2 = 2JAS^2$, $\phi = 2\Phi$

(ii) $H_{xy} = A \sum_n (S_n^z)^2, \; H_{sb} = -\mu B \sum_n S_n^x$ (4)

$m^2 = \mu B/JS, \; c^2 = 2JAS^2, \; \phi = \Phi$

(iii) $H_{xy} = J \sum_n S_n^z S_{n+1}^z, \; H_{sb} = -\gamma J \sum_n (S_n^x S_{n+1}^x - S_n^y S_{n+1}^y)$ (5)

$m^2 = 4\gamma, \; c^2 = 2J^2S^2, \; \phi = 2\Phi$.

In all of these examples, the soliton of the SG theory [4] is a possible kink mode of the corresponding 1d magnet. It describes the transition between the two equivalent equilibrium directions $\pm$ x in cases (i), (iii) and a complete turn in the xy plane in case (ii). Equivalent examples exist in 1d antiferromagnets and have been discussed in [5].

2. Pulse Modes

There are several well known examples of pulse modes: Nonlinear modes, which may be continuously deformed to localised small amplitude solutions have been shown to exist in the isotropic ferromagnetic chain [2], the xy like antiferromagnetic chain [5], and in the SG chain (breather solutions) [4]. An example not discussed so far is the xy like ferromagnetic chain (cases (i) and (ii) in section 1 with $H_{sb} = 0$). Using the method of [2] the following pulse mode is easily found to be a solution of the equations of motion (θ and Φ give the spin direction in polar coordinates)

$$\cos\theta(z,t) = \pm (1-u^2/c^2)^{1/2} \operatorname{sech}\gamma^{-1}(2A/J)^{1/2}(z-ut-z_0) \tag{6}$$

$$\Phi(z,t) = \Phi_0 \mp \left[\operatorname{arctg}\left(\frac{c}{u} \sinh\gamma^{-1}(2A/J)^{1/2}(z-ut-z_0)\right) + \frac{\pi}{2}\right] \tag{7}$$

$$E = 2cS\,(1-u^2/c^2)^{1/2}$$

with $\gamma^{-1} = (1-u^2/c^2)^{1/2}$ and $c^2 = 2AJS^2$. Thus, when a pulse modes moves through, the direction of the spin in the xy plane changes by π. The velocity u varies between 0 and c, values close to c being related to small amplitude, low energy solutions and small velocities being related to high energy solutions with the spin direction passing nearly through the pole, $\theta = 0$.

3. Crossover Between Kink and Pulse Modes

As is clear from the preceding discussion kink modes and pulse modes may evolve into each other as the anisotropy constants vary. E.g. the pulse mode carrying xy like ferromagnetic chain (section 2) becomes a kink carrying system (section 1) when H_{sb} is switched on. This crossover can be followed explicitly for the Hamiltonian (3) with general interaction constants, using solutions given in [6]. To discuss this crossover it is particularly instructive to consider the energy which is calculated to be

$$E = 2^{1/2} cS(1-u^2/c^2)^{1/2} \left[1 \pm \left(1 - \frac{1-(1+2C/A)^{-2}}{(1-u^2/c^2)^2}\right)^{1/2}\right]^{1/2} \tag{9}$$

with $c^2 = 2AJS\,(1+2C/A)$. There are always both pulse modes (upper sign) and kink modes (lower sign) present. In the Sine-Gordon limit (section 1) one has $A \to \infty$, which shifts the pulse mode energy to infinity, whereas the kink mode acquires the typical relativistic velocity dependence. In the isotropic limit C = 0 the kink mode energy vanishes and one obtains (8) for the pulse mode energy. The general behaviour is illustrated in Fig. 1.

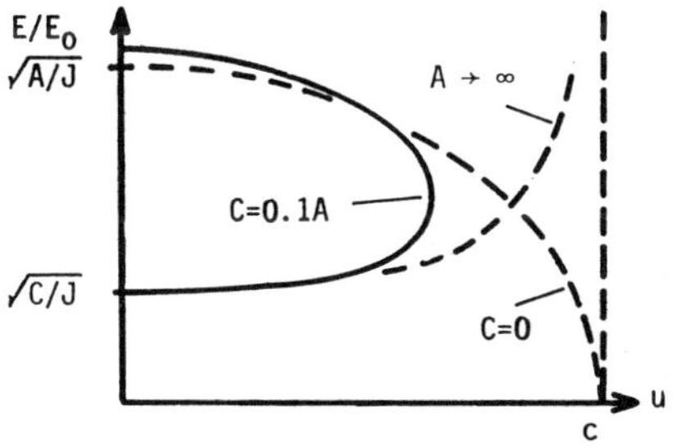

Fig.1 Energy of kink and pulse modes for the magnetic chain (3) $E_0 = 2\sqrt{2}\,JS^2$

In the xy ferromagnet with symmetry breaking owing to a magnetic field, see (4), a similar crossover is expected to occur.

4. Kink and Pulse Modes in Thermodynamic Equilibrium

For a comparison to experimental results we have to consider nonlinear modes in thermal equilibrium. In the case of kink modes the presence of an energy gap guarantees the existence of a low temperature region, in which kinks may be assumed as nonoverlapping. Then kinks can be treated as elementary excitations [7] and the kink density n [1,8] and the dynamic structure factor $S^\alpha(q,\omega)$ (α=x,y,z) [1], can be calculated. For SG kinks there results a central peak according to [1]

$$S^\alpha(q,\omega) = S_0^\alpha\,(q,\omega) \sim n\,\exp(-\,4m\omega^2/k_BTc^2q^2)\; f_\alpha(q) \quad . \tag{10}$$

For a realistic comparison to both neutron scattering experiments [3] and computer simulations [9] this simple result has to be modified to include higher order effects to be discussed in the following.

4.1 Kink-Magnon Interferences

In addition to the nonlinear modes discussed so far, magnetic chains allow for periodic, small amplitude modes, the conventional magnons. The presence of a SG kink modifies magnons according to [10]: the magnon decreases its amplitude inside the kink and acquires a phase shift after passing the kink. We have calculated the modification of the dynamic structure factor owing to the resulting interference effect, assuming nonoverlapping kinks. The kink induced central peak as given in (10) is reduced according to [11] (for α= x,y)

$$S^\alpha(q,\omega) = S_0^\alpha\,(q,\omega)\left[1 - \frac{k_BT}{2m}\left(1 - \frac{m^2+q^2}{12m^2}\right) + O(T^2)\right] \quad . \tag{11}$$

Thus the destructive interference is important for temperatures $T \simeq 2m$.

4.2 Pulse Modes

Attempts to include pulse modes, in particular the breathers of the SG chain, in the elementary excitation picture [12] suffer from the fact that it is not possible to incorporate correctly that low energy pulse modes are necessarily extended and overlapping. However, pulse mode contributions to thermodynamic quantities can be expanded in powers of the temperature. This follows from the absence of an energy gap and is also evident from exact (transfer matrix) results on static quantities [13]: The calculated nonanalytic contributions can be completely reproduced using the kink phenomenology, the remaining power law contributions have to be attributed to pulse modes or to multimagnon processes.

Power law contributions, however, can be calculated in a consistent low temperature expansion without using an elementary excitation picture. We have performed this calculation and obtained the following result [11]: there exists a second well-defined central peak at frequencies $|\omega|<cq$ with an integrated strength

$$\int_{-cq}^{cq} d\omega\, S^x(q,\omega) = \frac{(k_B T)^2}{32\pi m^3} \; \frac{1}{1+(q/2m)^2} \tag{12}$$

and with (classically) temperature independent width. Adding a second contribution of equal magnitude at higher frequencies, (12) agrees with the sum rule available from transfer matrix calculations [13].

Quantitatively, kink and pulse mode contributions are of comparable magnitude. Adding these contributions one obtains quantitative agreement with the measured strength of the central peak in $CsNiF_3$ (a 1d magnet described by (1), (4)) [3] up to T = 8K, when the scaling factor is determined from the one magnon integrated intensity in the long wavelength, low temperature limit. Computer simulations so far are at temperatures too high, to permit a quantitative comparison to the theory.

References

1. H.J. Mikeska: J. Phys. C11, L29 (1978)
2. K.A. Long, A.R. Bishop: J. Phys. A12, 1325 (1979)
3. J.K. Kjems, M. Steiner: Phys. Rev. Lett. 41, 1137 (1978)
4. A.R. Bishop, J.A. Krumhansl, S.E. Trullinger: Physica 1D, 1 (1980)
5. H.J. Mikeska: H. Phys. C13, 2913 (1980)
6. E.K. Sklyanin: preprint (1979)
7. J.A. Krumhansl, J.R. Schrieffer: Phys. Rev. B11, 3535 (1975)
8. J.F. Currie, et al.: Phys. Rev., in print
9. E. Stoll, T. Schneider, A.R. Bishop: Phys. Rev. Lett. 42, 937 (1979)
10. J. Rubinstein: J. Math. Phys. 11, 258 (1970)
11. E. Allroth, H.J. Mikeska: J. Phys. C, in print
12. J. Timonen, R.K. Bullogh: preprint 1980
13. W. Apel, H.U. Everts, H. Schulz: Z. Phys. B34, 183 (1979)

A Comparison of Static and Dynamic Properties of One-Dimensional Magnets and Corresponding sG Systems

J.M. Loveluck*, T. Schneider, E. Stoll, and H.R. Jauslin

IBM Zurich Research Laboratory
CH-8803 Rüschlikon, Switzerland

We examine the question of for what temperatures the mapping of an easy-plane ferromagnet in a field to the sG, the planar and the isotropic models are good approximations. We shall do this by comparing numerical transfer matrix calculations for these models. As far as dynamic properties are concerned, we shall present and compare dynamic form factors for the easy-plane ferromagnet in a field and sG models, as obtained from computer simulations, to unravel the physical origin of the excitation spectrum as observed in neutron-scattering experiments [1].

For these purposes, we consider the classical Hamiltonians defined by

$$\mathcal{H} = \sum_\ell \frac{\dot{\varphi}_\ell^2}{4A} + J\left(\frac{(\varphi_{\ell+1}-\varphi_\ell)^2}{2} - 1\right) - h\cos\varphi_\ell \quad : \mathrm{I} \qquad (1)$$

$$\mathcal{H} = -J\sum_\ell \cos(\varphi_\ell - \varphi_{\ell+1}) - h\sum_\ell \cos\varphi_\ell \quad : \mathrm{II} \qquad (2)$$

$$\mathcal{H} = -J\sum_\ell \vec{S}_\ell \cdot \vec{S}_{\ell+1} + A\sum_\ell (S_\ell^z)^2 - h\sum_\ell S_\ell^x \quad : \mathrm{III}. \qquad (3)$$

The results of numerous studies on $CsNiF_3$ have been reviewed by Steiner *et al.*, [2] the conclusion being that the system consists of ferromagnetic chains for $T > 3$ K. Below $T_N = 2.61$ K, the weak interchain interaction leads to a 3-d antiferromagnetic ordering. It was found that Hamiltonian (3), describing an easy-plane ferromagnet in a field, is appropriate for $T > 3$ K, with [3]

$$S = 1,\ \frac{J}{k_B} = 23.6\ \mathrm{K},\ \frac{A}{k_B} = 9\ \mathrm{K},\ g = 2.4, \qquad (4)$$

where $h = g\mu_B HS$, and H denotes the applied field. A classical description is expected to be valid if [4,5]

$$\frac{2A}{J} \ll 8\pi^2 S(S+1) \qquad (5)$$

$$k_B T \gg (2Ah)^{1/2} \qquad (6)$$

*Present address: Rutherford Laboratory, Chilton, Didcot, Oxfordshire OX11 0QX, England associated with a renormalization of A, reducing it to $A/k_B = 4.5$ [3]. For low temperatures but still consistent with condition (6), the out-of-plane fluctuations are expected to be small, and the planar model II (2) should be appropriate. It was pointed out by Mikeska [6] that, under similar conditions, the dynamics should be described by the sG Hamiltonian I. However, only detailed calculations can show under what conditions these mappings are reasonable, and can quantify the differences for a given temperature. Such results are crucial to calibrate the interpretation of experimental results.

Firstly, we discuss the comparison of the static properties of the three models, including the isotropic case [A = 0, (3)]. Using a numerical solution of the transfer-integral (TI) equation, we have calculated various thermodynamic properties. As an example, in Fig. 1 we show the temperature dependence of the magnetization at a fixed field, corresponding to 5 kGauss for the parameters given in (4). It is seen that the results of all three models coincide at low-enough temperatures, where quantum effects are crucial [7].

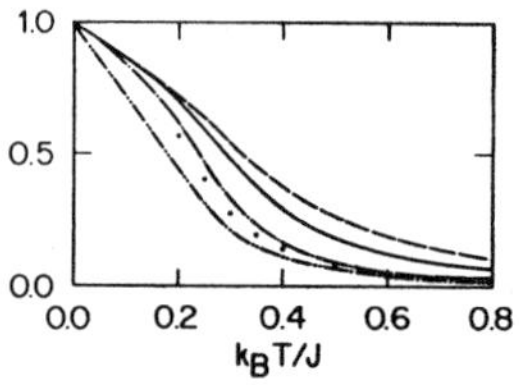

Fig.1 Transfer-integral results for the magnetization as a function of reduced temperature $k_B T/J$ for $h/J = 0.0342$. — — — Model I; ——— Model II; —·— Model III ($A/J = 0.38$); —··— Model III ($A/J = 0$); • Model III ($A/J = 0.19$). After Phys. Rev. Lett. 45. 1505 (1980)

For higher temperatures, the differences become important very rapidly for the planar and sG models. In fact, for $k_B T/J \gtrsim 0.4$, the isotropic model [A = 0, (3)] represents a much better approximation. These results clearly reveal that for $A/J = 0.38$ and $T/J = 0.428$, the sG model is inadequate. The above-mentioned values are just those realized in the neutron-scattering experiments of Steiner and Kjems [1], which have been interpreted in terms of the sG picture. The failure of the planar [8] and sG models in the temperature regime of interest demonstrates the importance of the out-of-plane spin fluctuations, leading to a crossover to the isotropic model at unexpectedly low temperatures.

The failure of the sG picture to describe the static properties leads us to conjecture that it is also inadequate for the dynamics.

To clarify this question, we next discuss some dynamic properties in terms of dynamic form factors defined by

$$S_{AA}(q,\omega) = \int_{-\infty}^{+\infty} dt\, e^{-i\omega t} \langle A(q,t)\, A(-q,0) \rangle , \tag{7}$$

$$A(q,t) = \frac{1}{\sqrt{N}} \sum_{\ell} A_{\ell} \exp iqa\ell , \tag{8}$$

a being the lattice constant, q the wave number, and A_ℓ denoting the variables: $S_\ell{}^x$, $S_\ell{}^y$ for Model III, and $\cos\varphi_\ell$, $\sin\varphi_\ell$ for Model I. In the neutron-scattering experiments reported [1], the measured spectrum corresponds to $S_{||}(q,\omega) = S_{xx}(q,\omega) + S_{yy}(q,\omega)$, or in the sG description to $S_{||}(q,\omega) =$

$S_{ss}(q,\omega) + S_{cc}(q,\omega)$. To estimate these dynamic form factors we used MD techniques [9-12]. We have taken 512 spins in Model III and 10^3 spins in Model I. Both systems were subjected to periodic boundary conditions. In Figs. 2 and 3 we show results for $S_{yy}(q,\omega)$ and $S_{xx}(q,\omega)$ for parameters corresponding to the above-mentioned experiments. Although the occurrence of spinwave resonance in $S_{yy}(q,\omega)$ and of the low-frequency peak in $S_{xx}(q,\omega)$ are in qualitative agreement with the neutron-scattering experiments, it is important to note that these experiments were performed at $q\ a/\pi = 0.1$. Here, we find in $S_{xx}(q,\omega)$ a very broad and weak resonance only.

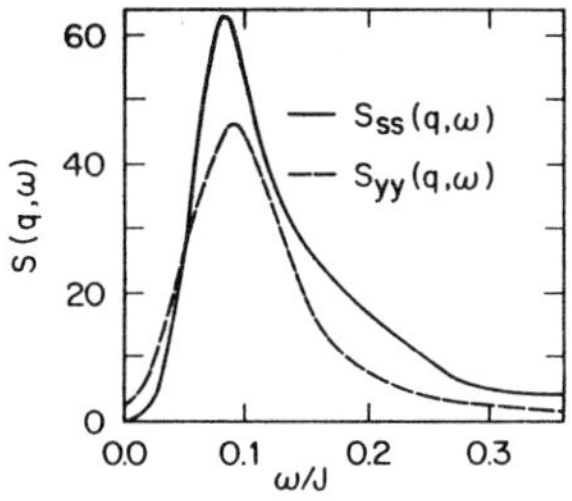

Fig. 2

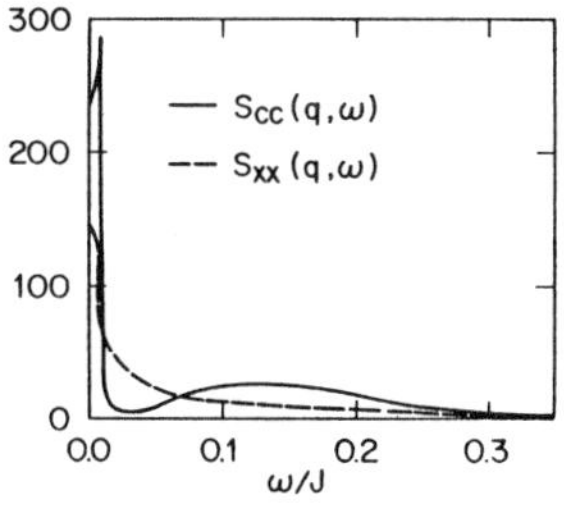

Fig. 3

Fig.2 MD results for $S_{yy}(q,\omega)$ at $k_BT/J = 0.428$ for Model III, and $S_{ss}(q,\omega)$ at $k_BT/J = 0.714$ for Model I, where $A/J = 0.38$, $h/J = 0.0342$ and $q\ a/\Pi = 0.004$. After Phys. Rev. Lett. 45, 1505 (1980)

Fig.3 MD results for $S_{ss}(q,\omega)$ at $k_BT/J = 0.714$ for Model I, and $S_{xx}(q,\omega)$ at $k_BT/J = 0.428$ for Model III with parameters as in Fig.2. After Phys. Rev. Lett. 45, 1505 (1980)

Let us now turn to the interpretation of these peaks. In doing so, it is useful to include for comparison the corresponding dynamic form factors of the sG model [$S_{ss}(q,\omega)$ and $S_{cc}(q,\omega)$] at a much higher temperature, chosen in such a way that the associated magnetization nearly equals that of Model III with $A/J = 0.38$. The resulting temperature renormalization for the sG Model from $k_BT/J = 0.428$ to 0.714 leads to a remarkable agreement between the positions of the spin-wave resonance in the two models (Fig. 2). As far as the physical origin of the resonance in $S_{ss}(q,\omega)$ is concerned, we refer to the recent study of the sG chain [11]. There, it was shown that for $k_BT/J = 0.714$ and higher temperatures, solitons are irrelevant and the spectrum can be described by a high-temperature self-consistent harmonic theory, because the field term in (1) acts as a perturbation. Thus, the peak in $S_{ss}(q,\omega)$ (Fig. 2) was traced back to a one spin-wave resonance, while that in $S_{cc}(q,\omega)$ (Fig. 3) was attributed to multimagnon difference and sum processes. In this view, it becomes clear that if one were to interpret the qualitative agreement between the spectra as evidence for the mapping to sG, it would necessitate a drastic temperature renormalization to a regime where the sG solitons, and in particular the kinks, are irrelevant. On the other hand, if the significant differences between the spectra seen in Fig. 3 were taken as signals for the failure of the temperature renormalization concept and, in turn, of the mapping to sG dynamics, consistency with the static properties would be maintained. In fact, the temperature dependence of the magnetization (Fig. 1) clearly revealed the crucial importance of the out-of-plane spin fluctuations for $k_BT/J = 0.428$, and, in turn, the failure of the sG approximation. Because these fluctuations are equally important in the dynamics, one is led to the conclusion that the sG mapping fails for both the static and dynamic properties.

It is for these reasons that the resonance in $S_{xx}(q,\omega)$ (Fig. 3) is attributable only to multimagnon difference processes or pulse-soliton [13] features of the nearly isotropic model. The former case appears to be more likely, because the specific heat for Model III ($h/J = 0.0342$ and $A/J = 0.38$) adopts its maximum close to $k_BT/J = 0.24$. Accordingly, $k_BT/J = 0.428$ belongs to the high-temperature regime where perturbative treatments are adequate.

In conclusion, we have demonstrated that the mapping of an easy-plane ferromagnet in a field to sG is inadequate for the parameters where the neutron-scattering experiments [1] in $CsNiF_3$ have been performed. This is because the out-of-plane spin fluctuations turned out to be of crucial importance, so that a nearly isotropic model [(3), $A \approx 0$] represents a much better approximation. Accordingly, the observed low-frequency resonance in $S_{xx}(q,\omega)$ must be attributed to multimagnon difference processes or pulse-soliton features of this model.

References

1. J.K. Kjems, M. Steiner: Phys. Rev. Lett. 41, 1137 (1978)
2. M. Steiner, J. Villain, C. Windsor: Adv. Phys. 25, 87 (1976)
3. M. Steiner, J.K. Kjems: J. Phys. C: Solid State Phys. 10, 2665 (1977)
4. H.K. Mikeska, E. Patzak: Z. Physik B26, 253 (1977)
5. A.R. Bishop: Unpublished work
6. H.J. Mikeska: J. Phys. C: Solid State Phys. 13, L29 (1978)
7. G. Reiter: Unpublished work
8. S. Sahni, J.V. José: Phys. Rev. B21, 5381 (1980)
9. T. Schneider, E. Stoll: Phys. Rev. Lett. 41, 1429 (1978)
10. E. Stoll, T. Schneider, A.R. Bishop: Phys. Rev. Lett. 42, 937 (1979)
11. T. Schneider, E. Stoll: Unpublished work
12. J.M. Loveluck, E. Balcar: Phys. Rev. Lett. 42, 1563 (1979)
13. A.R. Bishop: Z. Physik B37, 357 (1980)

Evidence for Soliton Excitations in the One-Dimensional Antiferromagnet TMMC

L.P. Regnault, J.P. Boucher, and J. Rossat-Mignod
Département de Recherche Fondamentale, Centre d'Etudes Nucléaires, 85 X
F-38041 Grenoble Cedex, France

J.P. Renard
Institut d'Electronique Fondamentale, Université Paris XI, Bat. 220
F-91405 Orsay, Cedex, France

J. Bouillot and W.G. Stirling
Institut Laue-Langevin, 156 X, F-38042 Grenoble Cedex, France

1. Introduction

The magnetic compound $(CD_3)_4NMnCl_3$(TMMC) represents a very good realization of a one-dimensional antiferromagnetic system with a spin S = 5/2 [1]. Calculations [2] have shown that dipolar interactions must induce a crossover transition from a Heisenberg to a X-Y behaviour. Recent quasi-elastic [3] and inelastic [4] neutron experiments have established that TMMC is actually in a X-Y state below T ≃ 5 K. Moreover a small in-plane anisotropy has been evidenced [5]. A magnetic field applied perpendicularly to the chain axis drives the spin system towards an Ising state [3,6]. Present theories [7,8,9] predict that a magnetic field induces sine-Gordon solitons. The aim of this study is to check the existence of these non-linear excitations [10,11].

2. Neutron scattering result

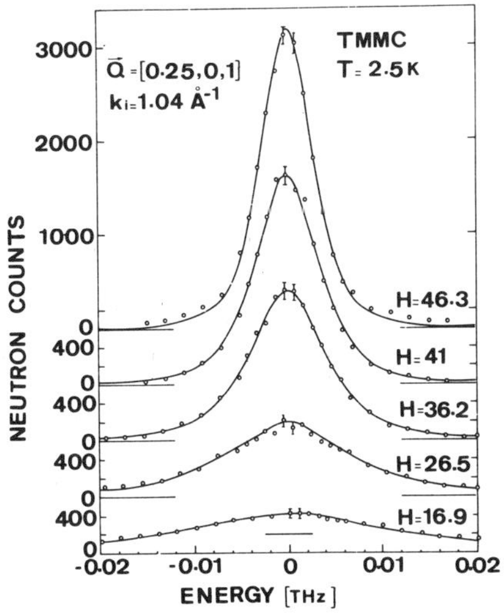

Fig. 1 Energy scans for T = 2.5 K at $\vec{Q}_0$ = [0.25,0,1].

Neutron experiments have been carried out using the three-axis spectrometer IN12 placed on a cold neutron guide tube of the H.F.R. of the Institut Laue-Langevin. To get a good instrumental resolution an incident wavelength was kept fixed at either 6.04 Å or 5.23 Å.

The crystal was oriented with the chain axis ($\vec{c}$-axis) in the horizontal plane such as scattering vectors $\vec{Q}$ = [h,0,ℓ] could be surveyed. The magnetic field was applied perpendicular to the chain-axis.

To obtain the scattering law $S(\vec{Q},\omega)$ an accurate determination of the resolution function of the spectrometer has been done which can be checked by the good agreement between measured and calculated energy widths ($\Delta\omega$ = 0.005 THz at ki = 1.04 Å^{-1}).

Typical energy scans are shown in Fig. 1. The magnetic scattering depends strongly on the magnetic field and the temperature. To deconvolute these data the analytic form of the scattering law must be known.

3. Scattering law [8]

The magnetic field being applied along the vertical y-direction, the magnetic cross-section can be written as :

$$\frac{d^2\sigma}{d\Omega d\omega}(\vec{Q}_0+q\vec{c}^*,\omega) = A(\vec{Q}_0)[S_{//}(q,\omega)+\cos^2\beta S_{\perp}(q,\omega)]$$

where β is the angle between the scattering vector $\vec{Q}$ and the $\vec{c}$-axis.

$$S_{\perp}(q,\omega) = \frac{1}{2\pi}\langle\sin^2\Theta\rangle \frac{a}{\Gamma_\omega\Gamma_q}\left[1+\left(\frac{\pi q}{\Gamma_{2q}}\right)^2+\left(\frac{\omega}{\Gamma_\omega}\right)^2\right]^{-3/2}$$

$$S_{//}(q,\omega) = \frac{D}{\Delta}\,\mathrm{sech}^2(Bq)\exp\left\{-\frac{\omega^2}{\Delta^2}\right\}$$

$$\Delta = \frac{4\sqrt{2}\pi J\ S}{\sqrt{\alpha}}\sqrt{\frac{T}{H}}|q| \ ;\ B = \frac{2\pi^2\ J\ S^2}{\alpha H}\ ;\ D = \frac{8\sqrt{2}\ J\ S^2}{\sqrt{\alpha}}\ a\sqrt{\frac{H}{T}}\frac{1}{H}\exp\left\{-\frac{\alpha H}{T}\right\}$$

α depends on the model because αH represents the energy of the soliton. The following units have been used : H in kOe, T and J in K, Δ and Γ_ω in THz, q in r.l.u. and Γ_q in inverse spin number.

4. Determination of Γ_q and Γ_ω

In the range of H/T value for which the soliton theory is valid (H/T>10 kOe/K) the convoluted values indicate that $S_{//}$ give a much smaller contribution than $S_{\perp}$. Therefore our data reflect mainly the behaviour of $S_{\perp}(q,\omega)$ which depends on two parameters Γ_q and Γ_ω. Γ_q is first determined by analysing the dependence on H/T of the intensity integrated over ω (Fig. 3). Then Γ_ω is obtained from the energy width of scans performed at q = 0 (Fig. 1).

An exponential dependence of Γ_q and Γ_ω on H/T is clearly demonstrated (Fig. 2).

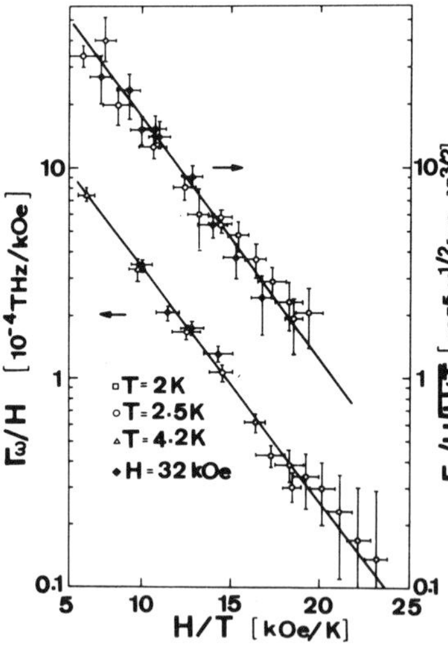

Fig. 2 Dependence of Γ_q and Γ_ω on H and T.

The following laws have been deduced :

$$\Gamma_q = (2.5\pm0.5).10^{-3}\ H\sqrt{\frac{H}{T}}\exp\left(-\ 0.26\ \frac{H}{T}\right)$$

$$\Gamma_\omega = (4.4\pm0.6).10^{-3}\ H\exp\left(-\ 0.26\ \frac{H}{T}\right)$$

$$\left(\frac{\Gamma_\omega}{\Gamma_q}\right)_{\exp} = (1.8\pm0.6)\sqrt{\frac{T}{H}}\ .$$

The dependence on H/T of the prefactor cannot be determined from this experiment, so the theoretical form has been used.

From the knowledge of Γ_q and Γ_ω, the maximum value of peak intensity at q = 0 and ω = 0 (Fig. 4) is well interpreted when the magnetic field or the temperature is varied. At T = 2 K the intensity drops for magnetic fields larger than H_c = 39.5 kOe, due to the build-up of the 3D ordering which consists in a ferromagnetic coupling of the antiferromagnetic chains.

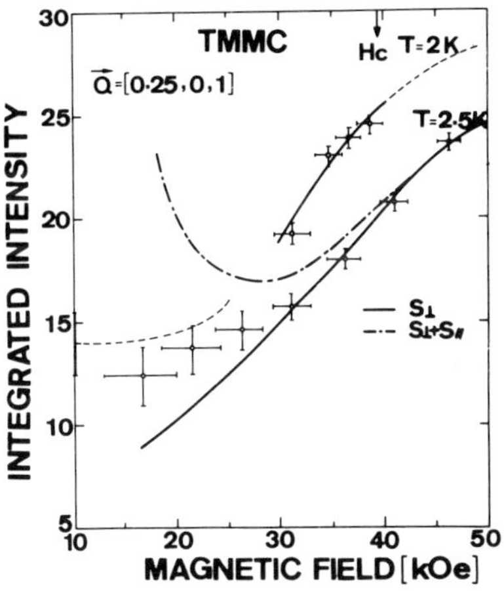

Fig. 3 Integrated intensity in energy versus magnetic field.

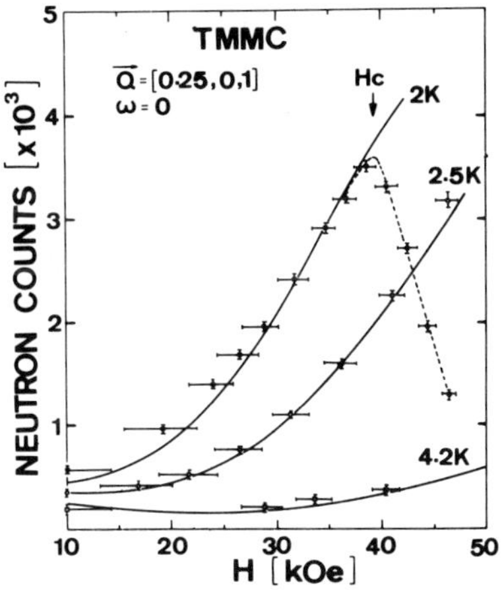

Fig. 4 Intensity at $\omega = 0$ and $q = 0$ versus magnetic field.

The soliton theory explains quite well the data for large H/T values (H/T > 15 kOe/K), i.e. when the contribution due to $S_{//}$ is small (Fig. 3). Whereas for H/T < 15 kOe/K the agreement is not good because the soliton theory gives a too large value for $S_{//}$ (dot-dashed line). The data are better explained by theories based on the small H/T approximation [6,8] (dashed line), where

$$S(q) = \frac{1}{\pi}\left[m_{//}\frac{\Gamma_{q//}}{\Gamma^2_{q//}+\pi^2q^2} + \cos^2\beta m_{\perp}\frac{\Gamma_{q\perp}}{\Gamma^2_{q\perp}+\pi^2q^2}\right]$$

$$\text{with } \Gamma_{q\perp,//}(H,T) \simeq \frac{kT}{4JS^2}\left[1 \pm \frac{\alpha^2}{16}\left(\frac{H}{T}\right)^2\right],\quad m_{\perp,//} \simeq \frac{1}{2}\left[1 \mp \frac{\alpha^2}{32}\left(\frac{H}{T}\right)^2\right] .$$

5. q-dependence of the energy width

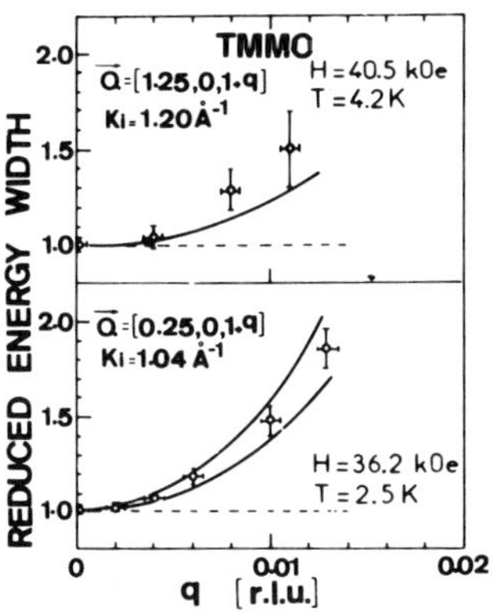

Fig. 5 $\Delta\omega(q)$.

Mikeska's approximation [7] leads to a scattering law in which the dependence q and ω is decoupled

$$S_{\perp}(q,\omega) \sim 1/\Gamma_{\omega}\Gamma_q[1+(\pi q/\Gamma_q)^2][1+(\omega/\Gamma_{\omega})^2]$$

$$\Delta\omega(q) = 2\,\Gamma_{\omega} .$$

In contrast Maki's approximation gives

$$S_{\perp}(q,\omega) \sim 1/\Gamma_{\omega}\Gamma_q[1+(\pi q/\Gamma_q)^2 + (\omega/\Gamma_{\omega})^2]^{3/2}$$

for which the energy width is q-dependent

$$\Delta\omega(q) = 1.533\,\Gamma_{\omega}\sqrt{1+(\pi q/\Gamma_q)^2} .$$

Energy scans performed for several q-values indicated that the energy width increases with 1 (Fig.5).

The agreement between experiments and the Maki's theory is quite good in particular for large H/T values because it represents a better approximation to the exact calculation which we have done numericaly [11].

6. Discussion

The obtained results can be compared with soliton theories [7-8] which give :

$$\Gamma_q = \frac{\alpha\sqrt{2\alpha/\pi}}{|J|S^2} H\sqrt{\frac{H}{T}} \exp\{-\alpha\frac{H}{T}\} \quad ; \quad \frac{\Gamma_\omega}{\Gamma_q} = \sqrt{\frac{2}{\pi\alpha}} \; 4JS\sqrt{1-\frac{D}{|J|}}\sqrt{\frac{T}{H}} \quad .$$

Using the numerical values J = 6.7 K, S = 5/2, D = 1.2 K and α = 0.26, the calculated values of the prefactor of Γ_q ($2.5.10^{-3}$) and the ratio Γ_ω/Γ_q(2.1) are in good agreement with the observed ones. However the experimental value of $\alpha = 0.26 \pm 0.03$ is smaller than the classical one $\alpha_c = g\mu_B S/k = 0.336$. This discrepency can be explained by quantum effects [12] which give a renormalization of the α coefficient :

$$\alpha_q = \alpha_c\left(1 - \frac{g^2}{8\pi}\right)\left(\frac{\alpha_c H}{8\pi|J|S^2}\right)^{\hat{g}^2/8\pi} \text{ with } \hat{g}^2 = g^2/(1-g^2/8\pi) \text{ and } g^2 = \frac{4}{S}\sqrt{1-\frac{D}{|J|}} \, .$$

As α_q is weakly H-dependent and in the magnetic field range used we calculate $\alpha_q = 0.24 \pm 0.01$ in excellent agreement with the experimental result.

The $S_{//}$ component arising from fluctuations parallel to H is difficult to study. In the soliton regime (H/T > 15 kOe) its contribution is too weak and superposed to $S_\perp$. A much higher q-resolution would be needed to separate the two contributions. For H/T < 15 kOe a $S_{//}$ contribution has been evidenced but there is no available theory appropriate to explain this regime.

References

[1] M.T. Hutchings, G. Shirane, R.J. Birgeneau and S.L. Holt, Phys. Rev. B5, 1999 (1972).
[2] D.W. Hone and A. Pires, Phys. Rev. B15, 1, 323 (1977).
[3] J.P. Boucher, L.P. Regnault, J. Rossat-Mignod, J. Villain and J.P. Renard, Sol. Stat. Comm. 31, 331 (1979).
[4] U. Heilmann, R.J. Birgeneau, Y. Endoh, G. Reiter, G. Shirane, S.L. Holt, Sol. Stat. Comm. 31, 607 (1979).
[5] K. Takeda, T. Koike and T. Tonegawa, J. Phys. Soc. Japan, 47, 1013 (1979).
[6] F. Borsa, J.P. Boucher and J. Villain, J. Appl. Phys., 49, 1326 (1978).
[7] H.J. Mikeska, J. Phys. C, 13, 2913 (1980).
[8] K. Maki, J. Phys. C (to be published).
[9] K.M. Leung, D.W. Hone, D.L. Mills, P.S. Riseborough and S.E. Trullinger, Phys. Rev. B21, 4017 (1980).
[10] J.P. Boucher, L.P. Regnault, J. Rossat-Mignod, J.P. Renard, J. Bouillot and W.G. Stirling, Sol. Stat. Comm. 33, 171 (1980).
[11] L.P. Regnault, J.P. Boucher, J. Rossat-Mignod, J.P. Renard, J. Bouillot and W.G. Stirling, J. Phys. C (to be published).
[12] K. Maki and H. Takayama, Phys. Rev. B20, 3223 (1979).

Quantum Effects in the Dynamics of the One-Dimensional Planar Antiferromagnet

Gerhard Müller and Harry Thomas
Institut für Physik der Universität Basel, CH-4056 Basel, Switzerland

Marcos W. Puga and Hans Beck
Institut de Physique de l'Université de Neuchâtel, CH-2000 Neuchâtel
Switzerland

In a recent publication /1/ we found that the T=o dynamics of the 1D s=1/2 Heisenberg antiferromagnet (HB AF) with nearest-neighbor exchange interaction J is almost completely dominated by a particular continuum of excitations (called spin-wave continuum, SWC) bounded by the dispersion branches $\varepsilon_1(q)=(\pi J/2)\sin q$ and $\varepsilon_2(q)=\pi J\sin(q/2)$. It differs markedly from the classical 1D HB AF where the spectral weight is concentrated on a single branch of spin-waves. The result for the structure function $S_{\mu\mu}(q,\omega)\equiv\langle S_\ell^\mu S_{\ell'}^\mu\rangle_{q,\omega}$, which is a special case of Eq.(6), is in good agreement with low-T neutron scattering data on CPC /2/ concerning excitation energies, lineshapes and integrated intensity.

In this paper we outline an extension to the planar s=1/2 AF with Hamiltonian

$$H=\sum_{\ell=1}^{N}\{J_\perp(S_\ell^xS_{\ell+1}^x+S_\ell^yS_{\ell+1}^y)+J_zS_\ell^zS_{\ell+1}^z\}\quad,\ J_\perp\equiv J>o\ ,\ o<J_z<J. \tag{1}$$

By using finite-chain calculations on systems with 4 to 10 spins, we again find a special class of excitations forming a two-parameter continuum (SWC) in (q,ω)-space which dominates $S_{zz}(q,\omega)$ at T=o. Further, we can identify each one of these excitations with a member of the "class C" of non-bound states defined in the framework of Bethe's formalism. By use of des Cloizeaux and Gaudin's method /3/, we obtain the SWC energies

$$\omega_m(q)\ =\ \frac{\pi J\sin\Theta}{\Theta}\sin\frac{q}{2}\cos\left(\frac{q}{2}-\frac{q_m}{2}\right);\ q\geq o\ ,\ o\leq q_m\leq q\ , \tag{2}$$

where $\Theta=\arccos(J_z/J_\perp)$, and q_m labels the different dispersion branches of the SWC. In particular, q_m=o and q_m=q yield the lower and the upper boundary, respectively,

$$\varepsilon_1(q)\ =\ \frac{\pi J\sin\Theta}{2\Theta}\sin q\ ,\ \varepsilon_2(q)\ =\ \frac{\pi J\sin\Theta}{\Theta}\sin\frac{q}{2}\quad. \tag{3}$$

From (2) we deduce the density of states in the SWC

$$D(q,\omega)=(N/2\pi)\{\varepsilon_2^2(q)-\omega^2\}^{-1/2}\quad;\quad \varepsilon_1(q)\leq\omega\leq\varepsilon_2(q). \tag{4}$$

In analogy to our result for the HB AF ($J_z=J_\perp$) and to exact results for the XY AF (J_z=o) /1/, we conjecture that $S_{zz}(q,\omega)$ can be written as a product $M(q,\omega)\cdot D(q,\omega)$ with

$$M(q,\omega)=\text{const}\cdot\{(\varepsilon_2^2(q)-\omega^2)/(\omega^2-\varepsilon_1^2(q))\}^{\alpha} \tag{5}$$

representing the squared matrix elements $|\langle G|S^z(q)|SWC\rangle|^2$ between the non-degenerate ground state and the SWC excitations. The exponent α can be related to exponents of the Baxter model /4/ yielding $\alpha(\Theta) = (\pi/2-\Theta)/(\pi-\Theta)$. Thus we are led to the analytic expression

$$S_{zz}(q,\omega) = \frac{2A}{B(1-\alpha,1/2+\alpha)} \frac{\theta(\omega-\varepsilon_1(q))\theta(\varepsilon_2(q)-\omega)}{(\omega^2-\varepsilon_1^2(q))^{\alpha}(\varepsilon_2^2(q)-\omega^2)^{1/2-\alpha}} \tag{6}$$

where the Beta function serves as a convenient normalizing factor, and the constant A will be determined below. In the limit $J_z/J_\perp = o$ $(\Theta=\pi/2)$, this result is exact /1/.

Fig. 1 shows $S_{zz}(q,\omega)$ at fixed $q=4\pi/5$ as a function of ω for various anisotropies. In the XY limit (α=o) $M(q,\omega)$ is a constant and $S_{zz}(q,\omega)$ reflects the SWC density of states (5) with a square-root divergence at $\varepsilon_2(q)$ and a finite step at $\varepsilon_1(q)$. For intermediate anisotropies ($o<\alpha<1/2$) $M(q,\omega)$ is no longer a constant, but diverges at $\varepsilon_1(q)$ and vanishes at $\varepsilon_2(q)$. This leads to a weaker singularity in $S_{zz}(q,\omega)$ at $\varepsilon_2(q)$ and gives rise to a second divergence at $\varepsilon_1(q)$. We therefore expect that inelastic neutron-scattering experiments will reveal a two-peak structure at low T. In the HB AF limit ($\alpha=1/2$) the matrix elements (5) vanish sufficiently rapidly at $\varepsilon_2(q)$ in order to overcome the density-of-states singularity. Most of the spectral weight is now concentrated at $\varepsilon_1(q)$. It should be remarked that the spectrum contributing to $S_{zz}(q,\omega)$ has no gap at q=o or $q=\pi$ in the whole range $o\leq J_z/J_\perp \leq 1$. This is a clear quantum effect, for it is in marked contrast to classical spin-wave theory. Here, the excitations contributing to $S_{zz}(q,\omega)$ consist of a single branch (in the extended zone) /5/

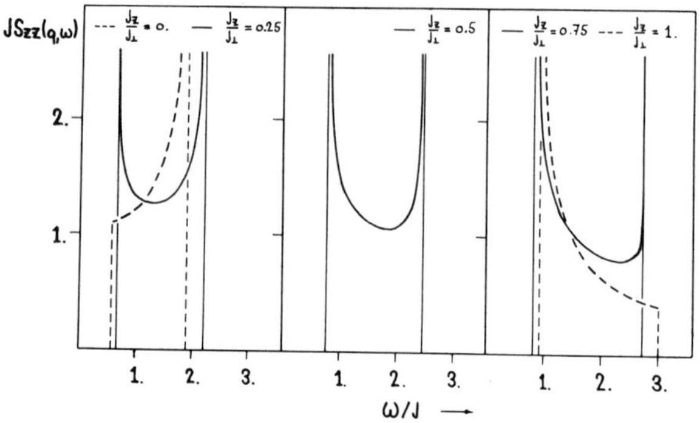

Fig. 1: Out-of-plane structure function $S_{zz}(q,\omega)$ of Eq. (6) versus ω at fixed $q=4\pi/5$ for various anisotropies $J_z/J_\perp$. The prefactor A has been determined for this figure via the sum rule.

$$\varepsilon_{cl}(q)=2Js\{(1-\cos q)[1+(J_z/J)\cos q])\}^{1/2} \quad , \tag{7}$$

which has a gap at $q=\pi$ increasing with planar anisotropy. Note that in the quantum case the spectral weight of $S_{zz}(\pi,\omega)$ is also redistributed, with increasing anisotropy, towards higher energies, but rather in terms of a varying exponent than in terms of an energy gap.

An important check for the validity of our result (6) is the application of the sum rule /6/

$$K_{zz}(q) \equiv \int_0^\infty \frac{d\omega}{2\pi} \omega S_{zz}(q,\omega) = -2JF_x(1-\cos q) , \qquad (8)$$

where the nearest-neighbor correlation function $F_x \equiv \langle S_\ell^x S_{\ell+1}^x \rangle$ has been calculated exactly by Yang and Yang /7/. The first frequency moment of (6)

$$K_{zz}(q) = (AJ\sin\Theta/4\Theta)(1-\cos q) \qquad (9)$$

reproduces the exact q-dependence irrespective of the anisotropy. The integrated intensity is obtained as

$$I_{zz}(q) \equiv \int_0^\infty \frac{d\omega}{2\pi} S_{zz}(q,\omega) = (A/2\pi)\tan(q/2)\,{}_2F_1[1/2,1-\alpha,3/2,-\tan^2(q/2)] \qquad (10)$$

$$I_{zz}(q) = (A/4\pi)q, \quad (q<<1) \;;\; I_{zz}(\pi)=(A/4\sqrt{\pi})\Gamma(1/2-\alpha)/\Gamma(1-\alpha). \qquad (11)$$

In the XY limit the linear term is already the exact result for all q. In the HB AF limit $I_{zz}(q)$ has a logarithmic divergence at $q=\pi$. The asymptotic behaviour of the correlations in real space is $\langle S_\ell^z S_{\ell+R}^z \rangle \sim (-1)^R/R^{2(1-\alpha)}$. Thus, the out-of-plane correlations are strongest in the isotropic limit. With increasing anisotropy the spins become more and more correlated in the easy plane, and hence the fluctuations out of the plane are increasingly suppressed. Another important static quantity is the q-dependent susceptibility

$$\chi_{zz}(q) \equiv 2\int_0^\infty \frac{d\omega}{2\pi}\omega^{-1} S_{zz}(q,\omega) = (A\Theta/\pi^2 J\sin\Theta)\,{}_2F_1[1,\alpha+1/2,3/2,\sin^2(q/2)] \qquad (12)$$

$$\chi_{zz}(o) = (A\Theta/\pi^2 J\sin\Theta) , \quad \chi_{zz}(q) \sim (\pi-q)^{-2\alpha} \quad (q\to\pi) \qquad (13)$$

The exact result for the uniform susceptibility is already available /7/: $\chi_{zz}=\Theta/[J\pi(\pi-\Theta)\sin\Theta]$. Hence, with the sum rule result (8), (9) and with the uniform susceptibility (13), we have two independent ways of determining A, yielding

$$A_K(\Theta) = -8\Theta F_x(\Theta)/\sin\Theta \quad , \quad A_\chi(\Theta) = \pi/(\pi-\Theta) . \qquad (14)$$

The difference between the two values is not greater than a few percent, except close to the HB AF limit, where it reaches 15%. This discrepancy in the isotropic limit is understood and extensively discussed in Ref. /1/.

We conclude by noting that the present approach also works for the in-plane component of the structure function $S_{xx}(q,\omega)$. In this case there are two dominating spin-wave continua, one of which is again confined between the branches $\varepsilon_1(q)$ and $\varepsilon_2(q)$ of Eq. (3) and the second one between the boundaries $\varepsilon_1(\pi-q) = \varepsilon_1(q)$ and $\varepsilon_2(\pi-q)$. We find that the SWC contributions to $S_{xx}(q,\omega)$ can be written as a sum of two terms of the form (6) with the boundaries in one of them replaced accordingly. The prefactors and the exponents in the two terms are different functions of the planar anisotropy. Hence the corresponding lineshapes are predicted to be even more complex than those of $S_{zz}(q,\omega)$. Furthermore we can show that an ansatz of the form (6) also holds for the isotropic $s>1/2$ HB AF with an exponent α which in this case depends on the spin quantum number s. Here we find that with increasing s the lineshape of $S_{\mu\mu}(q,\omega)$ becomes more symmetric. In the limit $s\to\infty$ the exact classical result is recovered. More details will be published elsewhere.

This work is supported by the Swiss National Science Foundation.

References

1. G. Müller, H. Thomas, H. Beck and J.C. Bonner, submitted to Phys.Rev. B

2. I.V. Heilmann, G. Shirane, Y. Endoh, R.J. Birgenau, and S.L. Holt, Phys. Rev. B18, 3530 (1978)

3. J. des Cloizeaux and M. Gaudin, J.Math.Phys. 7, 1384 (1966)

4. A. Luther and I. Peschel, Phys.Rev. B 12, 3908 (1975)

5. R.E. Dietz, L.R. Walker, F.S.L. Hsu, W.H. Haemmerle, B. Vis, C.K. Chau, and H. Weinstock, Sol.State Comm. 15, 1185 (1974)

6. P.C. Hohenberg and W.F. Brinkman, Phys.Rev. B 10, 128 (1974)

7. C.N. Yang and C.P. Yang, Phys.Rev. 151, 258 (1966)

Ground-State and Local Excitations in Co(PYR)$_2$Cl$_2$

W.J.M. de Jonge, C.H.W. Swüste, and K. Kopinga

Department of Physics, Eindhoven University of Technology
5600 MB Eindhoven, The Netherlands

Co(pyr)$_2$Cl$_2$ has been established as a reasonably good representation of a quasi-one-dimensional Ising model by specific heat, susceptibility, and magnetization measurements. The intrachain interaction was found to be ferromagnetic (J_0/k = 10 K). The interchain interactions J'/k amount to approximately two percent of J_0/k, and give rise to a three-dimensional long-range ordered state below T = 3.17 K [1].

The current interest in one-dimensional ferromagnetic systems and their excitation spectrum [2] made it seem worthwhile to investigate the ground state and excited states of the title compound, which is one of the very few model realizations.

In view of the limited size of this report and the general scope of these proceedings, we will, as far as a detailed presentation is concerned, emphasize the results of the experiments in the microwave range. For a complete and extensive account the reader is referred to papers published elsewhere [3,4].

The complex metamagnetic behavior [3] of the system at temperatures below T_N can be interpreted on the basis of a *non-collinear* array of ferromagnetic chains, which are weakly coupled. The spin configurations in the groundstate and the various field induced phases can be obtained from the application of the method of KUDŌ and KATSURA [5]. Starting from a basic element with two possible states (Ising) at each lattice point, the method yields all stable arrangements compatible with the structure. We will not go into details here butsimply refer to Fig. 1, where the resulting arrays are given together with the interaction parameters J_i/k defined by $\mathcal{H}_j = - \Sigma\, J_i\, S_i^z\, S_j^z$, in which the spin components refer to local coordinate systems. In this framework one may

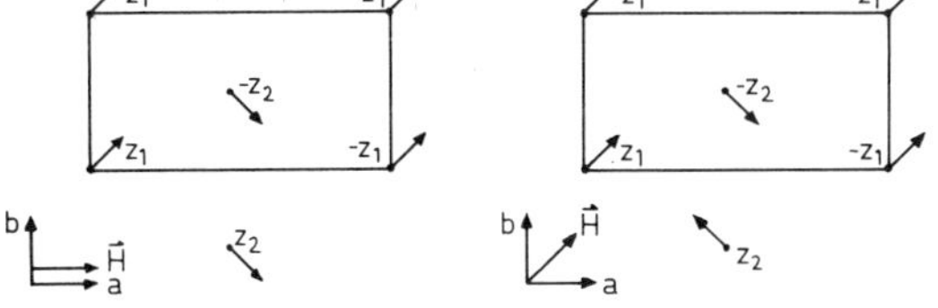

Fig.1 Ground-state and field induced arrays of CO(pyr)$_2$Cl$_2$. The arrows represent ferromagnetic chains perpendicular to the a-b plane.

confront the experimentally observed transtion fields [3] between the various phases with the theoretical expressions resulting from the model. This yields: $J_1/k = -3.9 \times 10^{-2}$ K; $J_2/k = -3.2 \times 10^{-2}$ K; $J_3/k = -14.7 \times 10^{-2}$ K and $J_4/k \simeq 0$.

In a pure Ising system (ferromagnetic or antiferromagnetic) the *intuitive* ground-state, i.e. spins completely parallel or completely antiparallel, is an eigenstate of the Ising Hamiltonian. The excitations in such a system are simple *local* spin reversals, in contrast to the familiar *collective* spin waves found in weakly anisotropic systems. These local spin reversals can be excited directly from the groundstate as has been demonstrated by TORRANCE and TINKHAM [6]. In the case of quasi-one-dimensional systems a typical phenomenon called spin cluster resonance can be observed [7,8]. We define a spin cluster $|m\rangle$ in a chainlike compound as a cluster of m adjacent spin reversals. Given a certain set of local interchain exchange parameters J_i, the excitation spectrum of a quasi-one-dimensional system, with intrachain exchange J_o, can be written in the Ising limit

$$E_m = 2|J_o| + m \sum_i \alpha_i J_i \mp 2m \bar{\mu} . \bar{H} , \qquad (1)$$

where α_i contains information on the specific local magnetic array, and m denotes the number of moments involved in the transition. Since the intra-chain exchange interaction in $Co(pyr)_2Cl_2$ amounts to $J_o/k \sim 10$ K, the excitation spectrum in this case lies at least 15 cm^{-1} above the groundstate and no transitions can be observed in the microwave range at low temperatures. However, when the temperature is increased, spin clusters will be thermally excited and transitions which change the size of such a spin cluster (spin cluster resonance SCR) may be observed by means of microwave irradiation. The transition frequency extrapolated to zero field is a measure for the small interchain interactions.

Fig. 2 shows the field dependence of absorption signals when the field was applied along ϕ. Comparable data was obtained for H//a and H//b which differ only in the slope $\partial\nu/\partial H$. Interpreting these data it may be useful to recall that from (1) the SCR frequencies can be found as

$$h\nu = n(\Delta E + 2 \bar{\mu} . \bar{H}) , \qquad (2)$$

where $\Delta E = 2J_3 - 2J_2 - 2J_1$; $2J_1 + 2J_2 + 2J_3$ or $-2J_1 + 2J_2 + 2J_3$ for H//a, H//b or H//ϕ, respectively, and n denotes the number of spins.

If we consider the results in Fig. 2, it is surpizing, to start with, that at this temperature, well above $T_N = 3.17$ K, any *local* excitation at non-zero frequency can be detected. The predicted linear dependence of ν - H is indeed apparent in the experimental data, indicating the Ising nature of the interactions. The experimental data also show that ΔE for H//a, H//b and H//ϕ, is the same within the experimental accuracy, indicating the same groundstate in all three cases.

This, at first instance, rather surprizing result is also indicated by the rotation diagrams shown in Fig. 3, taken at the same temperature T = 4.2 K. The rotation diagram is taken in the a-b plane, that is in the plane as is shown in Fig. 1, where, depending on the orientation of the field, three distinct high field configurations have to be distinguished at low temperatures. Now, accepting the fact that, as indicated by Fig. 2, local three-dimensional arrays, which allow SCR to be observed, can exist at this temperature, we might expect that different groundstates will show up in a rotation diagram. However, as can be seen in Fig. 3, the SCR pattern at 4.2 K varies rather smoothly in contrast with that observed at lower temperatures also shown in Fig. 3 where the various transitions between the different field induced states are clearly visible. The SCR at 4.2 K can be fitted to the expression $H_{\phi\nu} \simeq H_{\nu}/\cos\phi$ (see Fig. 3), a type of behavior typically for those phenomena where only the component of an external field along a fixed direction is relevant. In this case two branches of SCR can be distinguished with minima (corresponding to the directions of the magnetic moment) symmetrically around the a axis at 38° from that axis. This is in agreement with our model sketched in Fig. 1.

Now let us inspect the results obtained at lower temperatures. The effect of the change of ground-state array accompanying the rotation of the field in the region $H >> H_{critical}$ is clearly visible. The drawn lines represent the calculated SCR results based on the set of exchange parameters given above, which were obtained from the transition fields. Comparing the results one should realize that the data around 2 kOe originate from the SCR which cannot be uniquely identified (see the dashed line in Fig. 2). It is therefore not possible to obtain values for J_1 and J_2 from these data. However, the general agreement between the data and the calculated curves is certainly satisfying.

The results presented in Fig. 3 indicate that at 4.2 K indeed we are dealing with only one ground-state energy irrespective of the field direction. Equating ΔE to an exchange interaction or a sum of exchange interactions

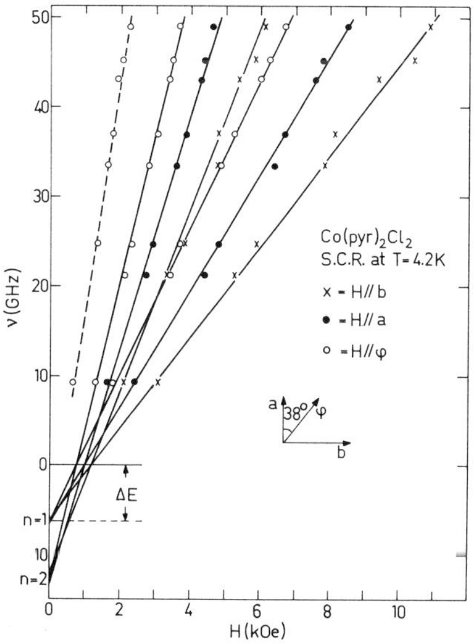

Fig.2 Field dependence of the n=1 and n=2 spin cluster resonances in $Co(pyr)_2Cl_2$. The dashed line connects data which presumably arise from closely spaced higher order excitations (n>2).

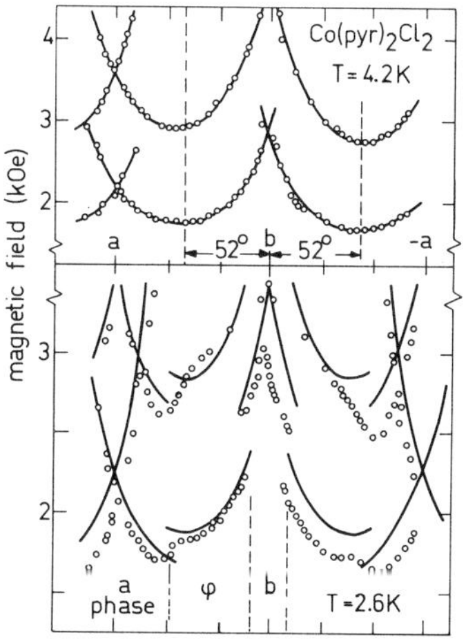

Fig.3 Angle dependence of a part of the spin cluster resonance spectrum at two temperatures. Note the "discontinuities" in $\nu(\phi)$ at 2.6 K, in contrast to the behavior at 4.2 K.

yields (see (2)) $\sum_i J_i/k = -16 \times 10^{-2}$ K. This value is very close to $J_3/k = -15 \times 10^{-2}$ K obtained from the phase transitions. It is clear that *qualitatively* any of the three-dimensional ground-state pictures in Fig. 1 can give rise to the observed behavior in Fig. 3. Physically however, it is hard to see why any of these states would be realized irrespective of the field direction. Moreover, comparing the $\sum_i J_i/k = -16 \times 10^{-2}$ K with the values for J_1/k, J_2/k and J_3/k given above we are led to the conjecture of a pseudo two-dimensional local ground-state in which only J_0/k and J_3/k can be locally distinguished. The much smaller J_1/k and J_2/k apparently average to zero, which may be due to their long ranged dipole character combined with the shrinking average size of the microscopic "ordered" regions. It can be easily shown from (2) that such a conjecture indeed leads to the single groundstate observed at 4.2 K and yields a quantitative agreement for ΔE.

If we now turn away from the local interactions, experienced by each individual spin, and consider the bulk properties, it appears that the values for the exchange parameters, found from the various experiments, show that in the ordered state the *effective* coupling between the b-c layers is rather weak. J_1/k, J_2/k and J_4/k are not only rather small, but their contribution to the "interplane" (a-c) interaction in the AF ground-state also averages approximately to zero. In view of this one might compare the experimental T_N with the theoretical prediction for the two-dimensional ordering temperature T_c for a rectangular Ising array given by $\sinh |J| / kT_c \sinh |J'| / kT_c = 1$. Taking for $T_c = 3.17$ K and $J/k = 10.6$ K, being the average between the reported values 11.7 K and 9.5 K, one obtains $|J'| / k = 0.21$ K, which is close to the actually measured interaction J_3/k. Also the behavior of the sublattice magnetization versus T in the ordered state can be compared with the theoretical behavior of a two-dimensional Ising rectangular model. Such a comparison, where we have taken the same set of parameters that yields the correct ordering temperature, gives a surprizingly good agreement. We may also compare our data with those obtained recently by PIRÉS and HONE [9] who solve the chains exactly within the Ising approximation and thereafter couple the chains by a molecular field mechanism. The resulting comparison with the susceptibility and specific heat data yields $J/k = 10.8$ K and $|J'| / k = 0.175$ K. In this case J'/k denotes the total interchain interaction which is comparable with J_3/k in our analyses.

Summarizing the present results one could state that spin cluster resonance, typically for (quasi-) one-dimensional Ising systems, has resulted in a consistent set of local interaction parameters. These parameters explain the rectangular Ising features observed in the ordered state. Spin cluster resonance also showed that outside the long ranged ordered state two-dimensional (rectangular) correlated regions do exist.

References

1. K. Takeda, S. Matsukawa and T. Haseda, J. Phys. Soc. Japan 30, 1330 (1971).
2. J.D. Johnson and J.C. Bonner, Phys. Rev. Lett. 44, 616 (1980).
3. W.J.M. de Jonge, Q.A.G. van Vlimmeren, J.P.A.M. Hijmans, C.H.W. Swüste, J.A.H.M. Buys and G.J.M. van Workum, J. Chem. Phys. 67, 751 (1977).
4. C.H.W. Swüste, A.C. Phaff and W.J.M. de Jonge, J. Chem. Phys., to be published.
5. T. Kudō and S. Katsura, Progr. Theor. Phys. 56, 435 (1976).
6. J.B. Torrance and M. Tinkham, Phys. Rev. 187, 587 (1969).
7. M. Date and M. Motokawa, Phys. Rev. Lett. 16, 1111 (1966).
8. Q.A.G. van Vlimmeren and W.J.M. de Jonge, Phys. Rev. B19, 1503 (1979).
9. A. Pires and D. Hone, J. Phys. Soc. Japan 44, 43 (1978).

New High-Field Phenomena in Spin-Peierls Systems*

I.S. Jacobs, J.W. Bray, and L.V. Interrante
General Electric Corporate Research and Development
Schenectady, NY 12301, USA

D. Bloch and J. Voiron
Laboratoire Louis Néel, CNRS, F-38042 Grenoble Cedex, France

J.C. Bonner
Physics Department, University of Rhode Island, Kingston, RI 02881, USA

The spin-Peierls (SP) transition in zero field is a spin-lattice dimerization phenomenon which occurs in a quasi-1-D Heisenberg S= ½ antiferromagnetic chains. The chains are coupled to the 3-D phonon field of the lattice, and a magneto-elastic transition results which is the insulating analogue of the well-known Peierls transition in a quasi-1-D conductor. For $T<T_{sp}$, the magnetic chains have a dimerized (alternating) character, resulting from an underlying progressive lattice distortion, with a non-magnetic singlet ground state and an excitation energy gap to magnetic excited states. The character of an alternating as opposed to uniform chain lattice, together with the corresponding magnetic excitations, is shown schematically in Fig. 1.

SP transitions are rare: they do not occur in the majority of quasi-1-D antiferromagnets. They were first observed in the donor-acceptor complexes $TTF^+MS_4C_4(CF_3)_4^-$ (abbreviated TTF-BDT), where M=Cu, Au; with T_{sp}=12K and 2.1K respectively. In the TTF-BDT systems, experimental measurements of susceptibility, EPR, NMR, specific heat and lattice dimerization are in good agreement with a simple mean-field theory [1]. The characteristic signature of the SP transition as illustrated in, say, the powder suscepti-

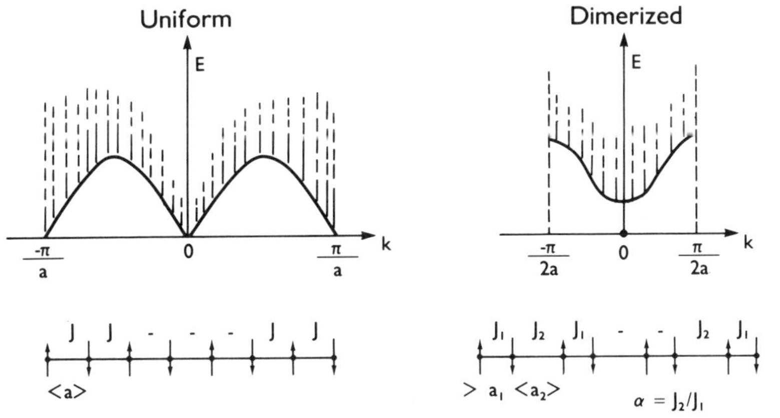

Fig. 1 Illustration of the conformation and elementary excitations of the uniform chain vs. the dimerized (alternating) chain.

*Work supported at GE by AFOSR and at URI by NSF and NATO.

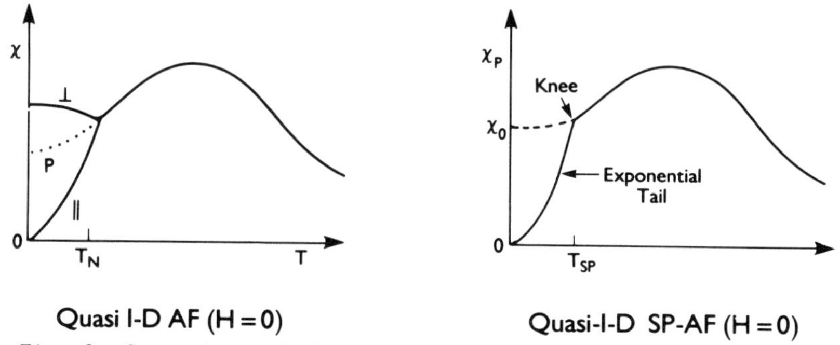

Fig. 2 Comparison of the ordering susceptibility phenomena in zero field of a quasi-1-D AF vs. a quasi-1-D SP-AF.

bility, is shown in Fig. 2. The ordering of a regular quasi-1-D antiferromagnet is also sketched for comparison.

The novelty of SP transition systems makes it of considerable interest to study their behavior in a magnetic field [2]. In zero field the SP system can be mapped into a half-filled pseudo-fermion band system where the dimerization wave-vector $q=2k_F=\pi/a$ (a=intrachain spacing; k_F=Fermi wave-vector). As the magnetic field increases, k_F decreases steadily towards zero. This effect is illustrated in Fig. 3. No corresponding "tuning" possibility exists for a Peierls conductor because k_F is fixed by the density of carriers. Initially, although k_F decreases, q sticks at π/a on account of Umklapp effects. Umklapp scattering produces a commensurability energy required for q to switch to a new (commensurate or incommensurate) value.

High magnetic field experiments and high-pressure studies are effective probes for exploring the extended phase diagram of TTF-CuBDT, e.g., out to magnetic fields of almost 200 kOe [3]. A number of interesting features have already been observed. For example, for a quasi-1-D antiferromagnet, an initial increase in T_N for H>0 is observed, whereas we have established

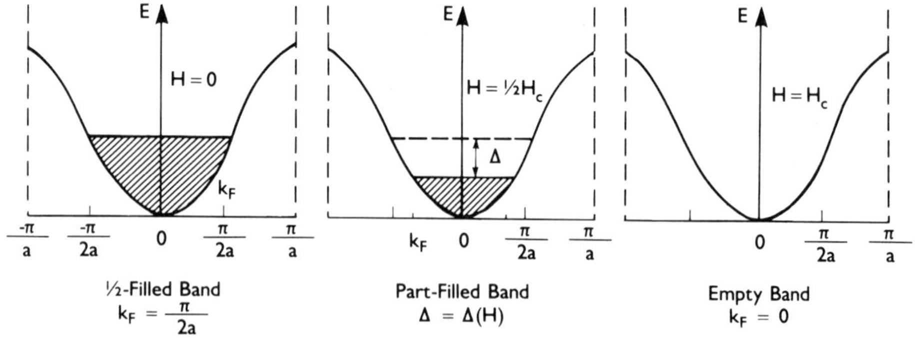

Fig. 3. Illustration of the effect of a magnetic field using the pseudo-fermion representation of the Heisenberg chain.

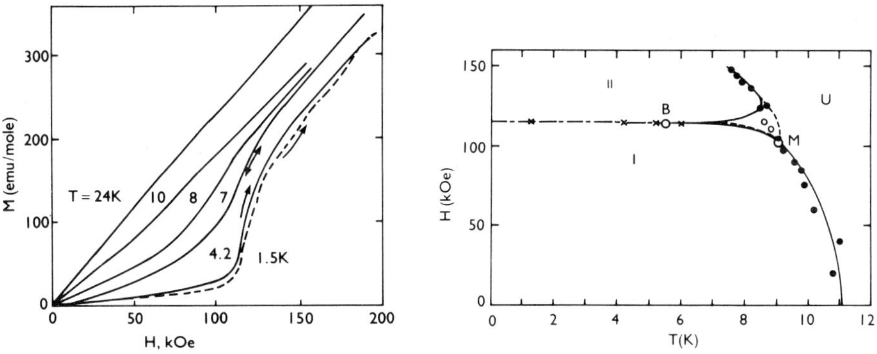

Fig. 4. The experimental magnetization isotherms out to high fields and the resulting field-temperature phase diagram.

both experimentally and theoretically that the opposite phenomenon happens for a quasi-1-D antiferromagnetic SP transition. This initial decrease in T_{sp} in a field has been studied by both neutron scattering and magnetization experiments, with results in reasonable agreement with theoretical predictions.

Recent experiments over an extended field range have demonstrated the existence of a new phase above about 115 kOe. This new phase is indicated in the magnetization isotherms of Fig. 4 (left) as inflection points at approximately 115 kOe at lower temperatures. Commensurability arguments predict that the dimerized SP phase will be stable only for a limited range of magnetic field, above which the distortion wave-vector q will switch to

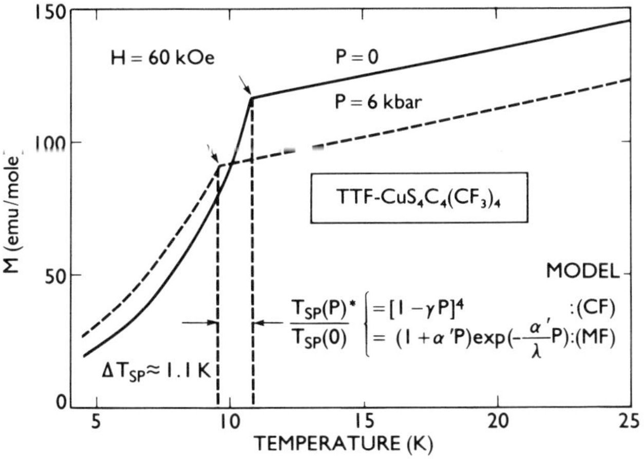

Fig. 5 Experimental results showing the depression of the SP transition by 6 kbar pressure.

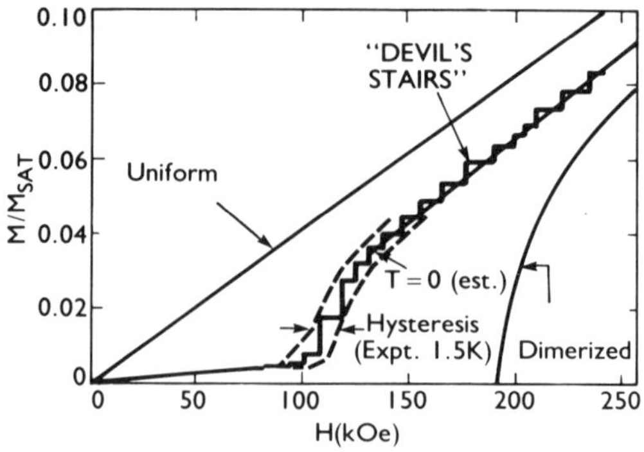

Fig. 6 A "Devil's Stairs"-like theoretical magnetization isotherm fitted to the experimentally observed hysteresis.

a different value. The values of field and temperature at which this occurs will correspond to a new type of multicritical point. The field-temperature phase diagram obtained experimentally is shown in Fig. 4 (right). Phase I is the SP phase, and phase II is the new high field phase corresponding, presumably, to this switch in q. A multicritical point, M, appears at approximately 9K and 105 kOe, and may be a Lifshitz point. Microscopic experimental studies to determine the precise nature of the new Phase II may be facilitated by the application of pressure. Initial results of high-pressure experiments show a depression of T_c (see Fig. 5). These results suggest that applying pressure may "shrink" the entire phase diagram of Fig. 4, and make the new phase accessible to experimental investigation. Relevant theoretical studies have been performed [4].

Finally, the high-field magnetization studies on TTF-CuBDT show a most unusual wide field range of hysteresis for temperatures below point B of Fig. 4 (right) [3]. Hysteresis is observed to begin at about 80 kOe and persist up to at least 200 kOe. Each magnetization curve is sheared in field as well as broadly curved. This may be the result of differential strain effects on the crystallites. However, the intriguing possibility exists that TTF-CuBDT approximates a Devil's staircase system at low temperature. This effect is shown in Fig. 6 as a multiple-step-function character of the very low temperature magnetization isotherms, and the extensive hysteresis results from the multiplicity of first-order phase transitions.

References

1. I.S. Jacobs, J.W. Bray, H.R. Hart, Jr., L.V. Interrante, J.S. Kasper, G.D. Watkins, D.E. Prober and J.C. Bonner, Phys. Rev. B14, 3036 (1976). L.V. Interrante, J.W. Bray, H.R. Hart, Jr., I.S. Jacobs, J.S. Kasper, P.A. Piacente and J.C. Bonner, "The Spin-Peierls Transition and Other Aspects of the Magnetic Behavior of the Compounds TTF.$MS_4C_4(CF_3)_4$(M=Pt,Cu,Au)" in *Quasi-One-Dimensional Conductors II*, ed. by S.Barišić, A. Bjeliš, J.R. Cooper and B. Leontić, Lecture Notes in Physics, Vol.96 (Springer, Berlin, Heidelberg, New York 1979) p.55
2. J.W. Bray, Solid State Comm. 26, 771 (1978). L.N. Bulaevskii et al., ibid, 27, 5 (1978). M.C. Cross, Phys. Rev. B20, 4046 (1979). M.C. Cross and D.S. Fisher, Phys. Rev. B19, 402 (1979).
3. D. Bloch, J. Voiron, J.C. Bonner, J.W. Bray, I.S. Jacobs and L.V. Interrante, Phys. Rev. Lett. 44, 294 (1980).
4. J.W. Bray, Solid State Comm., to appear.

Part IV

Polymers

Solitons in Polyacetylene: A Summary of Experimental Results

A.J. Heeger and A.G. MacDiarmid

Laboratory for Research on the Structure of Matter
University of Pennsylvania, Philadelphia, PA 19104, USA

Abstract

The experimental results relevant to the proposal that neutral and charged amplitude solitons dominate the electronic properties of polyacetylene, $(CH)_x$, are summarized.

I. Introduction

Interest in the study of the physics of polyacetylene, $(CH)_x$, has focussed on the proposal [1, 2] that neutral and charged amplitude solitons dominate the magnetic, electrical and optical properties. The soliton picture has received increasing experimental support through analysis of the magnetic, [3-6] infrared, [7, 8] transport, [9, 10] and photo-transport [11] properties of lightly doped samples.

In this paper we focus on the primary questions related to the existence of solitons in polyacetylene: a) Are the unpaired spins which appear on isomerization from cis-to-trans-$(CH)_x$ the result of mobile bond-alternation domain walls; i. e. neutral solitons? b) Does the doping of $(CH)_x$ proceed via the soliton mechanism?

The proposed soliton in $(CH)_x$ is a topological kink [1, 2, 12, 13, 14] in the electron lattice system; a bond alternation domain wall. Since there is complete degeneracy (the kink can be anywhere!), and since the mass is small, [2, 3] we expect the kink to be mobile. The competition of elastic and condensation energies spreads the domain wall over a region of about 15 a [2] where a is the carbon-carbon distance along the $(CH)_x$ chain. Associated with the kink is an electronic bound state which should appear near mid-gap. If this bound state contains one electron, the soliton is neutral with spin $\frac{1}{2}$ and therefore paramagnetic. If the bound state is empty then the soliton is charged, with spin 0, and non-magnetic. Similarly double occupancy would lead to a spin 0 charged state.

II. Neutral Solitons in Polyacetylene (B. R. WEINBERGER, E. EHRENFREUND, A. PRON, A. J. HEEGER and A. G. MAC DIARMID, J. Chem. Phys. 72, 4749 (1980))

It is well known [3, 15] that trans-$(CH)_x$ exhibits an intense spin resonance signal indicative of unpaired electrons at a level of a few hundred parts per million (i. e. about one per every 3000 carbon atoms). This signal is absent in pure cis-$(CH)_x$ and grows only on isomerization. [16] From analysis of the g-value and linewidth, GOLDBERG et al. [3] suggested that this spin resonance could arise from bond-alternation domain walls; i. e. defects induced during the isomerization. Such defects might arise, for example, by nucleation of the trans-$(CH)_x$ in the A phase (Fig. 1) on one end of a chain and the B phase on the other, from an originally perfect cis-chain. Alternatively, the defect might arise from re-construction of a chain with an odd number of carbon atoms, [17, 18] such odd length chains having been created by bond breaking during the complex isomerization process. With one kink per chain, the experimental density of unpaired electrons would imply a molecular weight of order a few thousand C-H units. This is consistent in magnitude with recent estimates obtained from radioactive labeling experiments. [19]

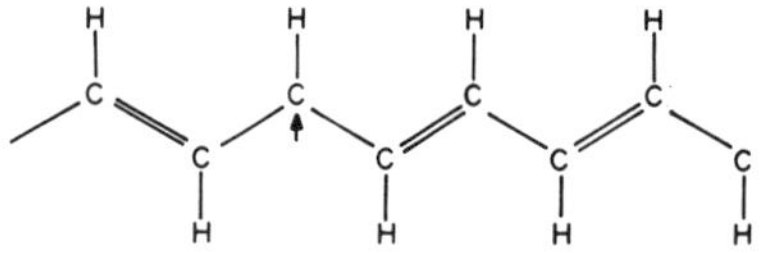

Fig. 1 Neutral domain wall kink in trans-$(CH)_x$.

In any case, the spin resonance in trans-$(CH)_x$ arises from defects induced during isomerization. The g-value precisely identifies their origin as the π-electron system, [3] consistent with the neutral soliton. There are two fundamental predictions of the neutral soliton picture which must be verified experimentally:

- The soliton should be highly mobile.
- The wave function of the electronic bound state should be spread over $\sim$ 15a.

The ESR spectra of the paramagnetic center in undoped polyacetylene have been studied in detail and the line shape and temperature dependence have been analyzed in both $(CH)_x$ and $(CD)_x$ in each of the two isomeric forms. [3] Both trans-$(CH)_x$ and trans-$(CD)_x$ display narrow resonances which broaden appreciably with decreasing temperature consistent with the picture of a mobile defect.

The mobility of the unpaired spins in trans-$(CH)_x$ has been confirmed by observation of the Overhauser effect in trans-$(CH)_x$ by NECHTSCHEIN et al. [5]. They found an enhancement of the proton NMR intensity by pumping the electronic system at $\hbar\omega_s$, the electronic Zeeman energy. This result implies a nuclear relaxation process due to the $(S_+I_- + S_-I_+)$ term in the hyperfine coupling Hamiltonian. The effectiveness of such a term requires mobility of the electronic spins so that the large difference in

Zeeman energy ($\hbar\omega_s - \hbar\omega_1$) implied by such mutual spin flips can be absorbed by the translational energy of the electronic spin. NECHTSCHEIN et al., [5] however, did not observe the Overhauser effect in cis-$(CH)_x$, but rather the "solid state" effect due to immobile electron spins. The results of the pumping experiments, implying mobile paramagnetic defects in trans-$(CH)_x$ and fixed defects in cis-$(CH)_x$ are consistent with ESR linewidths obtained by SHIRAKAWA et al. [15] for samples with varying cis-trans content.

The temperature dependence of the ESR linewidth of trans-$(CD)_x$ and -$(CH)_x$ are displayed in Fig. 2. Comparison of the data shows similar temperature dependences for the two isotopic forms. However, confirming the hypothesis that the proton-soliton hyperfine interaction is the major source of the $(CH)_x$ linewidth, the $(CD)_x$ resonance was found to be appreciably narrower throughout the entire temperature range studied. Both the trans-$(CH)_x$ and the trans-$(CD)_x$ line shapes had Lorentzian character. This, coupled with the broadening of the lines with decreasing temperature, is indicative of motional narrowing.

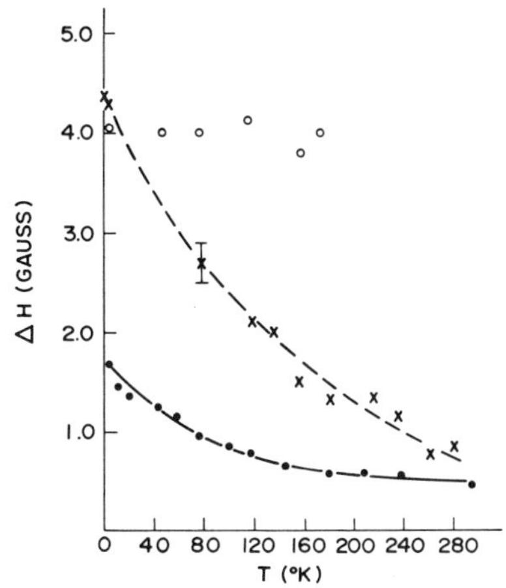

Fig. 2 First derivative peak-to-peak linewidths, ΔH vs temperature. (xxx) trans-$(CH)_x$; (•••) trans-$(CD)_x$; (ooo) cis-$(CD)_x$.

Also plotted in Fig. 2 is the temperature dependence of the ESR linewidth of cis-rich $(CD)_x$ over the temperature range 4.2 - 170 K. The Overhauser effect experiments of NECHTSCHEIN et al. [5] indicated that the defect is mobile in the trans and fixed in the cis isomer. The unnarrowed weakly temperature dependent resonance widths in both cis-rich $(CH)_x$ and $(CD)_x$ are consistent with that picture.

The immobile defects observed in the cis-rich films may be associated with small lengths of trans isomer locked in by regions of cis polymer. We, therefore, have attempted to fit the experimental cis-rich $(CH)_x$ and $(CD)_x$ resonance lines by numerical simulations of the unnarrowed hyperfine spectrum of a soliton of half-width seven lattice constants. Following SU et al. [2] we chose a wave function for the unpaired spin in the soliton of the form $\Psi(n) = (1/\sqrt{N})\mathrm{sech}(n/N)\cos[(\pi/2a)na]$, where $\Psi(n)$ is the wave function at the nth monomer unit away from the domain wall center and 2N defines the full width of the wall. The hyperfine constant at the nth monomer would be $a_{eff} = a_o|\Psi(n)|^2$, where a_o is the hyperfine constant for an unpaired spin localized to a single C—H unit. The results of the computer simulations are presented in Fig. 3. Each individual hyperfine line in the composite spectrum was assumed to be Lorentzian in shape having the high temperature asymptotic width of the experimental results (0.45 G). The best fit for soliton half-width N = 7 to the experimental cis-rich $(CH)_x$ absorption line (82 K) is shown in Fig. 4 (top) and corresponds to $a_o^H = 31$ G. Fig. 4 (bottom) depicts the best fit for soliton half-width of 7a to the cis-$(CD)_x$ line (113 K) and corresponds to hyperfine coupling constant for the deuteron $a_o^D = 5.4$ G. The value of a_o^H so determined is consistent with accepted values. Furthermore, the ratio a_o^H/a_o^D obtained in our simulations of 5.7 is in agreement with the ratio of proton and deuteron gyromagnetic ratios ($\gamma_H/\gamma_D = 6.5$). The assumption of a narrower soliton wave function leads to detailed structure in the lineshapes, inconsistent with the experimental results.

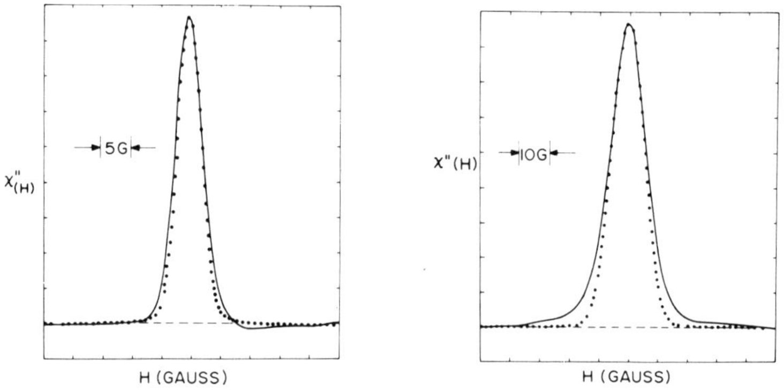

Fig. 3 Left: cis-$(CD)_x$ absorption line (113 K). (•••) Computer simulation assuming soliton half-width of 7a, $a_o^D = 5.4$ G, intrinsic hyperfine linewidth of 0.45 G. Right: cis-$(CH)_x$ absorption line (82 K). (•••) Computer simulation assuming soliton half-width of 7a, $a_o^H = 31$ G, intrinsic hyperfine line width of 0.45 G.

The topological nature of the soliton would imply a high degree of one dimensionality in its diffusive motion. Analysis of the esr lineshape [6] and the frequency dependence of the proton T_1 data [5] indicate an anisotropy in the spin diffusion in excess of 10^6. The observed anisotropy is much greater than would be expected on the basis of the anisotropic band structure of Grant and Batra. [20] The very clear one-dimensional character of the experimental spin diffusion in the presence of more moderate one-electron interchain coupling may be taken as direct evidence of the collective nature of the magnetic species.

III. Soliton Doping in $(CH)_x$

The soliton theory leads to specific predictions regarding the changes in physical properties to be expected upon doping.

- Charged solitons are spinless. [2] Thus if doping occurs via the generation of charged solitons, the charge carriers would be spinless, and the localized states formed by light doping would not show paramagnetism.

- A localized electronic bound state is predicted to appear at mid-gap; one such state for each soliton. [2]

- Since the bond lengths are altered near the center of the kink, and an antisymmetric displacement pattern results, the lattice dynamics will be locally perturbed, giving rise to intense infra-red absorption modes [21].

In the following paragraphs we summarize the experimental evidence related to these three key features of the soliton doping concept.

a. Solitons in Polyacetylene: Magnetic Susceptibility (S. IKEHATA, J. KAUFER, A. PRON, T. WOERNER, M. A. DRUY, A. SIVAK, A. J. HEEGER and A. G. MAC DIARMID, Phys. Rev. Lett. (in press)

TOMKIEWICZ et al. [22] have presented evidence against solitons in $(CH)_x$. They argue, from measurements of the magnetic susceptibility (χ) at a concentration near $y = 0.009$, that the properties of doped $(CH)_x$ are determined by an inhomogeneous mixture of metallic islands even at light doping levels, and that the semiconductor-metal (SM) transition is caused by the percolation of such metallic islands. The importance of the magnetic properties has caused us to re-examine the behavior of the susceptibility in a series of samples carefully prepared to achieve maximum uniformity.

$(CH)_x$ films were prepared at -78° C; [23] the doping utilized a technique developed to achieve a homogeneous distribution of dopant ions. A large flat trans-$(CH)_x$ film ($\sim$ 130 mg after doping) was placed in the center of a 500 ml bulb to which two cold fingers were attached, one entering at the top, the other at the bottom. The film rested on small glass supports thus permitting both sides to be exposed evenly to the AsF_5 vapor. A fixed amount of purified AsF_5, $\sim$ 10-20% in excess of that required for a given doping level, was transferred to the upper cold finger and the apparatus was evacuated and sealed. Doping was carried out by first holding the upper cold finger at -131° C for 12 hours. Independent mass spectroscopic studies showed that the vapor of AsF_5 (vapor pressure < 0.1 torr) over the AsF_5 contained no HF or AsF_3. After 12 hrs. the lower cold finger was held at -196° C for 12 hrs during which time the upper cold finger was gradually permitted to warm to approximately room temperature. After this time all the AsF_5 had been cryopumped into the lower cold finger. The lower cold finger was then held at -131° C and the upper cold finger at room temperature for 12 hours thus repeating the doping process a second time. This entire procedure was repeated a total of 4 times, and after final cryogenic pumping for 12 hours, the films were weighed and loaded into either an NMR tube (~ 100 mg of sample) or an EPR tube (~ 2 mg of sample). The large volume of the reaction bulb and the small vapor pressure of AsF_5 (which would decrease still further during the reaction) leads to very slow doping (~ 5 days). For the film having a 7% nominal AsF_5 content, an elemental analysis gave the composition $[CH_{1.03}(AsF_{5.96})_{0.067}]_x$. The sum of the analytical percentages of the constituent elements was 99.13%. It is therefore assumed that all films contained the AsF_6 species. Although different anions can be generated by AsF_5 doping, this method assures the presence of the AsF_6^- species [23].

Samples prepared in this manner have a more sharply defined SM transition as a function of y and show higher electrical conductivities in the transitional region. For example at $y = 0.002$ we found $\sigma = 0.5\ \Omega^{-1}\text{-cm}^{-1}$, and at $y = 0.02$ we found $\sigma = 30\ \Omega^{-1}\text{-cm}^{-1}$. These values are much larger than obtained earlier [10] using the more rapid doping (by more than three orders of magnitude for $y = 0.002$ and by more than one order of magnitude for $y = 0.02$). At higher (and lower) concentrations, the conductivities were typical of doped polyacetylene. The thermopower was measured as a function of y for the samples prepared for susceptibility measurements as described above. The transition to metallic behavior is more abrupt as a function of y and occurs at a lower concentration than obtained earlier [10], with standard doping procedures. For example, at $y = 0.0004$, $S = 840\ \mu V/^{\circ}K$ and at $y = 0.0025$, $S = 20\ \mu V/^{\circ}K$ whereas in earlier studies such small values were not achieved until $y > 0.01$. We conclude that for $0.001 < y < 0.02$, the electrical properties are particularly sensitive to the doping process. Inhomogeneous doping can occur resulting in lower quality electrical properties and a smeared SM transition. The doping technique described above appears to minimize the occurrence of such inhomogeneities.

The spin susceptibilities were obtained using quantitative spin resonance techniques (either at 10 GHz using a ruby standard or at 10 MHz using the Schumacher-Slichter method) and were fitted to

$$\chi(T) = \chi_p + N_c \mu_B^2 / k_B T \qquad (1)$$

where χ_p, N_c, μ_B and k_B are the Pauli susceptibility, number of unpaired spins, Bohr magneton and Boltzmann constant, respectively. The concentration dependence of N_c is illustrated in Fig. 4; N_c begins to decrease on doping for $y > 0.001$ and falls rapidly with continued doping in agreement with the earlier results of WEINBERGER et al. [4]. In the more heavily doped samples, the spin susceptibility is almost independent of temperature. Typical values for $y = 0.053$ are listed in Table 1, implying $N_c < 1$ ppm.

Table 1 Susceptibilities Measured for $[CH(AsF_5)_{0.053}]_x$

T	χ_s (10^{-8} emu/mole)
300 K	6 ± 3
77 K	5 ± 3
4.2 K	11 ± 3

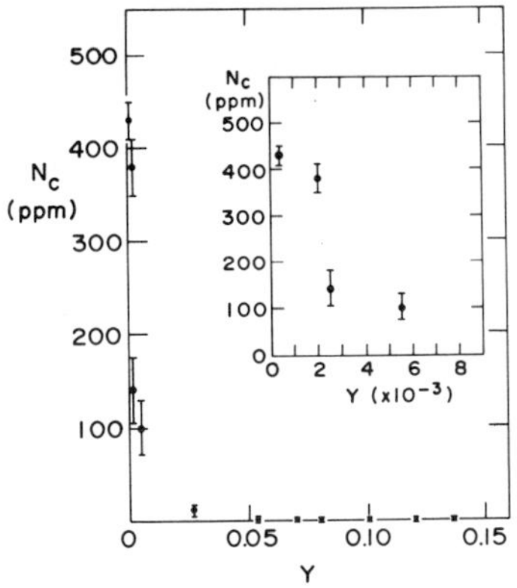

Fig. 4 Number of unpaired spins, N_c, as a function of dopant concentration.

The temperature independent Pauli susceptibility is shown in Fig. 5 as a function of y; χ_p is small at all concentrations below y = 0.053. At concentrations below y = 0.005 the accuracy is limited by the finite Curie-law contribution. However, for each of the low concentration data points with $y < 0.01$ $\chi_p < 5 \times 10^{-8}$ emu/mole. At higher concentration a finite χ_p emerges; e.g. at y = 0.053, $\chi_p = (5 \pm 3) \times 10^{-8}$ emu/mole. The 5.3% sample was cross-checked by comparing the independent results obtained using ^1H and ^{19}F as standards; values agreed to within the error. The Pauli term increases rapidly above y = 0.07 to a value of about 3×10^{-6} emu/mole at the highest doping levels, in agreement with the results of WEINBERGER et al. [3].

A heavily doped sample (10%) was prepared using cis-$(CH)_x$ starting material. The resulting $\chi_p \simeq 4.7 \times 10^{-6}$ emu/mole is in agreement with the high concentration data in Fig. 5; consistent with earlier observations [24, 25] of cis-trans isomerization during doping.

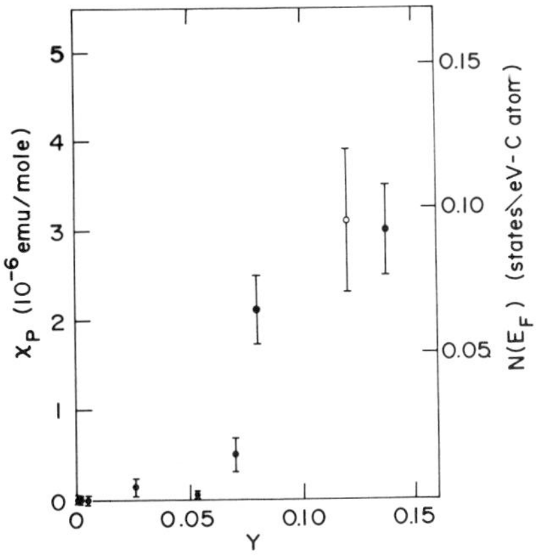

Fig. 5 Pauli susceptibility (χ_p) as a function of dopant concentration.

In the transition region χ_p appears to be particularly sensitive to the doping procedure; e.g. $y = 0.026$ gave $\chi_p \simeq 10^{-7}$ emu/mole whereas at $y = 0.053$, χ_p was considerably smaller. Given the abrupt increase by two orders of magnitude (Fig. 5), fluctuations in y due to inhomogeneous doping could easily result in an apparent continuous increase. Such fluctuations may well be the source of any residual χ_p for y below ~ 0.05.

Our results for lightly doped samples are consistent with the soliton doping mechanism; [2] the Curie-law contribution decreases upon doping, but the Pauli susceptibility is small and apparently zero in the limit of completely homogeneous doping. Thus the localized states generated by light doping are non-magnetic. In view of the dramatic increase in electrical conductivity, we conclude that the charge carriers are spinless.

χ_p remains small even in the high conductivity regime as shown for $y = 0.053$ in Table 1. The onset of "metallic" behavior is not observed in χ_p until $y > 0.07$, whereas the transport data imply a critical concentration more than an order of magnitude lower. This behavior is inconsistent with the proposal [22] that the SM transition is due to percolation through metallic islands. In that case, χ_p would turn on smoothly at y well below the percolation threshold indicated by electrical measurements, whereas the data show precisely the opposite behavior.

The small χ_p for $y < 0.05$ is not consistent with the results reported earlier by TOMKIEWICZ et al. [22]. Their higher values apparently result from inhomogeneous doping as suggested in their paper. However, our studies show that these inhomogeneities can be avoided (at least using trans-$(CH)_x$) leading to higher quality electrical properties and a sharp threshold for the onset of χ_p. Thus, our earlier conclusion [4] that the charge carriers are non-magnetic at low doping levels is confirmed.

The concentration range $0.005 < y < 0.05$ is of particular interest; χ_p has not turned on, whereas the transport data indicate an excellent conductor. The existence of spinless charge carriers seems to persist well into the highly conducting region where an independent soliton picture would be inadequate. The increase observed in χ_p near $y = 0.07$ may signify the closing of the energy gap [26, 27] and the onset of single-particle metallic behavior. At higher concentrations, the magnitude of χ_p coincides approximately with that expected from the single-particle band picture [20].

b. Solitons in Polyacetylene: Localized States near Mid-gap (N. SUZUKI, M. OZAKI, S. ETEMAD, A. J. HEEGER and A. G. MAC DIARMID, Phys. Rev. Letters (to be published))

In this section we examine the effects of soliton doping on the optical absorption spectra. Calculations of the absorption coefficient (α) show that a soliton kink on a chain strongly suppresses the interband transition, whereas transitions involving the mid-gap soliton level are found to have a significantly enhanced absorption cross-section. The results are in agreement with the experimental absorption spectra obtained from trans-$(CH)_x$ lightly doped with AsF_5 and subsequently compensated with NH_3.

TAKAYAMA, LIN-LIU and MAKI (TLM) [12] considered the continuum limit of the SU, SCHRIEFFER and HEEGER [2] (SSH) model of a one-dimensional (1-d) $(CH)_x$ chain. Their analytical results are in excellent agreement with the numerical results of SSH and allow explicit calculation of wave functions and matrix elements. In particular, SUZUKI et al. [28] carried out calculations of the matrix elements for optical transitions using the TLM theory [12]. They find that for a perfect dimerized chain ($\Delta(x) = \Delta_o$), the interband absorption coefficient (per carbon atom), for transitions from valence band (VB) to conduction band (CB), is

$$\alpha_i(\omega) = A f_i(\omega) \tag{2}$$

$$f_i(\omega) = \left(\frac{E_g}{\hbar\omega}\right)^2 \frac{E_g}{\sqrt{\hbar\omega^2 - E_g^2}} \tag{3}$$

$$A \equiv \frac{16\pi\hbar e^2 |M_x|^2}{m^2 n c W E_g} \tag{4}$$

where $E_g = 2\Delta_o$, $W = 4t_o$, m is the electron mass, e is the electron charge, c is the velocity of light, and n is the index of refraction. The energy dependence of $f_i(\omega)$ is shown in Fig. 6. Although $\alpha_i(\omega)$ diverges as $(\hbar\omega - E_g)^{-\frac{1}{2}}$, this square root singularity will be smeared out by disorder, interchain coupling [20], or fluctuations [13].

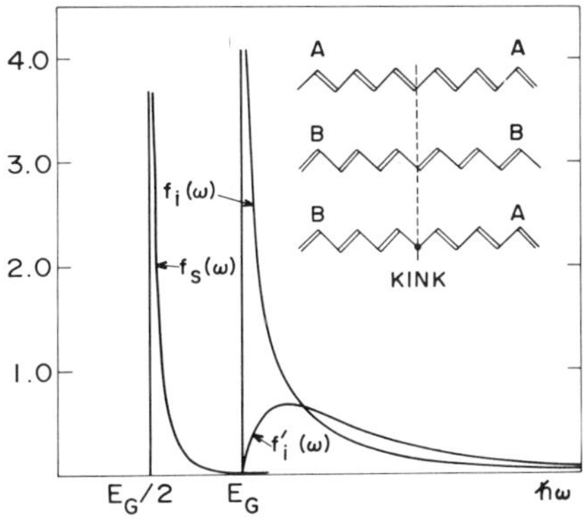

Fig. 6 The absorption function, $f_i(\omega)$ for the perfect dimerized chain is shown; $f_s(\omega)$ and $f_i'(\omega)$ describe the absorption in the presence of a soliton.

For a trans-$(CH)_x$ chain with a static kink ($\Delta(x) = \Delta_o \tanh(x/\xi)$), the soliton formation energy takes the minimum value $\frac{2}{\pi}\Delta_o$ with $\xi = \xi_o \equiv \hbar v_F/\Delta_o$, [12] and one bound state appears at the center of the gap. SUZUKI et al. [28] calculated α_s, for transitions between a soliton level (S) and the band states, and α_i', for the interband transition on a chain containing a kink. Neglecting Coulomb interactions the two transitions involving the soliton level are degenerate (VB → S and S → CB) with

$$\alpha_s(\omega) = A \frac{\pi^2 \xi_o}{a} f_s(\omega), \tag{5}$$

$$f_s(\omega) = \frac{E_g/2}{\sqrt{(\hbar\omega)^2 - E_g^2/4}} \operatorname{sech}^2\left(\frac{\pi}{2} \frac{\sqrt{(\hbar\omega)^2 - E_g^2/4}}{E_g/2}\right). \tag{6}$$

The factor $(\pi^2 \xi_o/a)$ in (5) indicates an enhancement of the soliton transition; for W = 10 eV and E_g = 1.4 eV, one finds $\xi_o \simeq 7a$ and $(\pi^2\xi_o/a) \simeq 70$. As a result, the transitions involving the soliton level should be observable even at extremely small concentrations. The interband $\alpha_i'(\omega)$ (per carbon atom) in the presence of a kink is

$$\alpha_i'(\omega) = A \frac{\xi_o}{L} f_i'(\omega), \tag{7}$$

$$f_i'(\omega) = 16 \frac{\hbar\omega}{\pi E_g} \int \frac{dz}{z'zK'K} \left[\pi \frac{\sinh\frac{\pi}{2}K \cosh\frac{\pi}{2}K'}{\cosh\pi K' - \cosh\pi K} - \frac{K'}{K'^2 - K^2}\right]^2 \tag{8}$$

where $z' = (\frac{2\hbar\omega}{E_g} - z)$, $K' = \sqrt{z'^2 - 1}$ and $K = \sqrt{z^2 - 1}$.

See note added in print.

The energy dependence of $f_s(\omega)$ and $f_i'(\omega)$ are shown in Fig. 6; $\alpha_s(\omega)$ diverges as $(\hbar\omega - E_g/2)^{-\frac{1}{2}}$ at half the energy of the interband transition. The singularity will be smeared in the real system, as described above. Note that (7) $\alpha_i'(\omega)$ is reduced by the factor ξ_o/L where L is the chain length. Thus we expect a uniform suppression of the entire interband transition due to the existence of a kink on the chain. The remarkable change in interband oscillator strength caused by the presence of a soliton can be understood on a simple physical basis. Since the soliton is a domain wall, the wavefunctions of the conduction (Ψ_k^c) and valence (Ψ_k^v) bands may be approximated by those of the A phase to the right and by those of the B phase to the left of the kink. Since the broken symmetry switches the "single and double bonds", $\langle\Psi_k^c(A)|p_x|\Psi_k^v(A)\rangle = -\langle\Psi_k^c(B)|p_x|\Psi_k^v(B)\rangle$ and the total matrix element $\langle\Psi_k^c|p_x|\Psi_k^v\rangle$ will vanish. Because of the coherent mixing within the domain wall, the matrix element is non-zero, but reduced by the fractional part of the chain which contributes; i.e. ξ_o/L.

The optical absorption spectra of <u>trans</u>-$(CH)_x$ are shown in Fig. 7. Curves 1 through 5 show the effects of doping; 1-undoped, 2-very lightly doped with AsF_5 (about 0.01%), 3-lightly doped with AsF_5 (about 0.1%), and 4-subsequently compensated with NH_3; all data are from the same $(CH)_x$ film. The doping was carried out <u>in situ</u> using extreme care so that the results could be directly and quantitatively compared. Curve 5 on Fig. 7 was obtained with a separate film doped to a somewhat higher level (~.5%); quantitative comparison was possible through normalization of the absorption curve prior to doping. The effect of temperature on undoped $(CH)_x$ is shown in the inset; data obtained at 77 K and 4.2 K are indistinguishable.

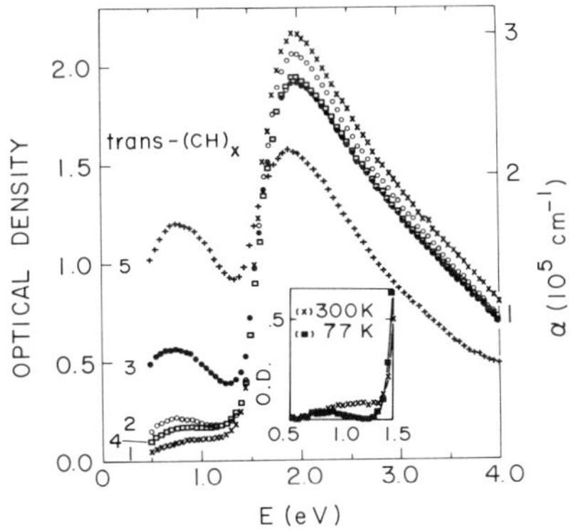

Fig. 7 Absorption spectra of trans-$(CH)_x$ (see text).

The strong absorption band with edge at 1.4 eV and peak at 1.95 eV has been attributed [26] to the direct interband transition in a 1-d band structure, and can be viewed as arising from a transition from the peak in the density of states in the valence band to that in the conduction band. The rounding appears to shift the position of the peaks in the VB and CB densities of states by about 0.2 eV. In addition to the main peak, a weak absorption is observed centered near 0.9 eV, corresponding to a transition between the peak in VB density of states and a level inside the gap. Relative to the VB edge, the gap state occurs at about 0.7 eV; i.e. near mid-gap.

On doping with AsF_5 the low energy absorption grows with increasing concentration and shifts toward lower energy; compensation decreases α almost to the same level as before doping. The strength of the main absorption decreases on doping with AsF_5, but it does not recover after subsequent compensation with NH_3.

The characteristic features of the low energy and main absorption bands can be explained if we assume that the doping proceeds through formation of positive charged solitons (S^+) and that the low energy α is associated with the transition from the valence band into the S^+ level to form a neutral soliton (S^o). As the number of S^+ increases with AsF_5 doping, the strength of the low energy transition ($S^+ + e \rightarrow S^o$) grows proportionally. Further, the calculations presented above predict that the interband transition will be suppressed with the introduction of soliton kinks, again in qualitative agreement with the experimental results. On reacting with NH_3, the S^+ level is compensated, and the low energy band decreases accordingly. On the other hand, the interband transition does not recover to its initial strength on compensation, since the π-electron kinks remain on the chain; only the charged center is compensated.

The integrated intensity of the low energy absorption band after light doping (curve 3) is about one-tenth of that of the main absorption band of the undoped sample. From this we estimate a doping concentration (using (5) and (6) of about 0.1%, in satisfactory agreement with the experimental value. In particular, the existence of an enhancement factor consistent with the theoretical parameter ($\pi^2 \xi_0/a$) is implied by the strength of the transition. The decrease in intensity of the interband transition is also evident in curve 3 and corresponds to about a 10% reduction relative to the undoped sample. If we assume each dopant induces a charged soliton kink, we have about one per 10^3 carbon atoms (0.1%) implying that each kink removes on the average 10^2 carbon atoms from the interband transition.

The data indicate a gap about 0.7 eV above the VB edge, precisely at the center of the 1.4 eV gap. This is consistent with recent studies [11] which showed that the enhanced photoconductivity in trans-$(CH)_x$ is directly related to the presence of localized states generated by isomerization and/or doping with energy near mid gap.

c. Solitons in Polyacetylene: Infrared Studies (C. R. FINCHER, JR., M. OZAKI, A. J. HEEGER and A. G. MAC DIARMID, Phys. Rev. B 19, 4140 (1979))

Several years ago FINCHER et al. [7] discovered that upon light doping, two new absorption modes appeared in the ir (1370 cm^{-1} and 900 cm^{-1}) with remarkable intensity. The strength of the induced modes grew in proportion to the dopant level, becoming comparable to any of the $(CH)_x$ ir lines at about 0.1%. Thus FINCHER et al. [7] concluded that the doping induced ir modes had oscillator strength enhanced by approximately 10^3 and that such a large enhancement must arise from coupling the new vibrational modes (induced by doping) to the electronic oscillator strength of the polyene chain. FINCHER et al. [7] emphasized the generality of their results; the same modes were observed for iodine and AsF_5 (p-type), and for Na doping (n-type). Thus these intense absorptions are not due to specific vibrations of the dopant molecules, nor to detailed interactions between the molecules and the polymer chain. The observed generality suggests that the intense ir absorption modes are intrinsic features of the doped $(CH)_x$ chain.

MELE and RICE [8] have recently shown that these phenomena can be understood as a direct result of the soliton model. They carried out theoretical studies of the vibrational excitations in $(CH)_x$ [29] and in $(CH)_x$ containing charged solitons. Two strongly infrared active modes, derived from bulk optical phonons, appear in the presence of a soliton kink. The calculated frequencies (1360 cm^{-1} and 980 cm^{-1}) and the calculated oscillator strengths are in excellent agreement with the data [7].

The origin of the intense ir modes associated with the bond alternation domain walls has a simple physical explanation [8]. The normal mode patterns associated with the two strong ir lines are shown in Fig. 8. The two displacement patterns correspond to an overall contraction of the weaker and stronger bonds, respectively, on one side of the defect, and an overall expansion on the other. These antisymmetric displacement patterns drive charge back and forth across the defect center leading to the large oscillator strength.

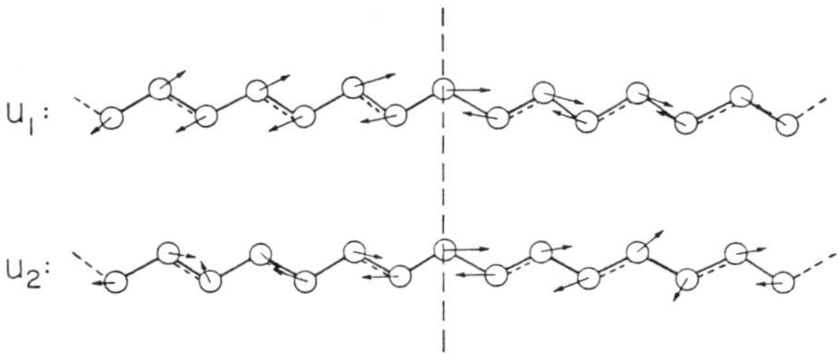

Fig. 8

IV. Summary and Conclusion

In summary, the paramagnetic center in undoped polyacetylene is a topological defect in the carbon $2p_{\pi}$ alternating bond structure; a bond-alternation-domain wall, approximately 14 lattice constants in extent. Experimental evidence shows the defect to be fixed in the cis-rich polymer but highly mobile in the trans-isomeric form.

The localized electronic bound state predicted to appear at mid-gap has been observed through optical absorption studies. Detailed calculations within the soliton model lead to two specific predictions regarding the optical absorption: 1) The intensity of transitions involving the mid-gap soliton level should be enhanced by about two orders of magnitude. 2) The existence of a soliton domain wall on a chain suppresses the interband transition. Both of these effects are observed experimentally with magnitudes in approximate agreement with the theoretical results.

An understanding of the intense infrared absorption modes, induced upon doping, has emerged in terms of the lattice dynamical changes associated with the charged soliton. MELE and RICE [8] calculate that ~ 20% of the total oscillator strength associated with the added charge appears in these "vibrational" excitations. That such a large fraction of the oscillator strength should be exhausted in these lattice modes is a remarkable consequence of the nature of the soliton defect.

Finally, the magnetic susceptibility results for AsF_5 doped $(CH)_x$ are fully consistent with the soliton doping mechanism and indicate that the SM transition is not due to percolation of metallic regions. For light doping, the decrease of the Curie-law contribution with no Pauli susceptibility has been demonstrated: the charge carriers are spinless consistent with the soliton doping mechanism. The Pauli susceptibility remains small well into the highly conducting regime and shows an abrupt increase near $y \simeq 0.07$ to values typical of simple metallic behavior. The intermediate range, $0.005 < y < 0.05$, is thus of particular interest, for it can be viewed as a high conductivity state generated by charged solitons at high density. The development of a theoretical understanding of the semiconductor-metal transition beginning from the soliton concept remains as a fascinating and important problem.

References

1. M. J. Rice, Phys. Lett. 71A, 152 (1979).
2. W. P. Su, J. R. Schrieffer and A. J. Heeger, Phys. Rev. Lett. 42, 1698 (1979), and Phys. Rev. B 22, 2099 (1980)
3. I. B. Goldberg, H. R. Crowe, P. R. Newman, A. J. Heeger and A. G. MacDiarmid, J. Chem. Phys. 70, 1132 (1979).
4. B. R. Weinberger, J. Kaufer, A. Pron, A. J. Heeger and A. G. MacDiarmid, Phys. Rev. B 20, 223 (1979).
5. M. Nechtschein, F. Devreux, R. L. Greene, T. C. Clarke and G. B. Street, Phys. Rev. Lett. 44, 356 (1980).
6. B. R. Weinberger, E. Ehrenfreund, A. Pron, A. J. Heeger and A. G. MacDiarmid, J. Chem. Phys. 72, 4749 (1980). More recent studies indicate even more rapid 1d spin diffusion; K. Holczer, J. P. Boucher, F. Devreux and M. Nechtschein (preprint).
7. C. R. Fincher, Jr., M. Ozaki, A. J. Heeger and A. G. MacDiarmid, Phys. Rev. B 19, 4140 (1979).
8. E. Melé and M. J. Rice, preprint.
9. Y. W. Park, A. Denenstein, C. K. Chiang, A. J. Heeger and A. G. MacDiarmid, Solid State Commun. 29, 747 (1979).
10. Y. W. Park, A. J. Heeger, M. A. Druy and A. G. MacDiarmid, J. Chem. Phys. (July, 1980); Y. W. Park, A. Denenstein, C. K. Chiang, A. J. Heeger and A. G. MacDiarmid, Solid State Commun. 29, 747 (1979).
11. S. Etemad, M. Ozaki, D. L. Peebles, A. J. Heeger and A. G. MacDiarmid, Phys. Rev. Lett. (Submitted).
12. H. Takayama, Y. R. Lin-Liu and K. Maki, Phys. Rev. B 21, 2388 (1980).
13. S. A. Brazovskiǐ, JETP Lett. 28, 656 (1978).
14. J. A. Krumhansl, B. Horovitz and A. J. Heeger, Solid State Commun. 34, 945 (1980).
15. H. Shirakawa and S. Ikeda, Polym. J. 2, 231 (1971); H. Shirakawa, T. Ito and S. Ikeda, ibid. 4, 460 (1973); T. Ito, H. Shirakawa and S. Ikeda, J. Polym. Sci., Polym. Chem. Ed. 13, 1943 (1975).
16. J. C. W. Chien, F. E. Karasz, G. E. Wnek, A. J. Heeger and A. G. MacDiarmid, J. Polym. Sci., Polym. Lett. Ed. 18, 45 (1980).
17. W. P. Su, Solid State Commun. (in press).
18. D. Vanderbilt and E. J. Melé, Phys. Rev. B (Submitted).
19. J. C. W. Chien and F. E. Karasz (private communication).
20. P. M. Grant and I. P. Batra, Solid State Commun. 29, 225 (1978).
21. E. J. Melé and M. J. Rice, Phys. Rev. Lett. (in press).
22. Y. Tomkiewicz, T. D. Schultz, H. B. Brown, T. C. Clarke and G. B. Street, Phys. Rev. Lett. 43, 1532 (1979).
23. T. Ito, H. Shirakawa and S. Ikeda, J. Polym. Sci., Polym. Chem. Ed. 13, 1943 (1975) and references therein. For details on the doping pro-

cedure and the dopant species, see A. Pron, Ph.D. Thesis, University of Pennsylvania, 1980.
24. L. Mihaly, S. Pekker and A. Jánossy, J. Synth. Metals (in press).
25. D. Moses, A. Denenstein, A. Pron, A. J. Heeger and A. G. MacDiarmid, Solid State Commun. (Submitted).
26. C. R. Fincher, Jr., M. Ozaki, M. Tanaka, D. Peebles, L. Lauchlan, A. J. Heeger and A. G. MacDiarmid, Phys. Rev. B 20, 1589 (1979).
27. J. Ritsko, E. J. Melé, A. J. Heeger, A. G. MacDiarmid and M. Ozaki, Phys. Rev. Lett. 44, 1351 (1980); J. Ritsko (to be published).
28. N. Suzuki, M. Ozaki, S. Etemad, A. J. Heeger and A. G. MacDiarmid, Phys. Rev. Lett. (Submitted).
29. E. Melé and M. J. Rice, Solid State Commun. 34, 339 (1980).

Note added in print

Equations (7) and (8) for the interband absorption strength of a $(CH)_x$ chain containing a soliton at the center are incorrect. Although the vertical matrix elements (k' = k) cancel as indicated in the text, the non-vertical (k' ≠ k) terms restore in part the strength of the transition. The correct expression is given by

$$\alpha_i'(\omega) = \alpha_i(\omega)\ [1 - \frac{\ell n2}{2N}\frac{\pi^2\xi_o}{a}]$$

for a chain containing one soliton where $\alpha_i(\omega)$ is the absorption coefficient for a perfect chain. This expression is in excellent agreement with the experimental results which show a decrease in the magnitude of α, but with no change in the shape of $\alpha(\omega)$. Moreover, the magnitude of the decrease implies a value for $\pi^2\xi_0/a$ consistent with the value obtained from the strength of the mid-gap transition and with theoretical estimates. We thank J. A. Krumhansl and J. Tinka Gammel of Cornell University for pointing out the restoration of oscillator strength due to non-vertical transitions and for several important discussions.

Magnetic Resonance Studies of Soliton Diffusion in Polyacetylene

F. Devreux[1], K. Holczer[2], and M. Nechtschein[1]
Département de Recherche Fondamentale, Centre d'Etudes Nucléaires
de Grenoble, 85 X, F-38041 Grenoble Cedex, France

T.C. Clarke and R.L. Greene
IBM Research Laboratory[3], 5600 Cottle Road
San Jose, CA 95193, USA

It has been known for many years that paramagnetic centers are present in conjugated polymers [1]. Recently, a renewal of interest in the properties of these centers has arisen from NMR [2] and ESR [3-9] experiments in undoped polyacetylene, $(CH)_x$, and a proposed explanation for the uncharged spins in terms of a bond alternation domain wall (topological soliton) model [10,11]. In this paper we will review our previous room temperature NMR results [2], present new results as a function of oxygen exposure and temperature variation, attempt to reconcile the apparent differences between the NMR and ESR data, and compare our results to the soliton theory.

In "pristine" trans-$(CH)_x$ the spin susceptibility (χ) obeys a Curie law [5] and the neutral spin 1/2 concentration (n) lies in the range of one spin per 1000-3000 carbon atoms, while in the cis-rich polymer the spin concentration is an order of magnitude smaller. Early observations [4] of a motionally narrowed ESR line suggested that the spins in trans-$(CH)_x$ were mobile. This was definitely confirmed by our dynamic nuclear polarization (DNP) experiments [2]. The DNP experiment consists of observing the NMR (at the nuclear Larmor frequency, ω_N) while pumping with microwave power near the electronic Larmor frequency (ω_e). Two limiting results may occur according to whether the electron-nuclear coupling is static or dynamic. In the static case, the electronic spin is fixed (at least its hopping frequency is less than ω_N) and forbidden transitions at $\omega_e \pm \omega_N$ can be induced. They give rise to a NMR enhancement which is negative, or positive, for pumping at $\omega_e+\omega_N$, or $\omega_e-\omega_N$, respectively. This is the so-called "solid-state effect" (SSE) [12]. On the other hand, if the electronic spin is mobile such that translational energy exists at $\hbar\omega_e$ one gets an enhancement of the NMR signal by pumping just at ω_e. This is the Overhauser effect (OE).

In Fig.1 we show recent DNP results which agree with our earlier work. We observe the OE in a trans-$(CH)_x$ sample and the SSE in a cis-rich $(CH)_x$ sample. This demonstrates conclusively that at room temperature the spins are mobile in trans-$(CH)_x$ and fixed in cis-$(CH)_x$. Mixed cis-trans samples exhibited both the OE and the SSE roughly in proportion to their cis-trans content.

To quantitatively determine the spin mobility and its dimensionality in trans-$(CH)_x$ we measured the frequency dependence of the proton nuclear relaxation time (T_1). It is well known [13] that

---

[1] ER CNRS no. 216.
[2] Permanent address: Central Research Institute for Physics, H-1525, Budapest, P.O.B. 49, Hungary
[3] Work at IBM partially supported by the ONR.

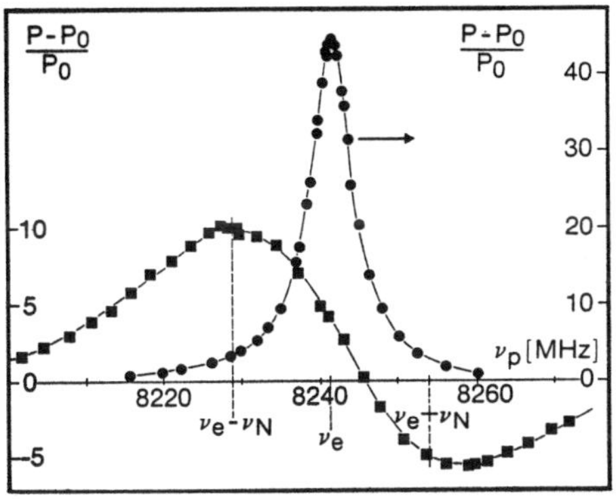

Fig.1 Enhancement of proton NMR amplitude (P) at 300 K as a function of the microwave pumping frequency ν_p near $\omega_e/2\pi$=8240 MHz for (■) cis-rich ($\Delta H_{pp}\approx$7.8G) and (●) trans ($\Delta H_{pp}\approx$0.8G) undoped $(CH)_x$. P_o is the NMR signal amplitude without pumping.

$$T_1^{-1} = kT\chi\left[\frac{3}{5}\, d^2 f(\omega_N) + \left(a^2 + \frac{7}{5}\, d^2\right) f(\omega_e)\right] \tag{1}$$

where a and d are the isotropic and dipolar electronic spin-proton hyperfine couplings, respectively, and $\chi=\chi_{molar}/N(g\mu_B)^2$. The spectral density function for the magnetic field fluctuations at the nucleus, $f(\omega)$, depends sensitively on the dimensionality of the spin motion (for ω<diffusion rate). For one dimensional (1D) diffusion $f(\omega)=(2D_{\parallel}\omega)^{-1/2}$, whereas in 2D $f(\omega)\propto\ell n(1/\omega)$ and in 3D it is frequency independent. Here $D_{\parallel}$ is the intrachain diffusion rate. One expects a cross-over between 1D and 2D or 3D regimes at a cutoff frequency $\omega_c\approx D_{\perp}$, $D_{\perp}$ being the interchain diffusion rate.

In Fig.2 we show our measurements of T_1^{-1} versus $\nu_N^{-1/2}$ for undoped trans-$(CH)_x$ over the frequency ($\nu_N=\omega_N/2\pi$) range 10-340 MHz. The room temperature data gives clear evidence for 1D diffusive behavior. Using (1), with values for a and d as explained previously [2] and the spin susceptibility corresponding to 1 spin per 2000 as measured by a Schumaker-Slichter method, we find $D_{\parallel}\approx 6\times10^{13}$ sec^{-1}. Since no deviation from 1D behavior is observed down to $\omega_N/2\pi$=10 MHz we obtain an upper limit for the cutoff frequency: $\omega_c<2\pi\times10^7$ sec^{-1}. A value for ω_c can be obtained by measuring the proton relaxation time in the rotating frame, $T_{1\rho}$, which is expressed [13] as

$$T_{1\rho}^{-1} = \frac{1}{2}\, kT\chi\left[\left(a^2 + \frac{4}{5}\, d^2\right) f(\omega_1) + \frac{3}{4}\, d^2 f(\omega_N) + \left(a^2 + \frac{13}{5}\, d^2\right) f(\omega_e)\right] \tag{2}$$

where ω_1 is the Larmor frequency in the rotating frame. Since $\omega_1\approx10^4$ sec^{-1}, we can definitely assume that $\omega_1\ll\omega_c$, and hence we have $f(\omega_1)\approx f(o)=1/\sqrt{2D_{\parallel}\omega_c}$. With the experimental value $T_{1\rho}^{-1}$=250 s^{-1}, Eq. (2) leads to ω_c=4.5×10^7 sec^{-1}. Noting that $\omega_c\approx D_{\perp}$ it then follows that the anisotropy in spin diffusion is of order $D_{\parallel}/D_{\perp}\approx10^6$. Our observation of highly one-dimensional spin diffusion supports the topological soliton theory [10] which describes the neutral magnetic defect as a collective entity, i.e., a domain wall defect would move easily along the chain but not easily between chains.

In principle one should be able to obtain similar results from an analysis of the ESR, i.e., evidence for 1D diffusion motion and a value for

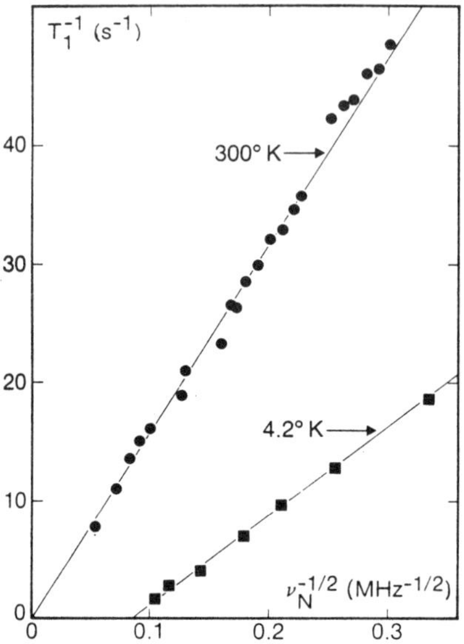

Fig.2 Nuclear relaxation rate T_1^{-1} versus $\nu_N^{-1/2}$ for undoped trans-$(CH)_x$ at 300 K and 4.2K.

the diffusion rate. In fact, the confrontation between NMR and ESR data raises some difficulties. First, the observed ESR line is broader than what one expects from motional narrowing with the value of $D_{\parallel}$ obtained from the NMR T_1 data. Using the observed ESR linewidth ($\Delta H \sim 1G$) one obtains a value of $D_{\parallel}$ which is one to three orders of magnitude less than the value found from T_1. Secondly, the ESR line does not display the shape [8] expected for narrowing by a 1D diffusive motion. Related to this is the observation [14] of ESR spin-echo signals which suggest the line has some inhomogeneous character.

We have done a detailed analysis of the ESR in trans-$(CH)_x$ which is reported elsewhere [15]. Here we briefly summarize our results. Our conclusion is that the analysis of the NMR data presented above and in Ref. 2 is basically correct. The apparent discrepancy with ESR is due to the fact that a small number ($\lesssim 10\%$) of fixed spins can completely mask the ESR line features of the majority of highly mobile spins, while they have only a negligible effect on the nuclear relaxation. Thus we suggest that there are two types of spins (diffusive and fixed) in all trans-$(CH)_x$ samples measured to date.

First, we discuss the ESR linewidth at 300K. If we had 3D motion we would expect $\Delta\omega \simeq (\Delta\omega_0)^2 D^{-1}$ where $(\Delta\omega_0)^2$ is the second moment of the unnarrowed line. We have estimated [15] $(\Delta\omega_0)^2$ by including both the hyperfine and dipolar spin-spin interactions (we find these contributions are of comparable magnitude). From the measured $\Delta\omega(\gamma_e^{-1}\Delta\omega \sim 1G)$ we then obtain $D \approx 10^{10}$ sec^{-1}, a value much less than found by NMR. For 1D diffusive motion we expect the resulting linewidth to be given by $\Delta\omega_{diff} \approx (\Delta\omega_0)^2/2\sqrt{D_{\parallel}\omega_c}$, from which we obtain $D_{\parallel} \approx 10^{12}$ sec^{-1}, still less than the NMR result. We conclude that an additional broadening mechanism must be present and this we attribute to the presence of some localized spins.

Using the values for $D_{\parallel}$ and ω_c obtained from T_1 and $T_{1\rho}$ data we estimate the linewidth for purely diffusive spins to be $\gamma_e^{-1}\Delta\omega_{diff}\approx 0.13G$. The diffusive spins and the fixed spins must be coupled otherwise two ESR lines of very different linewidth would be observed. To smear out the two lines an exchange frequency $\omega_{ex}>\Delta\omega_{diff}\approx 3\times 10^6$ sec^{-1} is required. This is a very small value and therefore the observation of one ESR line is quite reasonable. In fact, we may expect the exchange rate between the two kinds of spins to be of the order of the interchain exchange rate for the diffusive spins, which is also the cutoff frequency $\omega_c\approx 4\times 10^7$, as determined from $T_{1\rho}$. It can be shown [15] that the resulting linewidth may be approximated by

$$\Delta\omega \approx (1-c)\Delta\omega_{diff} + c\Delta\omega_{fixed} , \tag{3}$$

where c is the relative concentration of fixed spins. Since the fixed spin linewidth is not motional narrowed, we expect $\gamma_e^{-1}\Delta\omega_{fixed}\approx 5\text{-}10G$, and then $\Delta\omega_{fixed}/\Delta\omega_{diff}\approx 30\text{-}50$. It thus appears that for c as low as ~0.1 the observed linewidth is dominated by localized spins, roughly proportionally to their concentration.

We have performed [15] a memory function analysis of the ESR lineshape (exploring up to 100 times the linewidth) for several samples of different ΔH. The ESR line memory function can be decomposed and analyzed in terms of two contributions: (i) a diffusive contribution, which has the 1D diffusive characteristic frequency dependence $\propto(\omega D_{\parallel})^{-1/2}$; (ii) a narrow spectrum contribution ($\approx 5\times 10^8$ sec^{-1}), which is characteristic of fixed spins. It appears that the narrower the observed linewidth, the larger the proportion of the diffusive contribution. The extrapolated linewidth for a purely diffusive spin sample (c=0) is found to be 0.15G, in agreement with the value obtained independently from T_1 and $T_{1\rho}$ data. In addition, from the experimental diffusive contribution to the memory function, we have found the value for the diffusion rate $D_{\parallel}\approx 10^{13}$ sec^{-1}, which compares well with the NMR data.

Thus by assuming the coexistence of two types of spins--diffusive and fixed--the NMR and ESR data are fully consistent. It turns out that the nuclear relaxation is dominated by the diffusive spins and the ESR by a small number of localized spins. This is not unreasonable since each diffusive spin is able to make $\sim\sqrt{D_{\parallel}T_1}\sim 10^6$ protons relax, while the number of protons relaxing via the fixed spins is much smaller. Note that we have estimated the nuclear relaxation rate, via nuclear spin diffusion [12] to be between one and two orders of magnitude less than the measured T_1^{-1}. Another argument in favor of diffusive spin controlled nuclear relaxation is the following. For two samples with ESR linewidths as different as ΔH_{pp}=0.44G and 1.2G we have determined $D_{\parallel}$ from T_1 measurements and found values in agreement within 25%.

What is the origin of the fixed spins? One possibility is that imperfect isomerization leads to regions in the polymer chain where the spins cannot move. In addition, there exists experimental evidence for the localizing effect of oxygen. Prior work [4,9] showed that the ESR linewidth in trans-$(CH)_x$ increased upon exposure to air. This result, along with the wide variation in ESR linewidths reported in the literature for trans-$(CH)_x$ ($0.8G<\Delta H_{pp}<3G$) suggests that small amounts of residual oxygen could influence the ESR linewidth. With this in mind we have taken great care to avoid oxygen exposure and have been able to prepare trans-$(CH)_x$ samples with linewidths $\Delta H_{pp}\approx 0.44G$. This should be compared with our estimate of the pure diffusive linewidth of order $\Delta H_{pp} \approx 0.2G$. Upon exposure to air the linewidth rapidly broadens in agreement with prior results [4,9].

Other evidence for the presence of localized spins in the oxygen exposed samples is given in Figs.3 and 4. Figure 3 shows that T_1^{-1} drops rapidly upon air exposure. The broadening of the ESR line upon air exposure would imply that $D_{\parallel}$ decreases which should lead to an increase in T_1^{-1} [recall $T_1^{-1} \propto (D_{\parallel})^{-1/2} n_{diff}$]. The conclusion is that the number of diffusing spins must decrease. The evidence for the appearance of fixed spins is given in Fig.4 where we show the observation of a SSE in a DNP experiment after air exposure.

We now discuss our first NMR experiments designed to measure the temperature dependence of the spin motion. Figure 2 shows T_1^{-1} versus frequency for trans-$(CH)_x$ at 4.2°K. We see that we still have the straight line characteristic of 1D diffusion [the data cannot be fit by either Gaussian or Lorentzian forms of $f(\omega)$]. However, two changes from the 300K data are readily apparent; (i) decreased slope, and (ii) positive intercept on the ν axis.

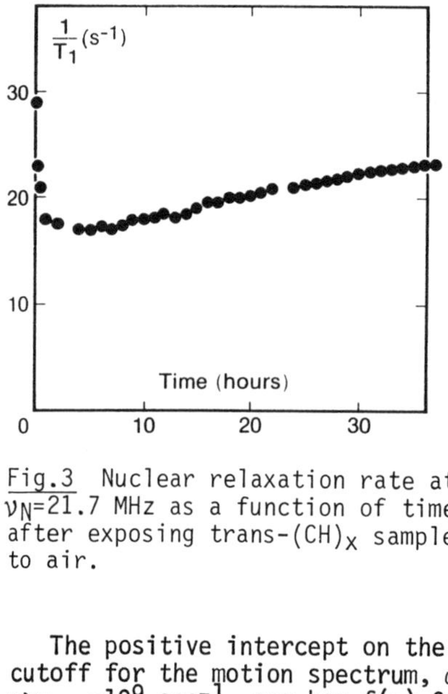

Fig.3 Nuclear relaxation rate at ν_N=21.7 MHz as a function of time after exposing trans-$(CH)_x$ sample to air.

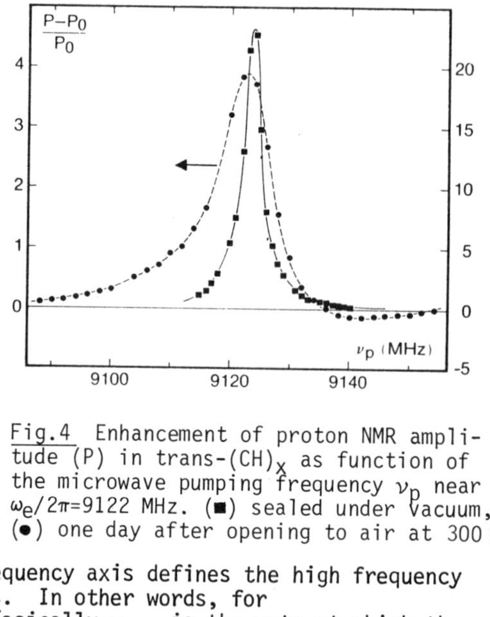

Fig.4 Enhancement of proton NMR amplitude (P) in trans-$(CH)_x$ as function of the microwave pumping frequency ν_p near $\omega_e/2\pi$=9122 MHz. (■) sealed under vacuum, (●) one day after opening to air at 300 K.

The positive intercept on the frequency axis defines the high frequency cutoff for the motion spectrum, ω_{max}. In other words, for $\omega > \omega_{max} \approx 10^9$ sec^{-1}, one has $f(\omega) \approx 0$. Basically ω_{max} is the rate at which the electronic magnetic field seen by a given proton fluctuates when a mobile spin is near by. If the spin wave function is peaked on only one C-H unit, this rate is nothing but the diffusion rate itself, and one has $\omega_{max} \sim D$. On the other hand if the spin is extended over $\ell >> 1$ lattice constants there will be no change of the induced magnetic field until the spin has moved by a distance of the order of its extension. Due to the random walk character of the motion this requires a time $\ell^2 D^{-1}$, which leads to $\omega_{max} \sim D/\ell^2$. The soliton theory [10] predicts such an extended spin defect domain with $\ell \sim 10$. This value is in agreement with an estimate based upon the hyperfine contribution to the ESR linewidth in cis-rich $(CH)_x$ [8]. Note that this estimate assumed that the same solitons exist in cis-rich and trans-$(CH)_x$, which must be proven. Taking $\ell \sim 10$, we obtain the qualitative estimate $D_{\parallel} \sim 10^{11}$ sec^{-1} from the intercept $\omega_{max} \sim 10^9$ sec^{-1}. Another way to determine the diffusion rate is from the slope of the data T_1^{-1} versus $\omega_N^{-1/2}$, i.e., Eq. (1). Since the slope at 4.2K is less than at 300K, we are led to $D_{4.2} > D_{300} \approx 6\times10^{13}$ sec^{-1}, which is in contradiction with the previous qualitative estimate $D_{4.2} \sim 10^{11}$ sec^{-1}. A way to overcome this contradiction is to assume that the number of diffusive spins n_{diff} has considerably decreased at 4.2K. Since we observe that the nuclear relaxation is controlled by 1D diffusion process, one should have $T_1^{-1} \sim n_{diff}(D_{\parallel}\nu_N)^{-1/2}$. A decrease of n_{diff} by 10-100 is necessary to overcome the contradiction. Since the total number of spins remain constant (Curie Law) one is led to the conclusion that 90-99% of the spins become trapped at 4.2K. More work is necessary to understand the origin of the spin localization at low T. Experiments to determine the temperature dependence of $D_{\parallel}$ and n_{diff} are in progress.

In summary our NMR and ESR results show that highly 1D spin diffusion occurs in trans-$(CH)_x$ with $D_{\parallel} \approx 6\times10^{13}$ sec^{-1} at room temperature and an anisotropy $D_{\parallel}/D_{\perp} \sim 10^6$. Oxygen exposure at 300K leads to a decrease in the number of diffusing spins but does not change $D_{\parallel}$ appreciably. The number of diffusing spins and their mobility decrease significantly at lower temperature. It appears that at all temperatures and conditions of sample preparation both fixed and diffusing spins are present, however, their ratio can vary significantly. We believe these results are in qualitative agreement with the description of the spins in terms of neutral domain wall solitons. It would be useful, however, to have a theoretical estimate for the magnitude and temperature dependence of the diffusion rate expected for these solitons. In addition elucidation of possible soliton trapping mechanisms (by oxygen and/or temperature decrease) and interchain soliton interactions would be useful.

REFERENCES

1. L. Salem, The Molecular Orbital Theory of Conjugated Systems (W. A. Benjamin, New York, 1966), pp. 516-523.

2. M. Nechtschein, F. Devreux, R. L. Greene, T. C. Clarke and G. B. Street, Phys. Rev. Lett. 44, 356 (1980).

3. H. Shirakawa, I. Ito and S. Ikeda, Die Macromolecular Chimie 179, 1565 (1978).

4. I. B. Goldberg, H. R. Crowe, P. R. Newman, A. J. Heeger and A. G. MacDiarmid, J. Chem. Phys. 70, 1132 (1979).

5. B. R. Weinberger, I. Kaufer, A. Pron, A. J. Heeger and A. G. MacDiarmid, Phys. Rev. B 20, 223 (1979).

6. A. Snow, P. Brant and E. Weber, Polymer Letters 17, 263 (1979).

7. Y. Tomkiewicz, T. D. Schultz, H. B. Broom, T. C. Clarke and G. B. Street, Phys. Rev. Lett. 43, 1532 (1979).

8. B. R. Weinberger, E. Ehrenfreund, A. Pron, A. J. Heeger and A. G. MacDiarmid, J. Chem. Phys. 72, 4749 (1980).

9. P. Bernier, et al., J. Phys. (Paris) Lett. 40, L297 (1979).

10. W. P. Su, J. R. Schreiffer and A. J. Heeger, Phys. Rev. Lett. 42, 1698 (1979).

11. M. J. Rice, Phys. Lett. 71A, 152 (1979).

12. A. Abragam, The Principles of Nuclear Magnetism (Oxford Press, Oxford, 1961), Chap. IX.

13. F. Devreux, J. P. Boucher and M. Nechtschein, J. Phys. (Paris) 35, 271 (1974).

14. Y. Tomkiewicz and N. Shiren, private communication.

15. K. Holczer, J. P. Boucher, F. Devreux and M. Nechtschein, submitted to Phys. Rev. B.

Theory of Polymers Having Broken Symmetry Ground States*

W.P. Su, S. Kivelson, and J.R. Schrieffer

Department of Physics, University of California
Santa Barbara, CA 93106, USA

Abstract

The theoretical basis for soliton excitations in polyacetylene is discussed. The unique charge-spin relations, light mass, and new electronic states of solitons lead to interesting experimental consequences. Models of the semiconductor metal transition are presented, and real time dynamics of $(CH)_x$ is illustrated.

1. Introduction

In quasi-one-dimensional metals such as KCP, TTF-TCNQ, or $(CH)_x$, a spontaneous symmetry breaking occurs in which a periodic distortion of the nuclear and electron density is set up with a wave vector $Q = \pm 2k_F$. The periodic potential generated by the distortion leads to an energy gap at the Fermi surface, $\pm k_F$. The mean field ground state is degenerate due to the invariance of the hamiltonian to a displacement of the charge density wave (CDW). For an incommensurate CDW, such as found in KCP and TTF-TCNQ, the degeneracy corresponds to a continuous symmetry, while in a commensurate system, such as $(CH)_x$, the symmetry is discrete. The formation of these CDW's is a consequence of the coupling of the electrons and phonons in 1D and is called the Peierls instability. A similar distortion occurs when the electron-electron Coulomb interaction U is sufficiently strong that a charge and/or spin density wave is set up, and a Mott-Hubbard type gap appears in the one-electron spectrum. We emphasize here the coupled electron-phonon model, treating U as a perturbation.

2. Broken Symmetry and Solitons

To be concrete, we consider a 1D trans $(CH)_x$ chain as shown in Fig.1. Since pure $(CH)_x$ has one π electron per CH group, the wave number of the Peierls distortion corresponds exactly to two lattice periods, $Q = 2\pi/2a$ and a commensurate (dimerization) distortion is set up. The σ electrons can be treated in the adiabatic approximation because of the large gap (~ 10 eV) between the bonding and antibonding σ states. Also, we treat only the coordinates u_n (see Fig.1). Since the change of bond length due to the distortion is small, we retain only second order terms in the σ energy about the undimerized state

$$E_\sigma = \frac{K}{2}\sum_n \left(u_{n+1} - u_n\right)^2 \qquad (1)$$

and assume the conventional linear electron-phonon coupling modulates the π electron hopping integral. The Hamiltonian is (in the notation of SSH) [1]

$$H = -\sum_{ns} t_o\left(c^+_{n+1,s}\, c_{n,s} + h.c.\right) + \sum_{ns}\left[\alpha\left(u_{n+1} - u_n\right)c^+_{n+1s}\, c_{ns} + \right.$$
$$\left. + h.c.\right] + E_\sigma + \frac{M}{2}\sum_n \dot{u}_n^2 \qquad (2)$$

where the undimerized bandwidth is $4t_o \simeq 10\,\mathrm{eV}$ and M is the CH mass.

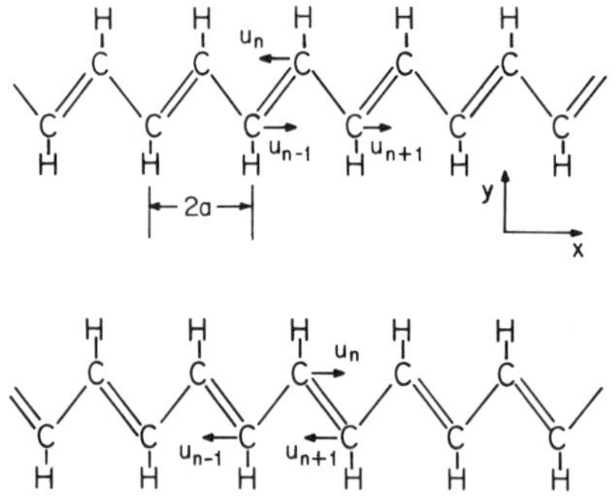

Fig.1 The two ground state configurations of trans-$(CH)_x$

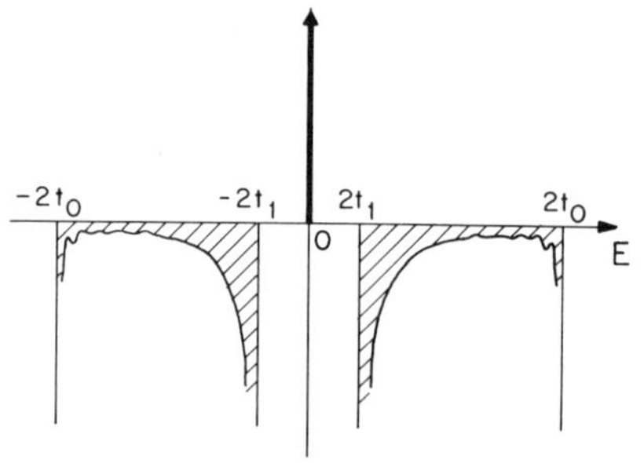

Fig.2 Change in the electronic density of states due to the presence of a soliton

It is useful to define the staggered displacement $\psi_n = (-1)^n u_n$ and evaluate the ground state energy E_o of H for uniform $\psi_n \equiv \psi$, neglecting $\dot{\psi}_n$. The adiabatic potential $E_o(\psi)$ for the dimerization mode has a double minimum at $\psi = \pm u_o \simeq 0.042\,\text{Å}$ for $K = 21\,\mathrm{eV/Å^2}$ and $\alpha = 4.1\,\mathrm{eV/Å}$ appropriate to $(CH)_x$, reflecting the Peierls instability about $\psi = 0$. Thus, if one treats the lattice coordinates classically, the ground state is twofold degenerate, corresponding to $\psi_n = u_o$ (A phase) and $\psi_n = -u_o$ (B phase), as in Fig.1. The electronic structure of the perfect A or B phase is that of a conventional semiconductor having a gap 2Δ. Since $2\Delta \gg \hbar\omega \simeq 0.1\,\mathrm{eV}$, the optical phonon energy, the adiabatic approximation for treating the nuclear dynamics is valid for small deviations about the ground state. However, if $|\psi|$ is reduced to a small fraction of u_o over distances larger than the coherence length $\xi_o = \frac{\hbar v_F}{\pi\Delta}$, nonadiabatic effects must be considered. We note that $E_o(\psi)$ is not of the Ginzburg-Landau form since it varies as $\psi^2 \ln \psi^2$ for small ψ and is strongly non-local, so that the conventional gradient expansion does not apply [2].

In conventional semiconductors, the elementary excitations are electrons, holes and phonons. Because of the twofold degenerate ground state in $(CH)_x$, an additional type of excitation occurs, namely, a topological soliton. To understand the origin of these excitations, consider finding the minimum energy configuration $\{\psi_n\}$ such that $\psi_n \rightarrow \pm u_o$ as $n \rightarrow \pm\infty$ respectively.

This problem was solved by SSH who found

$$\psi_n \cong u_o \tanh \frac{n}{\ell} \tag{3}$$

where $\ell \approx 7$ for $2\Delta = 1.4$ eV. The energy increase due to this kink or soliton is $E_s = 0.42$ eV $\cong 0.60\Delta$. TAKAYAMA, et al. [3] solved the corresponding problem for the continuum limit of (2) and found (3) to be an exact solution with $E_s = \frac{2}{\pi}\Delta \simeq 0.63\Delta$, in good agreement with the analysis of the discrete lattice model. It should be noted that while ψ_n changes sign as one passes through the soliton, it is the amplitude of the distortion which is changing (going to zero at the soliton's center) and not the phase of the wave, as in the phase soliton of conventional CDW theory. This distinction has important experimental consequences.

The electronic structure of a soliton in $(CH)_x$, as found by SSH, is illustrated in Fig.2, which shows a sharp state which is twofold spin degenerate located at the center of the gap $\epsilon_o = 0$. The wave function of this gap center state $\phi_o(n)$ is localized about the center of the soliton and has a width, $2\ell a$, the same as the soliton. ϕ_o is a nonbonding state split off symmetrically from the bonding (valence) and antibonding (conduction) bands. Since a soliton removes precisely one-half a state from the valence band for each spin, a total deficit of one electron occurs for a filled valence band in the presence of a soliton. Therefore, a neutral soliton, S^o, has charge 0 and spin ½. The charged solitons $S^{\pm}$ have the gap center state empty (+) or doubly occupied (-) so that $S^{\pm}$ has charge $\pm e$ and spin 0. These charge-spin relations as shown in Fig.3 are the reverse of electron and hole excitations in molecules and solids and are characteristic of this new type of excitation in the broken symmetry system. This striking result is related to a model in relativistic quantum field theory, where so-called "fermion number one-half" solitons appear [4].

We note that it is conventional to define as an antisoliton, $\bar{S}$, the solution (3) with u_o replaced by $-u_o$. Clearly, for topological reasons, the excitations must alternate between S and $\bar{S}$ as one moves along the chain.

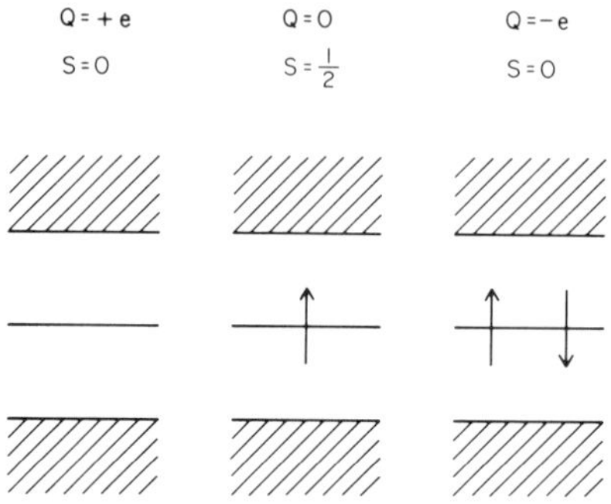

Fig.3 Spin and charge of neutral and charged solitons

The kinetic mass M_s of the soliton is defined as the coefficient of the $\frac{v_s^2}{2}$ terms in the system energy as a function of the soliton velocity v_s. One finds

$$M_s = \frac{4}{3\ell}\left(\frac{u_o}{a}\right)^2 M \simeq 6m_e \tag{4}$$

for $(CH)_x$, where m_e is the electron's mass. M_s is extremely small because of the small change of bond length on dimerization, and the soliton acts as a quantum particle.

Experimental evidence for neutral solitons of spin ½ come from measurements such as magnetic susceptibility, ESR and NMR, as discussed by A. J. HEEGER and R. L. GREENE in the previous papers. Although there are at present questions concerning procedures of sample preparation, there is strong support for solitons with $\ell \sim 7$ and very small mass. In addition, SU [5] has shown that the ground state of a chain having an odd number of CH groups corresponds to one soliton in the center of the chain. This may account for the number of free spins observed in undoped $(CH)_x$, i.e. one per several thousand CH groups.

Finally, while the breaking of charge-spin relations for S or $\bar{S}$ appears to violate Kramer's theorem (i.e. an even number of electrons must have spin 0,1... while an odd number must have spin $\frac{1}{2}, \frac{3}{2}, \frac{5}{2}, \ldots$), topological constraints require that S and $\bar{S}$ be created in pairs. As SSH showed, this removes the apparent difficulty and shows that the total state count for the valence or conduction bands remains an integer since half a state is split off, for S and for $\bar{S}$ from either the conduction or valence bands.

The phonon modes in the presence of a soliton have been worked out by MELE and RICE [6] using the SSH model. They find the soliton introduces two rather sharp phonon resonances localized in the domain wall which have strong infrared activity, in agreement with infrared absorption measurements on doped $(CH)_x$.

3. Lightly Doped $(CH)_x$: Soliton Doping

One of the reasons for the intense interest in $(CH)_x$ is its extreme dopability--its conductivity can be varied over more than seven orders of magnitude. At low impurity concentrations one can imagine that extra electrons (or holes) could be added to the system either as free carriers in the conduction (or valence) band or as charged solitons, depending on which leads to the lower energy state. Measuring energies relative to the gap center, to add two electrons requires an energy 2Δ, while to make two charged solitons (S^- and $\bar{S}^-$) requires $\simeq \frac{4}{\pi}\Delta \simeq 1.26\Delta$ so that the charged soliton is favored by an energy $\Delta(1-2/\pi) \simeq 0.25\,\mathrm{eV}$ per charge. Therefore, the electron-phonon model predicts that doping in $(CH)_x$ proceeds by creation of charged solitons ($S^\pm$ and $\bar{S}^\pm$).

This result ignores the presence of electron-electron interactions. If one treats these interactions within a Hubbard-like model (on-site repulsive U) in the Hartree Fock approximation, SCHRIEFFER [7] has shown that the energy to form a neutral S° or $\bar{S}^\circ$ is lowered by the self exchange energy of the unpaired electron in the φ_o orbital

$$E_{SU} = \frac{U}{4}\sum |\varphi_o(n)|^4 \simeq \frac{U}{12} \quad , \tag{5}$$

while the formation energy of the charged species $S^\pm$ or $\bar{S}^\pm$ is unaltered by U, since these excitations have spin zero. Because U is screened by the π electron dielectric constant, $\epsilon \simeq 10$, E_{SU} is quite small, of order 0.1 eV for the neutral soliton. The doping considerations above are not affected by U.

The condition of charge neutrality requires that there be the same number of charged impurities as charged solitons. We will assume these impurities are interstitial and randomly distributed. A charged soliton is normally bound to an oppositely charged impurity by their mutual Coulomb attraction. SSH estimate the binding energy to be $E_b \approx 0.3\,\mathrm{eV}$, although they assume that the interaction is screened by the full, in chain dielectric constant, so this estimate may be somewhat low.

Within the soliton model, two pictures of the charge transport have been discussed [8],[9] in the small concentration regime. The first, which is probably correct at sufficiently high temperatures or low impurity concentrations, is charged soliton conduction. Because the charged solitons are normally bound, this process is characterized by an activated conductivity and thermopower varying as 1/T,

$$\sigma \propto \exp(-E_b/kT) \quad , \qquad S \sim (e/k)(E_b/kT + \text{const.}) \quad . \tag{6}$$

The activation energy reflects the thermal population of unbound charged solitons. The conductivity depends weakly, at most linearly, on the concentration of impurities, y. Because of the large activation energy in (6), free soliton conduction is very inefficient at low temperatures, and another, more weakly temperature dependent process may take over: electron hopping between soliton midgap states. For concreteness we consider this process in an n type sample. An electron in a charged soliton bound state can make a phonon assisted transition to a nearby neutral soliton. If this neutral soliton is nowhere near a charged impurity, this process costs the same electostatic energy, E_b, as liberating a charged soliton. However, if the neutral soliton happens to be near another charged impurity, the energetic cost is much reduced. Thus, charge is transferred from impurity site to impurity site without a free soliton intermediary. This process has been studied in detail by KIVELSON [9]. The resulting conductivity and thermopower are of the form

$$\sigma \propto T^t \exp(-By^{-\frac{1}{3}}) \quad , \quad S \sim (e/k)\,\text{const.} \quad . \tag{7}$$

The hopping process is intrinsically three dimensional at low impurity concentrations. Although the hopping rate between a pair of solitons falls off more rapidly with distance in the out-of-chain direction, the typical in-chain separation between solitons varies roughly as y^{-1} while the inter-chain separation is much smaller, typically $\sim (4\pi y/3)^{-\frac{1}{3}}$. Thus, inter-chain hopping is more likely than intra-chain hopping so long as $y \ll (\ell_\perp/\ell)$, where $\ell_\perp$ is the decay length of the mid-gap state in the out-of-chain direction, $(\ell_\perp < \ell)$. The exponential dependence of σ on $y^{-\frac{1}{3}}$ arises from the overlap between the wave functions of electrons bound to a pair of solitons at neighboring impurity sites. The temperature dependence of the conductivity arises from the increase in the number of allowed hopping processes at higher temperatures. The details of the temperature variation depend on the nature of the electron-phonon coupling. The result in (7) was derived by KIVELSON from a simple model of the coupling. He found that t reflects the shape of the optical phonon spectrum, and is given approximately by the expression $t = (\omega_o/\Delta\omega)^2 \approx 9$ where ω_o and $\Delta\omega$ are respectively the center of gravity and half-width of the phonon spectrum.

The validity of the hopping picture rests on the assumption that there is a finite (although possibly very small) concentration of neutral solitons in the doped samples. The observed, roughly constant Curie law susceptibility up to doping concentrations of a couple tenths of a percent [10] apparently confirms the existence of neutral solitons in these samples, although a convincing argument as to why this should be so has not yet been made. Also associated with the spin hopping conduction, there is a degree of transverse spin diffusion. The transverse spin diffusion constant is related to the hopping conductivity by an Einstein relation, $D_\perp = (1/e^2)kT\sigma/c_{im}$ where c_{im} is the impurity concentration. For an impurity concentration of 0.04% and a typical room temperature conductivity of $10^{-5}\ \Omega^{-1}\ \text{cm}^{-1}$, $D_\perp/b^2 \approx 5 \times 10^7$ Hz where $b = 4.35\text{Å}$ is the interchain separation.

The experimentally observed [8] transport properties of polyacetylene are in qualitative agreement with (7). The conductivity varies more slowly with T than an exponential, but is consistent with power law behavior, and it increases very rapidly with increasing dopant concentration. Moreover, a large enhancement of the conductivity at 10 GHz has been reported [11], as would be expected of a hopping conductivity, although further experiments are needed to confirm this interpretation. The thermopower is roughly temperature independent and has a magnitude of order 10 (e/k), in excellent agreement with (7). However, (7) systematically overestimates the conductivity at any given impurity concentration. This may be partly a result of the complicated impurity chemistry which can lead to less than one charged impurity per dopant molecule.

Some of the most direct evidence for the existence of charged solitons comes from optical absorption experiments on lightly doped samples. The presence of a soliton is expected to introduce a peak into the absorption spectrum at half the interband gap, as analyzed by SUZUKI, et al. [12], and discussed by HEEGER.

4. Heavily Doped $(CH)_x$: The Metal Insulator Transition

As discussed above, for each electron (hole) introduced into the chain, one soliton is created. For concentration $y \ll 1/2\,\ell$, the solitons are widely spaced and aside from long-range Coulomb forces, they interact weakly. As y increases, the electronic structure of neighboring solitons mixes, e.g. hybridization of the gap center states on neighboring solitons occurs, thereby forming a broadened continuum of states centered about the gap center. KOTANI [13] studied this problem as a function of concentration for a periodic array of S and $\bar{S}$ in a cutoff linear band model. His results for the one electron spectrum E(k) are shown in Fig.4 for coupling strength α comparable to that for $(CH)_x$. For y = 0.5%, the gap center state $E_o(k)$ is nearly independent of k. As y increases to 5.0%, $E_o(k)$ broadens to roughly 1/3 the gap 2Δ, but sharp gaps of order $2/3\Delta$ remain between the $E_o(k)$ band the conduction and valence bands. Since the gap center band is either fully occupied for donor doping or completely empty for acceptor doping, the system remains a semiconductor having a gap of order 0.5 eV.

While a periodic soliton lattice minimizes the energy in KOTANI's model, Coulomb forces between charged solitons and charged impurities tend to randomize the soliton distribution in space. We have studied this problem by solving for the electronic spectrum of a chain of N CH groups having solitons centered at random sites $n_1, n_2, \ldots n_\nu$. In Fig.5 is shown a histogram of the gap center states for y = 0.08 and 2Δ = 1.9 eV. The average spacing between the solitons is about 4.3 times the half width, ℓ, of the soliton. As one might expect, the sharp KOTANI gaps have disappeared, but the degree of band tailing is surprisingly large, possibly due to an indirect interaction of gap center states via the conduction and valence band states. Preliminary results show that the states at the Fermi surface ($\sim\Delta$

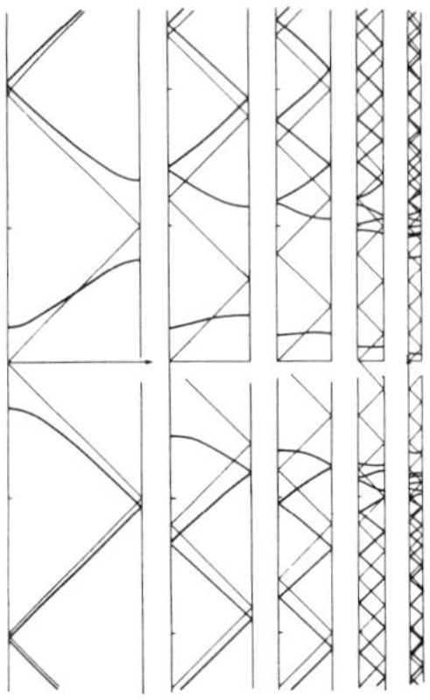

Fig.4 Soliton energies bands, E(k), for different concentrations: y = 0.005,0.01,0.02, 0.03,0.05 from right to left

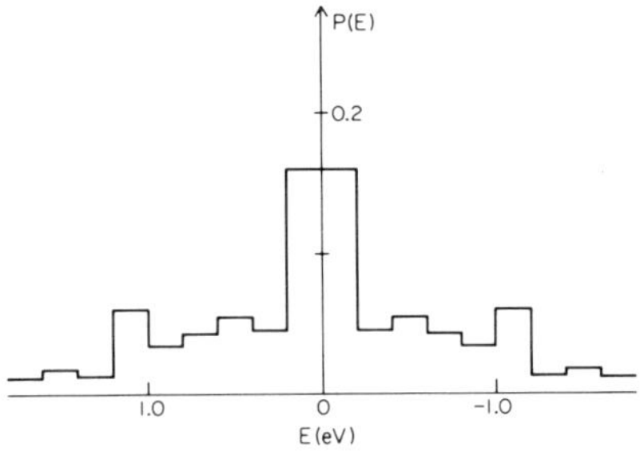

Fig.5 Histogram of the gap center states for a random array of solitons

and $-\Delta$ for electron and hole doping respectively) are localized, leading to hopping rather than band conductivity for electrons. Experimentally, a sharp increase of σ with doping occurs near $y_{c_1} \simeq 0.3\%$, while the Pauli susceptibility remains small until $y_{c_2} \sim 6\%$. We suggest that four regimes occur with increasing y: a) $0 < y < y_{c_1}$, the "soliton glass phase" in which $S^{\pm}$ are bound to the charged dopants; b) $y_{c_1} < y < y_{c_2}$, the "soliton liquid phase" in which cooperative screening effects enhanced by quantum fluctuations of the solitons causes the "melting of the glass" with the unpinned $S^{\pm}$ giving strong soliton transport; c) $y_{c_2} < y < y_{c_3}$, the soliton liquid becomes Pauli paramagnetic due to substantial tailing of the gap center state band, leading to a nonzero density of states at the Fermi surface; d) $y_{c_3} < y$, the mean square order parameter is strongly reduced, and the material approaches a conventional metal with strong polaronic effects.

Other models of the semiconductor metal transition in $(CH)_x$ have been proposed. Within a Landau-Ginzburg context, RICE and TIMONEN [14] have calculated the mean bond alternation amplitude versus carrier concentration for a periodic array of solitons, as in KOTANI's calculations. While they obtained a sharp transition at $y_c \sim 9\%$, BRAZOVSKII [15] has shown that for a periodic array, no transition occurs for any y if the Landau-Ginzburg approximation is not invoked. This result illustrates the importance of nonlocal effects in the theory. HOROWITZ [16] has developed a continuum model containing electron-phonon and zero range electron-electron interactions. Using amplitude and phase variables, he finds an effective attraction between solitons which leads to phase separation corresponding to regions of incommensurate CDW's between regions of commensurate CDW's. Which, if any, features of the above models relate to nature is unclear at present.

Turning to three-dimensional effects, still further hybridization of the gap center states occurs when interchain interactions are taken into account. The same considerations that led to the conclusion that the hopping process is three dimensional, lead to the conclusion that at low enough impurity concentrations, the amount of hybridization due to interactions between mid-gap states associated with solitons on different chains exceeds that due to purely in-chain hybridization. One possible mechanism for the metal-insulator transition is an Anderson transition in the mid-gap band, similar to that seen in doped semiconductors. If this mid-gap band is partially full, either due to overlap with the valence band tail, or due to the persistence of neutral solitons in heavily doped samples, metallic behavior will result [17]. The observed conductivity transition in $(CH)_x$ occurs at an impurity concentration at which in-chain and out-of-chain interactions are expected to produce comparable amounts of hybridization; thus the question of the dimensionality of the metal-insulator transition remains to be resolved.

5. Real Time Dynamics

It is interesting to study the time evolution of $\psi_n(t)$ to gain insight into the dynamics of solitons, phonons, electrons and holes, e.g. $S\bar{S}$ collisions, how an injected electron-hole pair becomes a charged soliton-antisoliton pair, etc. SU and SCHRIEFFER [18] have integrated the equations of motion for the $\psi_n(t)$ field for a finite length chain with forces given by adiabatic potential energy function $E_m(\{\psi_n\})$ for the many-electron state m. As an example, the time evolution of ψ following the sudden creation of an electron-hole pair at the gap edge (2Δ) at t = 0 is shown in Fig.6. ψ begins to drop near the chain

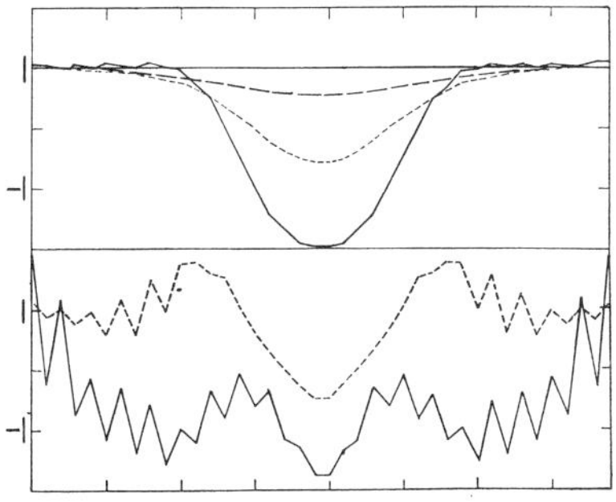

Fig.6 Evolution of the normalized $\psi(n)$ following the sudden creation of an electron-hole pair. The horizontal axis refers to site number n

center, localizing the electron and hole in that vicinity which feeds back further decreasing ψ until the depression separates and becomes an $S\bar{S}$ pair plus a cloud of phonons corresponding to the rapid wiggles of ψ_n or long wavelength phonons in u_n. This decay of eh → $S\bar{S}$ proceeds remarkably rapidly, i.e. in $\simeq 2 \times 10^{-13}$ sec without an activation barrier. Similar dynamics is obtained for dressing a single added electron except that the depression of ψ does not split into S + $\bar{S}$ but forms a strong polaron well, plus a cloud of shake-off phonons.

We are now developing methods to treat quantum effects in the statics and dynamics of the ψ field within the adiabatic approximation.

References

*This material is based upon work supported by the National Science Foundation under Grant No. DMR-8007432.

1. W. P. Su, J. R. Schrieffer, and A. J. Heeger, Phys. Rev. Lett. 42, 1698 (1979) and Phys. Rev. B 22, 2099 (1980).

2. M. J. Rice, Phys. Lett. 71A, 152 (1979).

3. M. Takayama, Y. R. Lin-Liu, and K. Maki, Phys. Rev. B 21, 2388 (1980).

4. R. Jackiw and C. Rebbi, Phys. Rev. D 13, 3398 (1976).

5. W. P. Su, Solid State Commun. (in press).

6. E. Mele and M. J. Rice (preprint).

7. J. R. Schrieffer (to be submitted).

8. Y. W. Park, A. J. Heeger, M. A. Druy, and A. G. MacDiarmid, Phys. Rev. B (to appear).

9. S. Kivelson (to be submitted).

10. B. R. Wienberger, J. Kaufer, A. J. Heeger, A. Pron, and A. G. MacDiarmid, Phys. Rev. B 20, 223 (1979); A. J. Heeger and A. G. MacDiarmid, in this volume.

11. G. Mihaly, G. Vancso, S. Pekker, and A. Janossy, J. Synthetic Metals (in press).

12. N. Suzuki, M. Ozaki, S. Etemad, A. J. Heeger, and A. G. MacDiarmid, Phys. Rev. Lett. (submitted).

13. A. Kotani, J. Phys. Soc. Japan 42, 416 (1977).

14. M. J. Rice and J. Timonen, Phys. Lett. 73A, 368 (1979).

15. S. A. Brazovskii, JETP Lett. 606 (1979); JETP 78, 667 (1980).

16. B. Horowitz, Solid State Commun. 34, 61 (1980).

17. This view of the metal-insulator transition is currently being investigated by Kivelson.

18. W. P. Su and J. R. Schrieffer, Proc. Natl. Acad. Sci. U.S. 77, No. 10 (1980).

Attracting Solitons and a First Order Lock-In Transition: Metallic Polyacetylene and the Spin-Peierls System

Baruch Horovitz

Department of Nuclear Physics, The Weizmann Institute of Science
Rehovot, Israel

The problem of a one-dimensional fermion system whose density is at or near to that of a half-filled band is of recent experimental and theoretical interest. For fermions with chemical potential μ and density ρ (measured from the half-filled band level) the relation $\rho(\mu)$ at zero temperature defines the lock-in phase transition. In the spin-Peierls problem [1] the magnetization is mapped into ρ of spinless fermions and the external magnetic field into μ. In a recent experiment [2] on a spin-Peierls system a first-order lock-in transition was observed. On the other hand, when doping polyacetylene [3], the experiment controls ρ - the doping level, and not the chemical potential. If the lock-in transition is first order it implies that for ρ smaller than some ρ^* the system will phase separate into regions with either $\rho=0$ or $\rho=\rho^*$. The system will then appear as metallic particles embedded in an insulating medium. This is indeed consistent with: a) infrared absorption data [4] which shows metallic absorption coexistent with the semiconducting gap, b) the appearance of metallic type magnetic susceptibility [5] below the threshold for the metallic d.c. behaviour and c) temperature dependence of the conductivity which is consistent with percolation between metallic particles [6].

The theoretical interest is in the existence of soliton solutions [7] - each soliton describes one electron charge in ρ or 1/2 fermion density for the spinless fermion problem. Define $2\Delta_0$ as the semiconducting gap of the commensurate system ($\rho=0$) and E_S as the soliton energy. Since $E_S<\Delta_0$ [7] fermions which are added to a half-filled band generate solitons, rather than occupying a state in the conduction band.

The lock-in transition is of first order if and only if solitons attract each other. To see this note that if solitons attract each other then for some sufficiently low soliton density the system is unstable and the curve $\rho(\mu)$ does not include these states, i.e. the lock-in transition is of first order. On the other hand, if the lock-in transition is of first order it means that a finite soliton density, ρ^*, has a lower energy than the same number of solitons with a density $\rho<\rho^*$, i.e. phase separation. Therefore solitons attract each other or attract collectively. If the attraction is at large separations it means that two solitons attract each other. However, the attraction may be a collective many-soliton interaction at short distance while the two-body interaction at large distance is repulsive. (This latter case indeed happens in the analogous binding of excitons into an electron-hole droplet. The corresponding first-order transition is the Mott transition [8].

We now show that if the single soliton energy E_S is sufficiently large the lock-in transition is of first order. The condensation energy of the dimerized ground state relative to the undistorted phase is [7,9]

$E_c = -\Delta_0^2/(2\pi v_F)$ where v_F is the Fermi velocity. In the incommensurate limit, where umklapp effects of the lattice reciprocal wavevector $2\pi/a$ are neglected, the gap is $\Delta_i \simeq \Delta_0^2/W$ where W is half the bandwidth. Thus, in the weak coupling limit, $\Delta_0 << W$, the incommensurate condensation energy can be neglected.

Now consider the Gibbs free energy $G(\mu) = E - \mu\rho$. In the incommensurate limit we have a metal with excess charge ρ whose energy E_m is

$$E_m = 4 \int_0^{\frac{1}{2}\pi\rho} v_F q \frac{dq}{2\pi} = \frac{\pi}{4} v_F \rho^2 \quad .$$

The chemical potential is $\mu = \frac{1}{2}\pi v_F$ and therefore the Gibbs energy for the metal is $G_m = -\mu^2/(\pi v_F)$. Now $G_m(\mu_c) = G_c(\mu_c)$ when $\mu_c = \Delta_0/\sqrt{2}$ so that for $\mu > \mu_c$ the incommensurate phase is favoured.

The final step in this argument is to show that if $E_s > \Delta_0/\sqrt{2}$ the lock-in transition must be of first order. Indeed, for $\mu_c < \mu < E_s$ a very low soliton density is not favoured compared to the commensurate phase (energy cost $E_s - \mu > 0$) while the commensurate phase is not favoured compared to the incommensurate phase. Thus at $\mu = \mu_c$ there is a lock-in transition which excludes low soliton density states, i.e. it is a first-order transition.

If one takes into account the finite bandwidth so that there is some gain from the incommensurate condensation energy the range of E_s for which the transition is first-order increases.

I have found [10] an exact solution for the soliton lattice in the continuum model [11] for solitons in polyacetylene, i.e. the adiabatic electron-phonon Hamiltonian in one dimension. Here [11] $E_s = 2\Delta_0/\pi$, rather close to the value of μ_c. In the limit of weak coupling ($W/\Delta_0 \to \infty$) the soliton-soliton interaction decays as a repulsive exponential while for a finite W/Δ_0 it decays as $(\text{distance})^{-1}$ and is attractive !

We have also solved this problem including a direct electron-electron interaction in a Hartree-Fock scheme [9,12]. We have found that solitons attract each other if the electron-electron coupling is not too strong compared with the electron phonon coupling.

The application to the experimentally studied systems shows the two faces of the same problem - attracting solitons and phase separation in doped polyacetylene and a first-order lock-in transition in the spin-Peierls system.

References

1. E. Pytte: Phys. Rev. B10, 4637 (1974)
2. D. Bloch et al.: Phys. Rev. Lett. 44, 294 (1980)
3. C.K. Chiang et al.: Phys. Rev. Lett. 39, 1098 (1977)
4. C.R. Fincher et al.: Phys. Rev. B20, 1589 (1979)
5. B.R. Weinberger et al.: Phys. Rev. B20, 223 (1979)
6. Y. Tomkiewicz et al.: Phys. Rev. Lett. 43, 1532 (1979)
7. W.P. Su, J.R. Schrieffer and A.J. Heeger: Phys. Rev. Lett. 42 , 1698 (1979), and Phys. Rev. B (to be published).
8. W.F. Brinkeman and T.M. Rice: Phys. Rev. B7, 1508 (1973)
9. B. Horovitz: Solid State Commun. 34, 61 (1980)
10. B. Horovitz (to be published)
11. H. Takayama, Y.R. Lin-Liu and K. Maki: Phys. Rev. B21, 2388 (1980
12. M. Grabowski, K.R. Subbaswamy and B. Horovitz: Solid State Commun. 34, 911 (1980)

Polyacetylene Revisited

Y. Tomkiewicz, N.S. Shiren, T.D. Schultz, K. Mortensen, M.L.W. Thewalt, J.D. Kuptsis, and R.G. Schad, IBM T.J. Watson Research Center
Yorktown Heights, NY 10598, USA

B.H. Robinson, IBM, Poughkeepsie, NY, USA

T.C. Clarke and G.B. Street, IBM, San Jose, CA, USA

H. Thomann and L.R. Dalton, State University of New York, Stony Brook, NY, USA

H.B. Brom, University of Leiden, The Netherlands

The conjugated organic polymer $(CH)_x$ has recently been the subject of extensive interest on two counts: 1) In pristine *trans*-$(CH)_x$, EPR and NMR evidence has been given for highly mobile spin carriers that carry no charge. These have been interpreted as arising from topological solitons (a localized reversal of the bond alternation pattern). The novelty of such spin carriers has stimulated much theoretical and experimental interest. 2) Upon doping, the conductivity of $(CH)_x$ increases by $\sim 10^{12}$ and undergoes an apparent insulator-to-metal transition at $\sim 1\%$ dopant concentration. The possibility that the doping produces charged solitons rather than band electrons and that with increasing density this soliton gas undergoes a phase transtion has also attracted much interest.

In this paper, these two aspects of $(CH)_x$ will be discussed. In the first part, spin-echo and CW EPR results for the undoped *trans* polymer will be presented. We shall show that both the magnitude and the temperature dependence of the diffusion rate as determined from the echo measurements should lead to a reconsideration of the soliton model. In the second part, magnetic and transport data obtained in the doped *cis* polymer will be presented showing that doping of $(CH)_x$ is very inhomogeneous. The doped and undoped parts of the polymer appear to be largely distinct, the doped regions being metallic. The conductivity changes that have previously been interpreted as arising from a transition *within* the doped region are attributed to a change in the topology of these regions, *i. e.*, the occurrence of a percolation threshold in which the metallic regions touch and form an essentially metallic continuum.

Spin Dynamics in *trans*-$(CH)_x$. Fig. 1 shows T_M of the electron spins in $(CD)_x$ as derived by spin-echo measurements[2] and T_2^* from CW EPR as functions of temperature. While T_2^* has a monotonic behavior as a function of temperature, T_M has a minimum value of 120 nsec at T_{min} = 130 K. For $T > T_{min}$, T_M is an increasing function of temperature and agrees well with T_2^*; for $T < T_{min}$, T_M is a decreasing function of temperature and has very different values from those of T_2^*, implying that the EPR line is inhomogeneously broadened for these temperatures. The monotonic temperature dependence of T_2^* was previously interpreted as an indication of a strong narrowing for all temperatures. The present measurements indicate that the strong narrowing regime is a proper description only for the temperature range $T > T_{min.}$ Thus, the data presented show[3] that the diffusion rate of the motion is temperature dependent and, if this temperature dependence can be represented by an activated process, then the activation energy is larger than 130 K.

At the position of the minimum in T_M, the determination of the diffusion rate D from the echo measurements has two advantages. First, the diffusion rate can be determined readily from the relationship

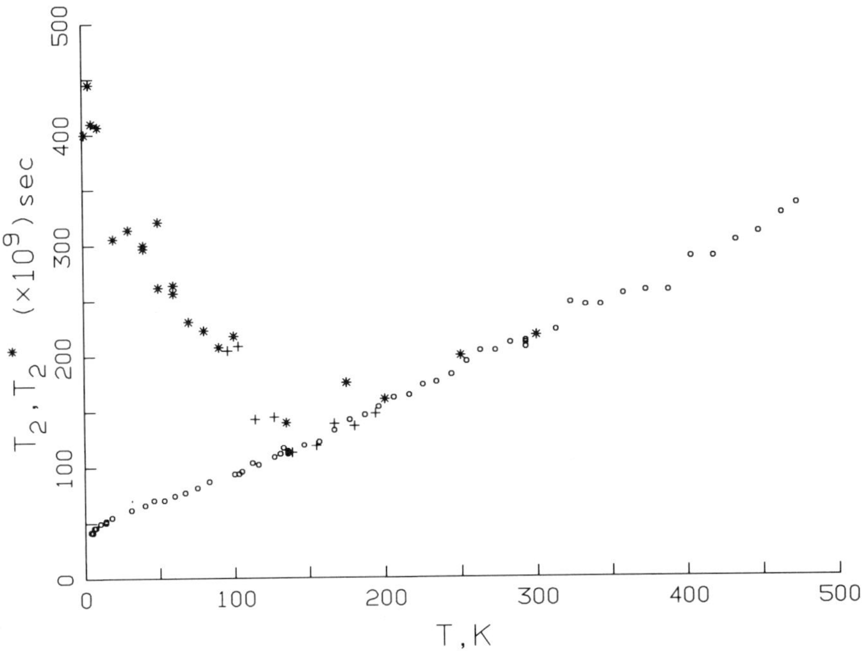

Fig. 1. T_M (as determined from spin-echo experiments) and T_2^* (as determined from CW EPR) as a function of temperature. The T_M data presented are on two different samples indicated by * and +. The T_2^* data are indicated by o.

$$D\tau = C \tag{1}$$

where τ is the time separating the 90° and 180° pulses. Although the value of C depends on the details of the particular model used, for the commonly accepted models of either 1d or quasi-1d diffusion C = 6. Thus D as determined from (1) is independent of the model, *i. e.* of whether the motion is 1d or quasi-1d. This should be contrasted with the strong dependence on dimension of D as determined from the EPR linewidth, a variation by a factor of ~10. Second, any other technique for determining D requires knowledge of either the linewidth of static spins (EPR) or the hyperfine coupling between the proton and electron spins (NMR). In determining D from echo measurements at the position of the minimum, no additional parameters are required.

The determination of D at any temperature other than that corresponding to the minimum of T_M requires a specific model. Since the value of the diffusion coefficient at 293 K is a subject of extensive interest, we have determined an upper bound for D to be 6 × 10^9 sec^{-1}. This upper bound comes from using a 1d diffusion model in analyzing the echo data. We know this overestimates D, because both the lineshape and the echo decay indicate that the diffusion is not one dimensional. If one further modifies this estimate by assuming a spin delocalization over 15 sites, as required by the soliton model, the upper bound for the diffusion rate becomes ~10^{11} sec^{-1}. Thus, even this upper bound, which must grossly overestimate the true rate, is lower by a factor of at least 100 than the rate of ~10^{13} sec^{-1} inferred from NMR.

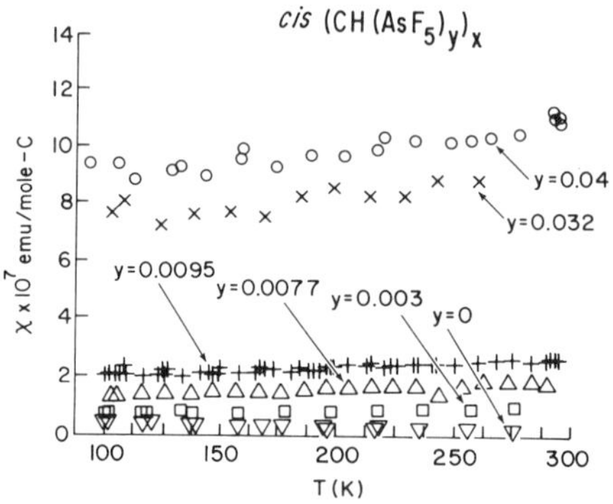

Fig. 2. Spin susceptibility as a function of temperature for *cis*-$(CH)_x$ doped with various levels of AsF_5.

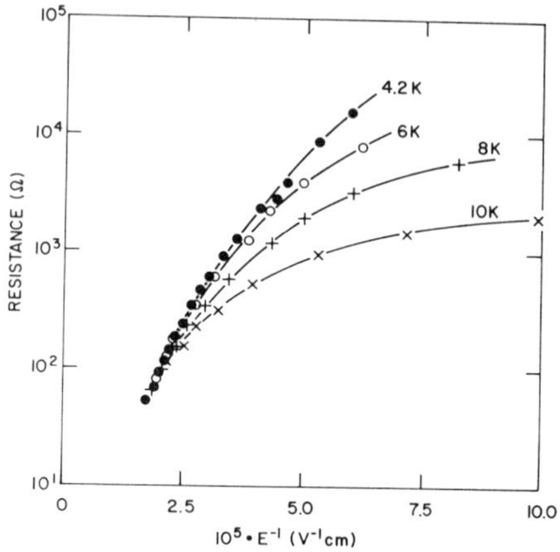

Fig. 3. Resistance as a function of 1/E for $[CH(AsF_5)_{0.005}]_x$ at several temperatures.

We may summarize the results of the spin-echo data experiments: 1) The spin diffusion rates are significantly lower than the values previously determined by NMR and expected by the soliton model. 2) The activation energy of the diffusion is of the order of hundreds of degrees at least, much higher than that expected from the soliton model. These results should lead to a reconsideration of the soliton model.

Insulator-to-Metal Transition of Doped $(CH)_x$. Fig. 2 shows the temperature dependence of the spin susceptibility of *cis*-$(CH)_x$ doped with various levels of AsF_5 in the temperature range $100 < T < 300$ K. For comparison, the susceptibility of pristine *cis*-$(CH)_x$ is also shown in this figure. It is clearly seen that doping *increases* the spin susceptibility and that the spin susceptibility generated by doping has hardly any temperature dependence in this temperature regime, even for the lowest concentration of dopant (0.3%). If doping were homogeneous, then at least for the lower concentrations the electrons would be expected to occupy localized, non-overlapping states, giving rise to a spin susceptibility having a Curie type of temperature dependence. Since the susceptibility is temperature-independent, banding apparently occurs. Thus, the impurity states must overlap even for the lowest dopant concentration, which would require inhomogeneity. Moreover, we find a linear relationship between the magnitude of the susceptibility and the dopant concentration over a broad range of concentrations (0.3% to 12%), suggesting that the effective transfer integral is concentration independent. This would indeed be the case if metallic domains were formed in the $(CH)_x$ matrix. The effect of increasing the concentration would be to increase the size of these domains and/or to increase their number so that eventually enough would be in essential contact to create percolation paths. While the transition to such a situation would be expected to affect the transport properties considerably, no similar effect is expected for the magnetic susceptibility. And indeed, neither the temperature dependence of the susceptibility nor its magnitude shows any indication of the semiconductor-to-metal transition apparently seen in the conductivity at the level of about 1% doping.

As for the conductivity, in the low-concentration regime the low-field (ohmic) resistance has the temperature dependence[4] $\exp(T_0/T)^{1/2}$. At high electric fields, namely when the voltage drop between neighboring grains is much larger than the thermal energy, the resistance should have a field dependence given by the equation[5]

$$R = R_o \exp(E_o/E) \cdot \quad (2)$$

Figure 3 shows the field dependence of the resistance in the temperature range 4.2–10 K. The field dependence follows equation (2) with a temperature-independent $E_0 \simeq 2 \times 10^5$ V/cm. By applying a similar analysis to that used for granular metals, we have determined the average separation of the metallic grains for the studied sample to be 60 Å. Thus, our studies on doped $(CH)_x$ for concentrations below the "insulator-to-metal transition" indicate the existence of metallic domains with high tunneling barriers between them, analogous to granular metals below the percolation threshold. Measurements on concentrations above this transition show a much weaker temperature and field dependence, indicating that barriers still exist but that they are much reduced and that both the temperature dependence and the field dependence of the conductivity can be understood by fluctuation-induced tunneling.

We may summarize the conclusions about doped materials. 1) The doping of $(CH)_x$ is inhomogeneous for all concentrations of dopant. A phase separation exists between the doped and undoped parts of the polymer and metallic clusters exhibiting Pauli-like susceptibility are formed in the doped parts. 2) The conductivity results, which were previously interpreted as indicative of an insulator-to-metal transition, can be understood in terms of the existence of two structural regimes: a) a dielectric regime in which isolated metallic particles are dispersed in the undoped polymer, yielding a low conductivity involving thermal activation, and b) a metallic regime where there are percolation paths *via* metallic domains that are very close and separated by only small barriers.

References

1. M. J. Rice, Phys. Lett. **71A**, 152 (1979); W. P. Su, J. R. Schrieffer and A. J. Heeger, Phys. Rev. Lett. **42**, 1698 (1979).
2. The evaluation of T_M was done by fitting the echo amplitude data with a function of the form $Ae^{-T_M/2\tau}$
3. Even if two kinds of spins, localized and fast-moving, were to exist in the sample, as suggested by Holczer *et al.*, the only way the minimum in T_M would occur would be if above 130 K only the fast spins were observed. Thus, even in this situation our conclusions are still applicable.
4. Y. Tomkiewicz, T. D. Schultz, H. B. Brom, T. C. Clarke and G. B. Street, Phys. Rev. Lett. **43**, 1532 (1979).
5. B. Abeles, Appl. Solid State Sci. **6**, 1 (1976).

Magnetic Behavior of Polyacetylene, Polyparaphenylene and Polypyrrole

M. Peo, S. Roth, and K. Dransfeld
Max Planck-Institut für Festkörperforschung, Heisenbergstr.1
D-7000 Stuttgart 80, Fed. Rep. of Germany

J. Hocker
Wiss. Hauptlabor, Bayer AG, D-5090 Leverkusen-Bayerwerk, Fed. Rep. of Germany

It is well known that doping of polymers with conjugated double bonds leads to metallic conductivity [1], but the mechanism of the insulator-to-metal transition and of the electric conductivity is not yet understood. The present work is an attempt to study the electronic properties of metallic polymers by measuring the Pauli spin susceptibility.

Magnetic measurements of AsF_5-doped polyacetylene have already been carried out by WEINBERGER et al. [2] and by TOMKIEWICZ et al. [3]. Our work is a comparative study of metallic polyacetylene doped by AsF_5 and by iodine, respectively and it shows that there is a qualitatively different magnetic behaviour between iodinated and AsF_5-doped polyacetylene samples. In addition we present magnetic data on polyparaphenylene doped by SbF_5 and on polypyrrole containing BF_4^--ions.

Polyacetylene was prepared by the standard procedure described by SHIRAKAWA et al. [4], polyparaphenylene was obtained from the polymerization of benzene following IVORY et al. [5], and polypyrrole was produced electrochemically applying the method of KANAZAWA et al. [6]. Doping by iodine or AsF_5 was carried out by exposing the samples to iodine vapour or AsF_5-gas, respectively. For doping by SbF_5 the samples were treated with a solution of SbF_5 in pentane and the BF_4^--ions were incorporated into polypyrrole during the electrochemical synthesis. The doping level has been determined after the magnetic measurements by weight uptake or by chemical analysis. The magnetic susceptibility was studied by means of a Faraday balance in the temperature range from 10 to 300 K.

Figure 1 shows the raw data for polyparaphenylene at various doping levels. The measured molar [7] susceptibility is plotted as a function of temperature. To analyze our data we assume that the measured susceptibility is composed of three terms

$$\chi = \chi_{core} + \chi_{Pauli} + C/T.$$

[χ_{core} is the diamagnetic contribution of the electrons in closed shells. As seen in Fig.1 it predominates in all samples at room temperature. χ_{core} can be calculated from the diamagnetic increments. To do this we have used the compilation of HABERDITZL (6.61 for CH, 126 for I_3, 61 for AsF_5, 69 for SbF_5 and 39 for BF_4^-; all in units of 10^{-6} emu cm^3/mole] [8]. C/T is the Curie susceptibility originating from localized magnetic moments (impurities or localized charge carriers) and χ_{Pauli} is the Pauli spin susceptibility of the conduction electrons, the term of principal interest here. χ_{Pauli} is proportional to $N(E_F)$, the density of electronic states at the Fermi surface. If $\chi - \chi_{core}$ is plotted versus the reciprocal temperature we obtain straight lines, the slope of which corresponds to the Curie constant and the intercept with the vertical axis to χ_{Pauli}.

Figures 2, 3, and 4 show this analysis for polyacetylene, polyparaphenylene, and polypyrrole, respectively. Our most important result is the

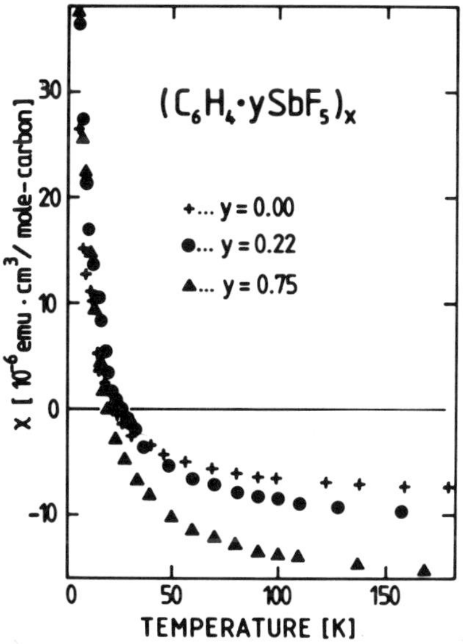

Fig.1 Magnetic susceptibility of polyparaphenylene for various doping levels versus temperature.

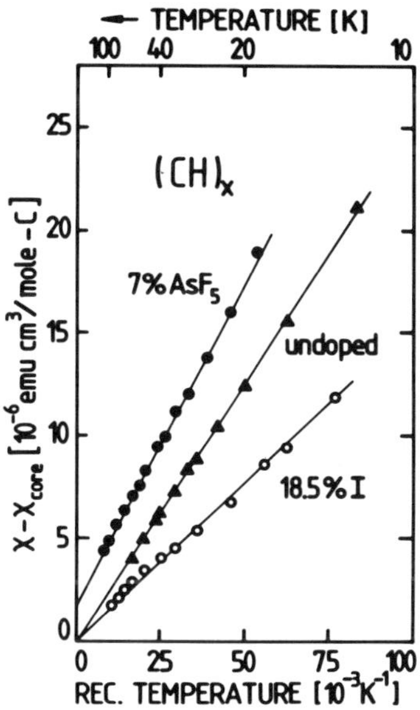

Fig.2 χ - $χ_{core}$ for polyacetylene versus reciprocal temperature.

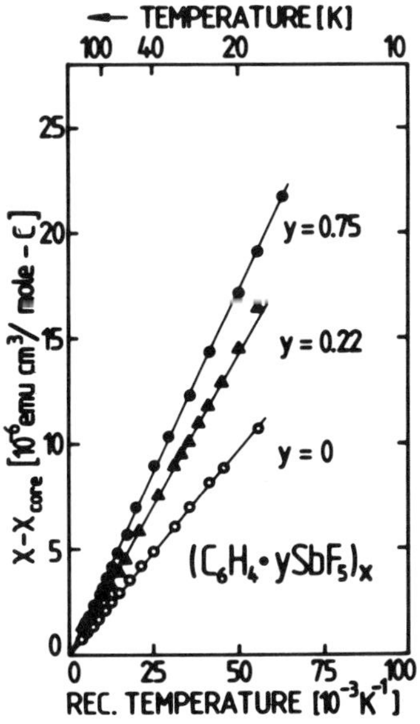

Fig.3 χ - $χ_{core}$ for polyparaphenylene versus reciprocal temperature.

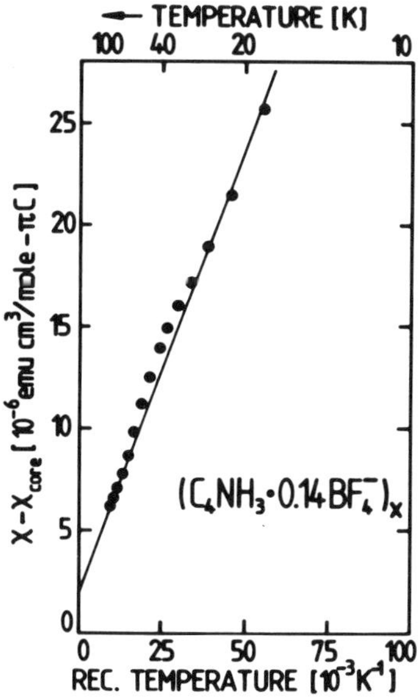

Fig.4 χ - $χ_{core}$ for polypyrrole versus reciprocal temperature.

qualitatively different behaviour of iodinated and of AsF_5-doped polyacetylene (Fig.2). The dopant concentrations (7 at % AsF_5 and 18.5 at % I) correspond roughly to the same doping level because three iodine atoms are needed to remove one electron from the polymer π-system thereby forming an I_3^--complex [9], whereas each incorporated AsF_5-group takes one electron. Both concentrations are well above the insulator-to-metal transition as seen in measurements of the electric conductivity and of the thermoelectric power [10]. We see no simple explanation why there exists a finite Pauli susceptibility in the AsF_5-doped sample ($\chi_{Pauli} = 1.9 \cdot 10^{-6}$ emu cm^3/mole) but no detectable Pauli term in iodinated polyacetylene ($\chi_{Pauli} < 2 \cdot 10^{-7}$ emu cm^3/mole). Furthermore the absence of a Pauli susceptibility in the iodinated sample is not consistent with a conventional mechanism of electronic band conductivity and normal density of states at the Fermi energy.

Figure 3 shows the analysis of our polyparaphenylene data. Here we find no Pauli paramagnetism at doping levels even as high as 75 at % of SbF_5. Polypyrrole, on the other hand, shows a Pauli paramagnetism of $2 \cdot 10^{-6}$ emu cm^3/mole (Fig.4).

The data presented here show that - as far as the magnetic susceptibility of the conduction electrons is concerned - different metallic polymers do not behave alike. A Pauli paramagnetism is found in AsF_5-doped polyacetylene and polypyrrole with BF_4-ions which is of the same order of magnitude as has been measured in $(SN)_x$ ($5 \cdot 10^{-6}$ emu cm^3/mole [11]). In iodinated polyacetylene and in SbF_5-doped polyparaphenylene, on the other hand, which both show metallic conductivity as well, χ_{Pauli} is equal to zero within our experimental limits of $2 \cdot 10^{-7}$ emu cm^3/mole.

We want to thank A. Stark for her assistance during the magnetic measurements.

References

1. C.K. Chiang, C.R. Fincher, Jr., Y.W. Park, A.J. Heeger, H. Shirakawa, E.J. Louis, S.C. Gau, A.G. MacDiarmid: Phys. Rev. Lett. 39, 1098 (1977)
2. B.R. Weinberger, J. Kaufer, A.J. Heeger, A. Pron, A.G. MacDiarmid: Phys. Rev. B20, 223 (1979)
3. Y. Tomkiewicz, T.D. Schultz, H.B. Broom, T.C. Clarke, G.B. Street: Phys. Rev. Lett. 43, 1532 (1979); Y. Tomkiewicz, T.D. Schultz, H.B. Broom, T.C. Clarke, G.B. Street: Bull. Am. Phys. Soc. 25, 161 (1980)
4. H. Shirakawa, T. Ito, S. Ikeda: Polym. J. 4, 460 (1973)
5. D.M. Ivory, G.G. Miller, J.M. Sowa, L.W. Shacklette, R.R. Chance, R.H. Baughman: J. Chem. Phys. 71, 1506 (1979)
6. K.K. Kanazawa, A.F. Diaz, W.D. Gill, P.M. Grant, G.B. Street, G.P. Gardini, J.F. Kwak: Synthetic Metals 1, 329 (1979/80)
7. The term "molar" refers to a "carbon mole", for which the true molar weight had to be divided by the number of C-atoms per molecule.
8. W. Haberditzl: "Magnetochemie", Akademie-Verlag, Berlin (1968) (Vieweg & Sohn 6050, Pergamon Press 080134637)
9. S.L. Hsu, A.J. Signorelli, G.P. Pez, R.H. Baughman: J. Chem. Phys. 69, 106 (1978)
10. Y.-W. Park, A.J. Heeger, M.A. Druy, A.G. MacDiarmid: J. Chem. Phys., to be published
11. J.C. Scott, J.D. Kulick, G.B. Street: Solid State Commun. 28, 723 (1978)

Static and Dynamic Susceptibilities of Magnetic Polymers*

J.C. Scott, R.E. Stahlbush, P.D. Krasicky, and R.H. Silsbee

Laboratory of Atomic and Solid State Physics, Cornell University
Ithaca, NY 14853, USA

1. Introduction

The inorganic polymers known as poly(metal phosphinates) contain transition metal ions as an integral part of the linear molecule, and therefore provide examples of one-dimensional magnets [1]. In this brief report, we summarize data obtained on a triply bridged chromium phosphinate (see Fig.1). Complete details are given in [2,3].

2. Magnetic Susceptibility

Susceptibility data, obtained with Faraday balance and with an ac mutual inductance bridge, show that, at temperatures above about 10K, PCrP-C is well described as a one-dimensional Heisenberg antiferromagnet, with spin S=3/2 and exchange J=-2.5K [4]. Below 6K an anisotropy develops (see Fig.2). X-ray diffraction reveals that the polymer molecules are preferentially aligned within the plane of the sample sheet. Hence we conclude that there is either an easy-plane or an easy-direction perpendicular to the molecular axis.

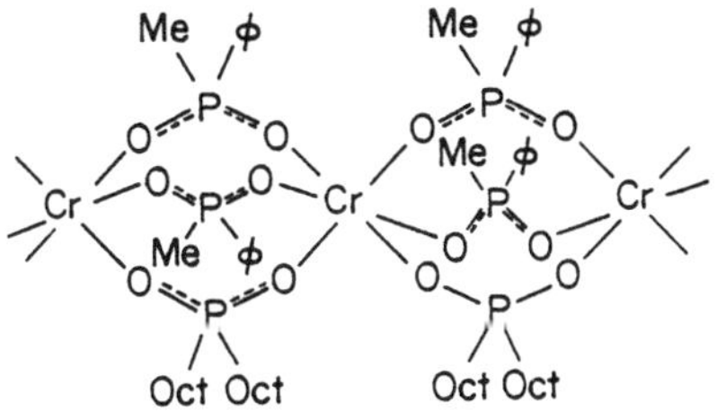

Fig.1 Molecular structure of the chromium tris-phosphinate PCrP-C

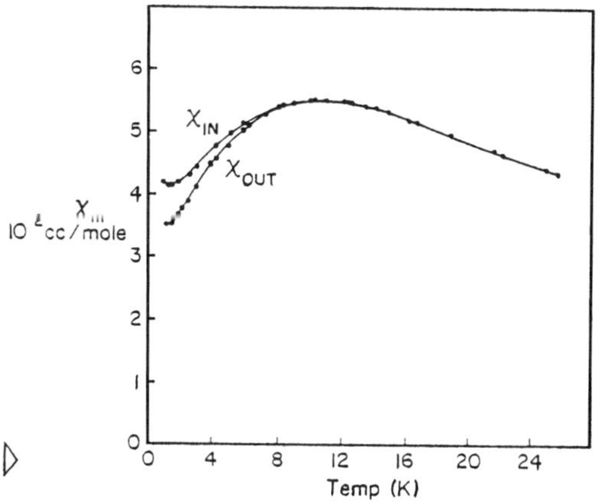

Fig.2 Initial susceptibility. χ_{in} is measured with the ac field in the plane of the sample sheet; χ_{out} perpendicular to it.

The dependence of the differential (low frequency ac) susceptibility on applied magnetic field is shown in Fig. 3. The increase and subsequent saturation of the susceptibility arises because the field tends to align the exchange correlated spins into a flopped configuration. Thus at high field the response is that of a perpendicular antiferromagnet. We have performed detailed calculations (see Section 4) which indicate that this observed behaviour is that of an easy-plane system, for two reasons: first, the onset of the spin-flopped alignment occurs at a field which is proportional to the temperature; second, there is no experimental evidence of the peak predicted by our calculations of $\chi(H)$ for the easy-axis case.

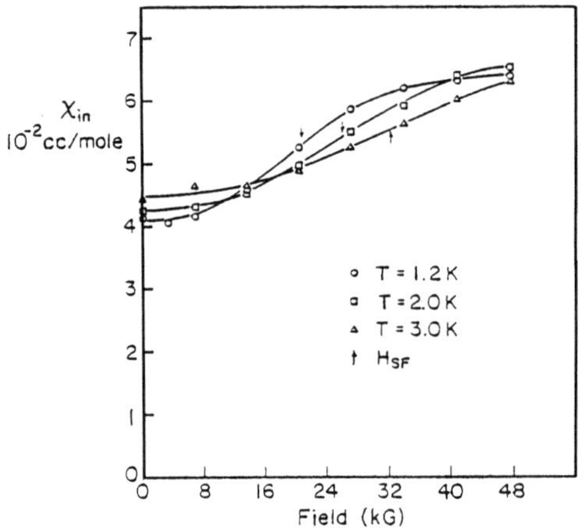

Fig.3 Field dependence of the susceptibility of PCrP-C. Data are taken with both the ac and dc magnetic fields in the plane of the sample sheet. Behaviour of χ_{out} is similar.

3. Specific Heat

The specific heat of PCrP-C has been measured by a quasi-adiabatic, heat pulse method. Careful comparison of the experimental data with numerical results of Blöte [5], and using a linear spin wave extrapolation to lower temperature, reveal an excess specific heat at temperatures below about 1K (see Fig.4). This excess specific heat appears in the temperature range where the susceptibility, χ_{in} which has mostly $\chi_{\perp}$ character, is approaching the value expected for a linear XY antiferromagnet. The entropy difference is due to the extra spin degree of freedom of the Heisenberg system as compared to the XY chain.

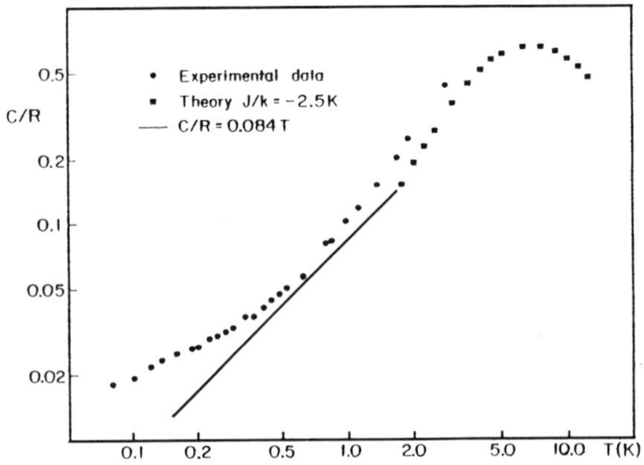

Fig. 4 Specific heat of PCrP-C. Numerical data are from calculation of Blöte [5].

4. Calculation of Thermodynamic Functions

A classical Heisenberg antiferromagnetic chain Hamiltonian, including both anisotropy and Zeeman energy, has been treated using the transfer matrix technique. The susceptibility and specific heat have been calculated as functions of temperature and magnetic field for both easy-axis and easy-plane anisotropy. Details and results are given in [2]. Using an easy-plane anisotropy parameter D=-0.03K gives semi-quantitative agreement with the experiments described above. The discrepancies, on the order of 10%, are attributed to the classical nature of the calculation.

A phenomenological description of the magnetic properties can be obtained using a "Correlated Cluster Model" in which, below the mean field temperature the magnetic element is taken to be the block of spins within a correlation length, $\xi \sim J/T$. Competition among the thermal, Zeeman, anisotropy and exchange energies of this cluster give rise to the following phenomena: (i) locking of the spin directions to the easy-axis or easy-plane around the crossover temperature, $T_A \sim \sqrt{|JD|}$. Below T_A the susceptibility is anisotropic; (ii) development of perpendicular, or spin-flopped, alignment of the individual moments in an applied field. For a field applied parallel to an easy-axis the crossover field is $\mu_B H_{SF} \sim \sqrt{|JD|}$. For fields in an easy-plane $\mu_B H_{SF} \sim kT$.

5. Electron Spin Resonance

The temperature dependence of the ESR linewidth, as observed with an X-band spectrometer, is shown in Fig. 5. At room temperature the lineshape is Lorentzian over ± 4 linewidths and the width can be analysed in terms of anisotropic three-dimensional diffusion. The interchain interaction, which cuts-off one-dimensional diffusion along the molecular chain, is of order 70mK, comparable to a sum of contributions from interchain dipolar coupling and chain branching/crosslinking at spacings greater than 20 units.

A memory-function analysis [3,6] of the linewidth, using static correlation functions for the classical antiferromagnetic chain, yields the temperature dependence depicted in Fig. 5. The parameters used were the room temperature

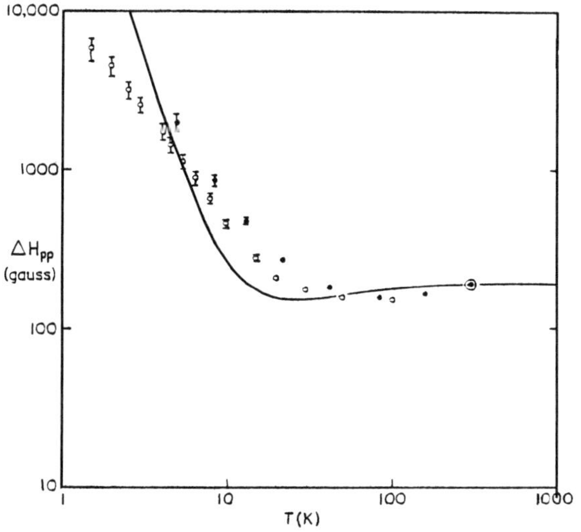

Fig.5 ESR linewidth of PCrP-C showing data on two separate samples. Solid line is result of memory function calculation.

linewidth and the exchange, J=-2.5K, found from the static properties. The theory yields a low temperature linewidth proportional to T^{-3} with a minimum near 50K. The experimental data, on the other hand, have a minimum near 100K and show a much slower low temperature increase. In an attempt to understand the origin of this discrepancy between theory and experiment we have constructed a model which considers the dynamics of correlated spin clusters.

6. Correlated Cluster Model: Dynamics [3]

The dynamical unit to be considered is the cluster of spins within the correlation length $\xi \sim J/T$. The individual spins within this cluster then experience anisotropy torques, due to their local environments, which must be summed to give the total broadening torque for the correlated cluster: $\Gamma_A \sim |JD|/T$. The resonance is "exchange narrowed" by the interaction across the ends of neighbouring clusters. The correlation time, i.e. the length of time each cluster may be considered to maintain its identity is found to be $\tau_e \sim 1/T$. Standard theory then yields a linewidth: $\Gamma_a{}^2 \tau_e \sim J^2D^2/T^3$.

This description is valid within the temperature range $S^2\sqrt{|JD|} \lesssim kT \lesssim |J|S^2$. At low temperature the linewidth divergence is cut-off at the antiferromagnetic resonance frequency $\omega_{AFMR} = \sqrt{|JD|}$, and near the mean field temperature, the cluster size approaches one spin. It may be that in PCrP-C this temperature window is too narrow to observe the T^{-3} behaviour.

7. Conclusion

The chromium polymer, PCrP-C, is an example of a one-dimensional Heisenberg antiferromagnet, J=-2.5K. Anisotropy, due to dipolar interactions, and to distortion of the chromium ligand field, in conjunction with the collective effects of exchange, induces a crossover from Heisenberg to XY spin symmetry. The low temperature ESR linewidth increases dramatically as a result of the same interactions.

*Work supported by NSF/MRL through Cornell Materials Sciences Center, DMR76-81083.

References

1. J.C. Scott, T.S. Wei, A.F. Garito, A.J. Heeger, H.D. Gillman and P. Nanelli, A.I.P. Conf. Proc. 29, 506 (1975).
2. R.E. Stahlbush, C.M. Bastuscheck, A.D. Raychaudhuri, J.C. Scott, D. Grubb and H.D. Gillman (to be published); R.E. Stahlbush, Ph.D. thesis, Cornell, 1980.
3. P.D. Krasicky, R.H. Silsbee and J.C. Scott (to be published); P.D. Krasicky, Ph. D. thesis, Cornell, 1980.
4. H.W.J. Blöte (private communication).
5. H.W.J. Blöte Physica (Utr) 78, 302.
6. T.T.P. Cheung, Z.G. Soos, R.E. Dietz and F.R. Merritt, Phys. Rev. B17, 1266 (1978); F. Lado, J.D. Memory and G.W. Parker, Phys. Rev. B4, 1406 (1971).

Part V

Quasi-One Dimensional Conductors

Structural Ordering in Quasi-One-Dimensional Systems*

J.D. Axe
Brookhaven National Laboratory
Upton, NY 11973, USA

This talk reviews what has been learned about nearly one-dimensional (1-D) systems, and in particular about the growth of long-range order in such systems, using thermal neutron probes. We discuss in turn

i) structural transformations coupled to electronic charge density wave (CDW) instabilities, i.e. Peierls transformations;
ii) transformations induced by coupling of 1-D ion (or mass) density waves.

Nearly 1-D conductors necessarily have large, almost flat portions of Fermi surface which efficiently screen lattice distortions of wave vector $\vec{q} = 2\vec{k}_F$. This leads to giant phonon Kohn anomalies and (in mean-field theories) to "soft mode" phase transformations involving simultaneous condensation of $2k_F$ CDW's and lattice displacements. In general, πk_F^{-1} is incommensurate with the spacing of the underlying lattice and, if the wavelength of the distortion is not forced by nonlinear effects to "lock in" to a commensurate value, a new gapless "phason" mode should result. The predicted giant Kohn anomalies have been seen best in $K_2Pt(CN)_4Br_{0.3} \cdot xH_2O$ (KCP). However, KCP never achieves long-range order probably because of impurity pinning. In TTF-TCNQ, by contrast, there is a true CDW transformation and interesting "lock-in" effects. However, although the transformation is anticipated weakly in the phonons, a simple soft mode description is inadequate. Much of the neutron and X-ray structural and dynamical studies of KCP, TTF-TCNQ and related compounds has been reviewed [1], [2]. Structural work on TTF-TCNQ under hydrostatic pressure, too recent to have been previously reviewed, reveals new CDW re-arrangements characterized by changing transverse wavevectors [3], [4].

$Hg_{3-\delta}AsF_6$ is composed of linear Hg chains which interpenetrate on ordered 3-D AsF_6 sublattice. At high temperature, there is no interchain positional correlation of the Hg atoms. Independent 1-D chains cannot positionally order and the resultant Hg intrachain correlations closely approximate a 1-D harmonic liquid [5], [6]. At lower temperatures, interchain interactions cause a "freezing" of the 1-D liquid, initially forming a sinusoidal mass density wave, analogous to the smectic liquid crystalline state. Neutron studies of both the statics and dynamics will be reviewed [7], [8].

* Research supported by the Division of Basic Energy Sciences, U. S. Department of Energy, under Contract No. DE-AC02-76CH00016.

References

1. R. Comès, G. Shirane in J. Devreese et al. (eds): Highly Conducting One-Dimensional Solids (Plenum, New York 1979) p. 17
2. J. D. Axe: Proceedings of International Conference on Neutron Inelastic Scattering (International Atomic Energy Agency, Vienna 1978) p. 101
3. S. Megtert, R. Comès, C. Vettier, R. Pynn, and A. F. Garito: Solid State Comm. 31, 977 (1979)
4. C. R. Fincher, G. Shirane, R. Comès and A. F. Garito: Phys. Rev. 21, 5424 (1980)
5. V. J. Emery and J. D. Axe: Phys. Rev. Lett. 40, 1507 (1978)
6. J. D. Axe in T. Riste (ed): Ordering in Strongly Fluctuating Systems (Plenum, New York 1980) p. 399
7. J. M. Hastings, J. P. Pouget, G. Shirane, A. J. Heeger, N. D. Miro, and A. G. MacDiarmid: Phys. Rev. Lett. 39, 1484 (1977)
8. I. U. Heilmann, J. D. Axe, J. M. Hastings, A. J. Heeger, and A. G. MacDiarmid: Phys. Rev. B 20, 751 (1979)

One-Dimensional Metals: Theory versus Experiment

T.M. Rice

Max-Planck-Institut für Festkörperforschung
D-7000 Stuttgart 80, Fed. Rep. of Germany, and

Bell Laboratories*, Murray Hill, NJ 07974, USA

Attention is focussed on one-dimensional metals which do not have a Peierls transition. Such materials are possible theoretically but only within a narrow range of values of the coupling constants. Two examples are discussed, $Hg_{2.88}AsF_6$ which has a transition to a superconducting state and $(TMTSF)_2PF_6$ where recent experiments by WALSH and coworkers favor a spin density wave ground state. It is suggested that the superconducting phase of $Hg_{2.88}AsF_6$ could be triplet rather than singlet state superconductivity.

1. Introduction

In the past decade many examples of one-dimensional (1-D) metals have been discovered and at the same time the theory of such one-dimensional metals has been developed and shown to be quite different to three-dimensional (3-D) metals. Yet one cannot say at this point that the comparison between theory and experiment has gone beyond the stage of a qualitative comparison. A detailed check on theory where functional forms and parameters are compared in a number of experiments on the same materials is still lacking. In part this deficiency can be traced to the complexity of the organic conductors which have not proven as easy to parametrize as hoped at first. The many degrees of freedom of the large molecules have made it difficult to give a simple description over the whole temperature regime.

For a theorist, the attraction of 1-D metals is not that they are simpler than 3-D metals but rather that they have an epidemic of instabilities which compete and interact with each other in a unique way. Already in the nineteen fifties, Peierls [1] and FROEHLICH [2] realized that the electron phonon interaction would lead to a formation of a charge density wave (CDW) state accompanied by a periodic distortion of background lattice. This state has now been seen in many 1-D metals [3]. The two materials that have been studied most are KCP[$K_2Pt(CN)_4Br_{0.3}\cdot 3H_2O$] and the organic conductor TTF-TCNQ. The former represents probably the best case for a comparison of theory and experiment [4],but because the Peierls transition has essentially occurred at all temperatures at which this material can be studied and because the material has an intrinsic randomness, the comparison between the theory and experiment is limited. The very recent discovery of a new inorganic material with a Peierls transition, reported at this conference [5], will hopefully lead to a better test of the theory.

The organic conductor TTF-TCNQ is more complex and a lot of different models were generated to explain its properties. Although it is perhaps premature to say that a consensus has emerged, I think it is now generally agreed

*Permanent address

that both Coulomb and electron-phonon effects are important, especially the Coulomb effects on the TTF chains, and that a combination of both effects plus librons, interchain coupling etc. are needed to adequately explain experiments. The comparison of theory and experiment on this material alone could take up this whole lecture and rather than do this I refer the interested reader to the recent summary by T.D. SCHULTZ [6].

The interest in the Peierls-Fröhlich (or Charge Density Wave (CDW)) state is also because of its special properties. For example, the electrical conductivity in the CDW state has some intriguing possibilities. One is the possibility of drifting the CDW as a whole in an applied electric field as FROEHLICH proposed a long time ago [2]. This requires, of course, an incommensurate CDW system so that the only force opposing the motion comes from the pinning of the CDW to randomly placed impurities. This effect has now been seen in $NbSe_3$, which, however, is not a 1-D metal, and since this is the topic of R.M. FLEMING's lecture [7] I will not discuss it further here. In 1-D metals the nearest one has come to a drifting CDW is in the fluctuation regime at temperatures just above Peierls transition. Some years ago it was hotly debated whether a fluctuating and drifting CDW could enhance the conductivity over the value in the absence of the CDW and the final consensus was that this would not occur [8]. However, it was clear that such collective contributions to the conductivity will be reduced if the system is commensurate [9] and recent experiments which show this effect will be discussed later in this conference by D. JEROME and H. SCHULZ [10].

Another intriguing possibility is conduction by mobile charged solitons. M.J. RICE and coworkers [11] were the first to introduce this idea for a CDW system which was weakly commensurate with the underlying lattice. Unfortunately no such experimental system is at hand at the moment, but there is no reason why such a system cannot exist. More recently these ideas have been developed for the special case of 2x1 commensurability and applied to the polymer polyacetylene. As this is the topic of a session later on, I will refrain from further comment here.

The CDW state proposed by PEIERLS and FROEHLICH is certainly the ground state of a system of electrons interacting with phonons, but if one includes also the interaction between electrons then there are more possibilities. This will be the main topic of my lecture. Can one obtain 1-D metals which do not have a Peierls transition and what should be expected for their behavior? I will discuss first the theory and secondly review the experimental situation in two classes of materials which, on the one hand, appear to be 1-D metals but on the other, appear not to make transitions to the CDW state.

2. 1-D Metals Without a Peierls Instability; Theory

As will be discussed below, several materials seem to be 1-D metals and yet do not have a Peierls transition. It is therefore of interest to examine more closely how this may occur. To discuss this question, both the electron-electron interaction and electron-phonon interaction on a single chain must be discussed and also the coupling between electrons on different chains. The single-chain problem with only electron-electron interactions has been studied extensively. An excellent review of this field by J. SOLYOM [12] has recently appeared. The essential result is the phase diagram shown in Fig. 1. Clearly our interest is in the upper half where $g_1>0$(g_1 is the coupling constant for backward scattering i.e. momentum transfer $\approx 2k_F$). In this region the value of g_1 scales to the weak-coupling limit ($g_1=0$), and already in first-order renormalization theory all singularities are reduced to power law divergences as $(\omega,T)\to 0$. In this region then, this method will give reliable results. The four key response functions which characterize the behavior are the charge density response at $q=2k_F$, $N(2k_F,\omega,T)$; the spin density response at $q=2k_F$, $\chi(2k_F,\omega,T)$; and the singlet and triplet superconducting response functions $\Delta_s(\omega,T)$ and $\Delta_t(\omega,T)$ respectively. The behavior of these response functions is most conveniently studied through their derivatives, $\overline{R}(\omega)=[\pi v\partial R(\omega)/\partial \ln\omega]$ and the result of the first-order renormalization group calculation for these derivatives is

$$\overline{N}(2k_F,\omega,T) = [1-(g_1/\pi v)\ln(\tilde{\omega}/E_F)]^{-3/2}(\tilde{\omega}/E_F)^{\alpha} \tag{2.1}$$

$$\overline{\chi}(2k_F,\omega,T) = [1-(g_1/\pi v)\ln(\tilde{\omega}/E_F)]^{1/2}(\tilde{\omega}/E_F)^{\alpha} \tag{2.2}$$

$$\overline{\Delta}_s(\omega,T) = [1-(g_1/\pi v)\ln(\tilde{\omega}/E_F)]^{-3/2}(\tilde{\omega}/E_F)^{-\alpha} \tag{2.3}$$

$$\overline{\Delta}_t(\omega,T) = [1-(g_1/\pi v)\ln(\tilde{\omega}/E_F)]^{1/2}(\tilde{\omega}/E_F)^{-\alpha} \tag{2.4}$$

where v and E_F are the Fermi velocity and energy, and $\alpha = (g_1-2g_2)/2\pi v$ with g_2 as the forward scattering coupling constant. The symbol $\tilde{\omega}$ denotes max.(ω,T).

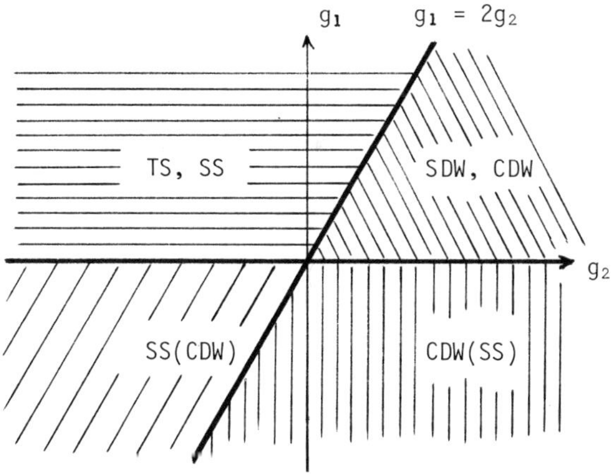

Fig.1 The phase diagram for a single chain showing the response functions which are divergent for different values of the coupling constants g_1, g_2

The critical parameter in the region with $g_1 > 0$, is the value of α. For $\alpha > 0$ the superconducting response functions diverge, but N and χ do not and are reduced rather than enhanced by the interactions. This effect is very important if we are to suppress the Peierls transition. To attain the condition $\alpha > 0$ is not easy. It requires that g_2, the screened forward scattering coupling constant be less than half g_1, the unscreened backward scattering coupling constant which demands very strong screening of Coulomb interactions as $q \to 0$.

In practice one must include also the electron-phonon interaction. These two interactions are not simply additive because of the cutoff problem. The relevant cutoff for the Coulomb interaction is E_F, but for the electron-phonon interaction it is ω_D, the Debye energy. The problem is simplest if one looks at the region $\tilde{\omega} \gtrsim \omega_D$, then the phonon-mediated graphs are not in general divergent and the only graphs one must include are those which are completely reducible with respect to a single phonon line shown in Fig. 2 [13,14]. Summing these graphs leads to the result for the phonon propagator, $D(2k_F,\omega,T)$ that

$$D^{-1}(2k_F,\omega,T) \sim 1 + \pi v\lambda N(2k_F,\tilde{\omega}) \tag{2.5}$$

where λ is dimensionless electron-phonon coupling constant.

Fig.2 The wavy line denotes the bare phonon propagator and the bubbles the $2k_F$ density-density response function including all electron-electron interactions but with no virtual exchange of phonons

In the absence of electron-electron interaction

$$N^o(2k_F,\tilde{\omega}) = (\pi v)^{-1}\ln(\tilde{\omega}/E_F) \tag{2.6}$$

and the condition $D^{-1}(2k_F,\omega=0,T_P) = 0$ determines the mean-field transition temperature for the Peierls transition

$$T^o{}_P = E_F\exp(-1/\lambda). \tag{2.7}$$

However, the electron-electron interaction strongly modifies the behavior of N, and in particular, if $g_1 > 0$ and $\alpha > 0$ then $N(2k_F,\omega=0,T)$ does not diverge as $T \rightarrow 0$, rather $N \sim (T/E_F)^\alpha$. As a result for small λ the condition $D^{-1} = 0$ may never be satisfied. On the other hand if $\alpha > 0$ then the divergence of N is stronger than that of N^o and the mean-field Peierls temperature T_p is enhanced. The criterion, therefore, within this approximation is simply the sign of α. Note that unlike the case for a 3-D CDW state this criterion, for the absence of a higher temperature Peierls transition, does not involve the comparison of electron-electron and electron-phonon strengths but is determined by the electron-electron interaction alone.

The above discussion applies to the case of $T_p > \omega_D$. If one is to find a 1-D metal without a Peierls transition, then the opposite limit $T_P \ll \omega_D$ must also be considered. In this limit one must first reduce all interactions to the single cut off ω_D. The Coulomb interaction reduced to this limit requires the replacement

$$g_1 \rightarrow g_1/(1+(g_1/\pi v)\ln(E_F/\omega_D)) \quad \text{and} \tag{2.8}$$

$$g_2 \rightarrow g_2 - 1/2\frac{(g_1{}^2/\pi v)\ \ln(E_F/\omega_D)}{1+(g_1/\pi v)\ln(E_F/\omega_D)} \ . \tag{2.9}$$

If one then simply adds the backward scattering electron-phonon scattering (forward scattering by phonons in 1-D metals is greatly reduced by kinematic restrictions) then the two critical parameters are the total values g_1^{tot} and α^{tot} given by [14]

$$g_1^{tot} = \frac{g_1}{1+(g_1/\pi v)\ln(E_F/\omega_D)} - \pi v\lambda \tag{2.10}$$

and

$$\alpha^{tot} = (g_1 - 2g_2 - \pi v\lambda)/2\pi v \quad . \tag{2.11}$$

The criterion then that $g_1^{tot} > 0$ and $\alpha^{tot} > 0$ becomes harder to satisfy. Clearly a small value of λ is required. Note that this criterion which leads to the absence of a Peierls transition at all temperatures, depends on the relative strengths of the electron-phonon and electron-electron interaction.

When interchain effects are included the criteria that determine the boundary region between the different instabilities do not shift greatly. There has been considerable effort made on the problem of combined interchain and intrachain interactions [14,15]. First if only the interchain Coulomb interaction is considered, i.e. no interchain hopping is allowed, then the key result is that this interchain Coulomb interaction stabilizes a 3-D ordered CDW state in any region where $N(2k_F,\tilde{\omega})$ of a single chain diverges. Thus this interaction will cooperate with the electron-phonon interaction and stabilize a 3-D Peierls transition in this region. The actual criterion for N to diverge differs somewhat from that quoted in (2.10-11) but if the interchain interactions are much weaker than the intrachain interactions this change is not substantial.

Interchain hopping has also been studied. The regime of most interest to us here is when $g_1 > 0$ and this has been studied by SUZUMARA and FUKUYAMA [16]. They find that the effect of the interchain hopping is to stabilize a 3-D ordered state corresponding to the most strongly divergent response function. Thus they find for $\alpha < 0$ the SDW phase is more stable while for $\alpha > 0$ it is the triplet superconductor phase which is most stable. Their results raise the possibility of stabilizing these ground states. In particular if one focuses on the SDW state which, as discussed below, has recently been proposed, then it requires that $\alpha < 0$ and obviously a fairly strong interchain hopping, since both the intrachain electron-phonon interaction and the interchain Coulomb interaction act together to enhance the 3-D ordering temperature of the rival CDW state. These various effects are in competition in this region, and while a detailed calculation incorporating all three effects should be made, it is clear that only with strong interchain hopping and weak intrachain electron-phonon and weak interchain Coulomb interaction can one hope to stabilize the SDW phase.

It would also be of great interest to stabilize the triplet superconducting state. Similar conclusions apply here namely that one must be in the region where $g_1^{tot} > 0$ and $\alpha^{tot} > 0$ and have some interchain hopping to stabilize the 3-D ordered state. At least in this case there is not a competing effect from the interchain Coulomb interaction.

In this regime the CDW response function $N(2k_F,\tilde{\omega})$ is going to zero and this has some interesting consequences. First if one looks at the Kohn anomaly in the phonon spectrum at $q = 2k_F$. Normally one expects a strong Kohn anomaly but in this case the anomaly should be greatly weakened by the Coulomb interaction. Since the relevant energy is the maximum of T and $\Omega(2k_F)$ (the phonon energy at $2k_F$) one may use the expansion in Fig. 2, This leads to a value of

$$\Omega(2k_F) = \omega_0(2k_F)[1 + \lambda\pi v N(2k_F,\Omega,T)]^{1/2}$$

$$\text{where } N \sim -(\pi v)^{-1}([\max(\Omega,T)]/E_F)^{\alpha} \quad . \tag{2.12}$$

Since $\alpha > 0$ the Kohn anomaly weakens as $T \to \Omega$ and is small and temperature independent for $T < \Omega$.

A second interesting consequence concerns the resistivity from impurity scattering. Several authors have pointed up the relation between the scattering rate, Γ_i for backward scattering by impurities and N. Using the Born approximation to describe Γ_i one obtains |15|

$$\Gamma_i = cV_0^2\left(\frac{2}{\pi}\right)^2\overline{N}(2k_F,\omega{=}0,T) \tag{2.13}$$

where c and V_0 are the impurity concentration and potential strength respectively. When the form (2.1) is substituted for $\overline{N}$ one obtains that $\Gamma_i \propto (T/E_F)^{\alpha}$. This result is obtained only in the Born approximation for the impurity scattering and the more subtle effects due to localization of all electrons in a random potential are ignored. It does, however, point to a suppression of the impurity resistance by the interactions. The full theoretical treatment must involve both localization and electron-electron interactions and the outcome is unclear.

3. 1-D Metals Without a Peierls Instability; Experiment

a) $Hg_{2.88}AsF_6$

There are two canditates that I am aware of which may qualify as 1-D metals without a Peierls transition. The first is the compound $Hg_{2.88}AsF_6$. This material has one-dimensional Hg chains which are incommensurate with the host AsF_6 lattice. This leads to a fascinating ordering transition among the Hg chains and a unique phonon structure that was described in the previous lecture by J. AXE [17]. The electronic properties of this material are also very unusual. The groups at McMaster and Pennsylvania have studied the electronic structure extensively in the past few years and their results will be briefly reviewed.

First the d.c.conductivity and the optical properties are highly anisotropic with a ratio of 10^2 between the d.c.conductivity in the ab plane along the Hg chains and the conductivity along the c axis, perpendicular to the Hg chains [18,19]. The ordering transition between the Hg chains shows that the coupling between Hg ions on adjacent chains is very weak ($\sim 10^{-4}$eV). At low temperatures the samples are good enough for de Haas-van Alphen measurements to be made and these show a Fermi surface which has very little variation along the c axis and whose cross-sectional area in ab plane agrees well with that deduced from a simple one-dimensional band structure [20]. The magnetoresistance is large in low fields and has been explained also on the basis of an essentially one-dimensional Fermi surface with interchain mixing of the wavefunction confined to a small region near the corners of the square Fermi surface [21]. So all the evidence is in favor of an essentially 1-D band structure.

The AsF_6 molecular units each take up one electron leaving each Hg atom with 1.65 [= (2x2.88 - 1)/2.88] conducting electrons. The lattice constant of the Hg chains a_{Hg} = 2.67 Å leading to Fermi surface planes located at $\pm k_F$ (= 0.97 $Å^{-1}$). The Fermi surface in the extended zone is shown in Fig. 3. The zone boundaries from the Hg chains are at $\pm\pi/a_{Hg}$ but are tilted due to the phase slip between the Hg ions on parallel chains. The pseudopotential for Hg has a node very near to a momentum transfer of $2\pi/a_{Hg}$ (= 2.35 $Å^{-1}$)[22] and so the energy gap at this zone boundary may be quite small. This suggests that it is better to describe the electronic structure as a 1-D electron metal rather than a 1-D hole metal. This model also agrees with the optical data which show no observable interband absorption across this gap [19]. There are also energy gaps which have the periodicity of the AsF_6 lattice. These gaps which arise from the AsF_6 lattice and from the perpendicular Hg chains are relevant in determining the number of carriers in transport measurements since, when these gaps are included and the Fermi surface is mapped back into the first Brillouin zone of the AsF_6 lattice, there are two filled zones which do not contribute to the conductivity. The resulting Fermi surface is also shown in Fig. 3 and it agrees very well with that deduced from the de Haas-van Alphen measurements [20]. The gaps induced by the AsF_6 lattice should also be small and therefore may be ignored in discussing the interaction effects.

The electrical conductivity of this compound continues to increase as the temperature is lowered with no sign of the Peierls transition. Indeed such a transition would truncate the Fermi surface and is totally ruled out by the Fermi surface observations. Nor is there any neutron scattering evidence for such a transition. There is prima facie evidence that $Hg_{2.88}AsF_6$ is essentially a 1-D metal but there is no Peierls transition. If viewed as a 3-D metal the Fermi surface must be classified as having close to perfect nesting yet there is no CDW or SDW transition! Clearly it is a good candidate for a 1-D metal which scales to the Tomonaga-Luttinger line (i.e. $g_1 \equiv 0$). The crystal structure of this compound is sufficiently simple that accurate a priori calculations could be made of the Fermi surface and of the coupling constants g_1, g_2 and λ. It would be of great interest to see if these coupling strengths obey the relations $g_1^{tot} > 0$ and $\alpha^{tot} > 0$.

The low temperature d.c.conductivity and magnetoconductivity of $Hg_{2.88}AsF_6$ are highly anomalous [21]. The d.c.resistivity fits the form $\rho(T) = \rho_0 + aT^3$ but ρ_0 is so small it cannot be reliably measured. The limit on ρ_0 puts a lower limit of 100 μm on the mean-free path($\sim 4\times10^5$ a_{Hg}) [21]. Further the anomalous magnetoconductivity is also consistant with a negligible scattering rate from impurities between the flat portions of the Fermi surface (electron-phonon scattering requires a $2k_F$ phonon and the number of these is exponentially small at $T \sim 4K$). It is hard to believe that these materials are so well ordered that this large mean-free path (>4×10^5 Hg atoms) does not arise from a suppression of electron-impurity scattering. Since such a suppression is explicitly predicted in (2.13) it is further evidence in favor of this model. A key experiment would be to examine the Kohn anomaly at $2k_F$ in the extended zone (1.94 $Å^{-1}$) in the phonon branches of the Hg chains.

Finally at a temperature T_c = 0.4 K the compound becomes a 3-D superconductor [23]. This transition can be interpreted if g_1^{tot} is small and negative and g_2 is also negative so that at low temperatures one scales to the superconducting state rather than the CDW state. Alternatively it is tempting to speculate that $Hg_{2.88}AsF_6$ has become a triplet superconductor with $g_1^{tot} > 0$ and $\alpha^{tot} > 0$ and the three-dimensionally ordered superconducting state is stabilized by interchain hopping. The usual objection that such a triplet state is suppressed by impurities would not apply here since the

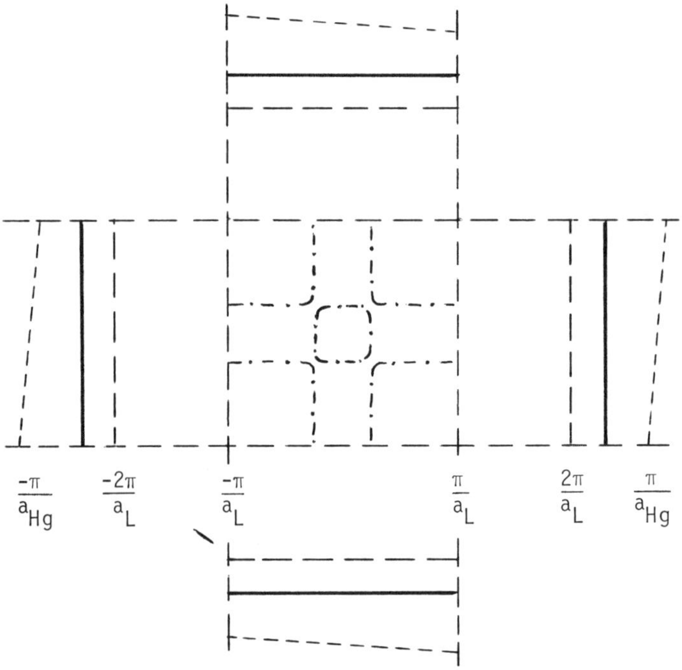

Fig.3 The Fermi surface of $Hg_{2.88}AsF_6$ in the extended zone scheme (solid lines) and in the reduced zone scheme (dot-dashed lines) of the background lattice. The small dashed line denotes the position of Brillouin zone due to the Hg chain periodicity

mean-free path ℓ is so long. In fact a simple estimate gives $\ell > \xi_{BCS}$, the BCS correlation length! As remarked earlier the large value of ℓ is also an argument in favor of assinging $Hg_{2.88}AsF_6$ to the region $g_1^{tot} > 0$ and $\alpha^{tot} > 0$ since in this region impurity scattering is suppressed. If this assignment is correct then the low temperature state is a triplet superconductor.

b) $(TMTSF)_2PF_6$ and $(TMTSF)_2AsF_6$

The second group of compounds are the organic compounds $(TMTSF)_2PF_6$ and $(TMTSF)_2AsF_6$. These compounds consist of planar stacks of TMTSF molecules with PF_6 or AsF_6 complexes stacked separately. These compounds were synthesized by BECHGAARD and coworkers [24] who found them to be highly conducting down to a temperature ∿ 18 K at which point the conductivity dropped rapidly. They interpreted this drop as a Peierls transition. JEROME and coworkers [25] found that under a pressure of 12 kbar this transition was suppressed and a superconducting transition was observed with a value of $T_c = 0.9$ K - the first organic superconductor. Very recently WALSH and coworkers [26] have examined the transition at zero pressure and found that the drop in conductivity could be totally suppressed by passing a modest current (∿10 mA) corresponding to a quite small electric field (∿10 mV/cm). This behavior is reminiscent of the non-linear conductivity of $NbSe_3$ [7]. However the explanation of a depinning of the CDW to obtain a drifting CDW does not apply because WALSH and coworkers [26] found a dramatic increase in the conduction electron resonance signal when the microwave electric field increased beyond a critical threshold. At low microwave fields the CESR signal is absent but above the critical field strength the CESR signal is intense and narrow. They tentatively interpret as the low temperature state a SDW state and that the non-linear conductivity and CESR response is due to depinning of the SDW state leading to motional narrowing of the CESR line. Neutron scattering experiments should decide the matter. From the discussion given above a SDW state is possible if the interchain hopping is substantial - a reasonable hypothesis for this material - and if $g_1^{tot} > 0$ and $\alpha^{tot} < 0$. In the case of an organic conductor reliable a priori estimates of these parameters are difficult.

4. Conclusions

In this comparison between theory and experiment I have chosen to focus not on the 1-D metals which show a Peierls transition but on those which do not. The possibility of having a 1-D metal which does not have a Peierls transition exists in principle and now it appears that two such systems have been found. Further study especially by neutron scattering should prove very valuable in establishing whether there are materials not just in the lower half of the phase diagram in Fig. 1 but in the upper half too. Unusual behavior is predicted for the Kohn anomaly in the phonon spectrum of such materials and this behavior should be observable in the phonon spectrum of the Hg chains in $Hg_{2.88}AsF_6$.

Acknowledgements

The author is grateful to Profs. P. Fulde and O.K. Anderson for useful discussions and wishes to thank them and the Max Planck Institut für Festkörperforschung for their hospitality. The support of the Alexander von Humboldt Foundation through a fellowship is gratefully acknowledged.

References

1. R.E.Peierls, Quantum Theory of Solids, Oxford Press (1955) p. 108
2. H.Fröhlich, Proc. Roy. Soc. A223, 296 (1954)
3. See the conference proceedings *Organic Conductors and Semiconductors*, ed. by L. Pál, G. Grüner, A. Jánossy and J. Sólyom, Lecture Notes in Physics, Vol. 65 (Springer, Berlin, Heidelberg, New York 1977); . S. Barišić, A. Bjeliš, J.R. Cooper and B. Leontić, eds., *Quasi-One-Dimensional Conductors I, II*, Lecture Notes in Physics, Vols. 95, 96 (Springer, Berlin, Heidelberg, New York 1979); and the review article by G.A. Toombs, Phys. Rep. 40, 181 (1978)

4. P. Brüesch, S. Strässler and H.R. Zeller, Phys. Rev. B 12, 219 (1975)
5. G. Travaglini, P. Wachter, P. Marcos and C. Schlenker (Paper P1.17 this conference)
6. T.D.Schultz in Festkörperprobleme (Advances in Solid State Physics), Volume XX, J.Treusch (ed.) Vieweg, Braunschweig, 1980
7. R.M. Fleming (Paper I.06 this conference)
8. H. Fukuyama, T.M. Rice and C.M. Varma, Phys.Rev.Lett. 33, 305(1974)
B.R.Patton and L.J. Sham, Phys. Rev. Lett. 33, 608 (1974)
[Note the latter authors obtained a conductivity enhancement but only in the limit of weak impurity scattering and strong electron-phonon coupling]
9. B.R. Patton and L.J.Sham, Phys.Rev. Lett. 31,631 (1973)
10. D.Jerome and H.J.Schulz (Paper I.04 this conference)
11. M.J.Rice, A.R.Bishop, J.A. Krumhansl, and S.E.Trullinger Phys.Rev.Lett. 36, 432(1976)
12. J.Sólyom, Adv. in Phys. 28, 201 (1979)
13. S.T.Chui, T.M.Rice and C.M.Varma, Sol.State Comm. 15, 155(1974)
14. L.P.Gor̆kov and I.E.Dzyaloshinskiĭ, Zh.Eksp.Teor.Fiz. 67, 397 (1974) [Eng.Trans.Sov.Phys. JETP 40, 198 (1974)]
15. P.A.Lee, T.M.Rice and R.A.Klemm, Phys.Rev.B 15, 2984(1977)
16. Y.Suzumara and H. Fukuyama, J.Low Temp.Phys. 31, 273(1978)
17. J.Axe(Paper I.02 this conference)
18. C.K.Chiang, R.Spal, A.Denenstein, A.J.Heeger, N.D.Miro and A.G. MacDiarmid, Solid State Comm. 22, 293(1977)
19. D.L.Peebles, C.K. Chiang, M.J.Cohen, A.J.Heeger, N.D. Miro and A.G. Mac Diarmid, Phys.Rev. B 15, 4607 (1977)
20. F.S.Razavi, W.R.Datars, D.Chartier and R.J. Gillispie Phys.Rev.Lett. 42, 1182(1979)
21. D.K.Chakraborty, R.Spal, A.M.Denenstein, K.-B.Lee, A.J. Heeger and M.Ya. Azbel, Phys.Rev.Lett. 43, 1832 (1979)
22. M.L.Cohen and V.Heine, Sol.State Phys. 24, p.38(Acad.Press,N.Y. 1970)
23. R.Spal, C.-E.Chen, A.Denenstein, A.R.McGhie, A.J.Heeger and A.G.MacDiarmid, Sol.State Comm. 32, 641(1979)
24. K.Bechgaard, C.S.Jacobsen, K.Mortensen, H.J.Pedersen and N.Thorup Sol. State Comm. 33, 119(1980)
25. D.Jérome, A.Mazaud, M.Ribault and K.Bechgaard, J.Physique Letters 41, L 95(1980)
26. W.M.Walsh, F. Wudl, G.A.Thomas, D.Nalewajek, J.J.Hauser, P.A.Lee and T.Poehler (to be published)

Electronic Instabilities in Quasi-One-Dimensional Conductors: Insulator or Superconductor?

D. Jérome and H.J. Schulz*

Laboratoire de Physique des Solides, Université Paris Sud
F-91405 Orsay, France

1. Introduction

A strictly one-dimensional metal shows instabilities at low temperatures. Depending on the type of interaction between the electrons these may be either of the (spin or charge) density wave type or superconducting [1].

Here we present two examples of organic quasi-one-dimensional metals showing such instabilities : (i) in TTF-TCNQ at low temperatures a charge density wave (CDW) state is realized (Peierls instability, sec. 2), (ii) in $(TMTSF)_2 PF_6$ a superconducting state is realized (sec.4). Though not strictly one-dimensional, metallic-like properties appear in these substances (as well as in many related compounds) only along one crystallographic axis ; the conductivity anisotropy is of the order 10^2-10^3 [2,3] and a plasma edge is only found for light polarized along that axis [3,4,5].

One-dimensional systems are characterized by large thermodynamic fluctuations. In sections 3 and 5 we give experimental evidence for strong fluctuation effects in the above substances : CDW fluctuation conductivity in TTF-TCNQ and superconducting paraconductivity in $(TMTSF)_2 PF_6$.

2. Peierls transition

The best-studied example of a quasi-one-dimensional conductor is TTF-TCNQ. In this compound both the TTF and the TCNQ chains contribute to the metallic conductivity. Phase transitions at ambient pressure have first been discovered by anomalies in the longitudinal (b-axis) dc conductivity at two different temperatures : T_H (high) = 54K [6,7] and T_L (low) = 38K [8,9]. Subsequent structural investigations by X-ray [10,11] and neutron [12] scattering have shown that at T_H the lattice becomes modulated with period 2a x 3.4b x c (a and c are the transverse lattice constants). Below T_I = 49K the period in the a-direction increases and jumps discontinously to 4a at T_L. The decrease of conductivity and the lattice modulation are well explained by the Peierls mechanism [13] which leads to the appearance of a CDW and an associated lattice modulation of wavelength π/k_F in a one-dimensional electron gas. At the transition temperature a gap opens at the Fermi level so that a sharp maximum appears in $\partial \log \rho/\partial T$, ($\rho$ = resistivity). This maximum is usually considered as the sign of a phase transition [14]. In its original form the Peierls mechanism is based on the electron-phonon coupling. However, direct repulsive electron-electron-interactions can lead to the same kind of instability. In fact, the existence of strong diffuse X-ray scattering at $4k_F$ far above T_H [15] as well as a large enhancement of the spin susceptibility indicate that electron-electron interactions play a major role in the mechanism [16,17]. Below the T_H transition the conducti-

*Now at: I. Institut für Theoretische Physik, Universität Hamburg
D-2000 Hamburg, Fed. Rep. of Germany

vity is still quite high ($\sigma > 10^3\ \Omega\ cm^{-1}$) and it is only below T_L that real semiconducting behaviour is found. This suggests that the two sets of chains undergo Peierls transition separately at T_H and T_L. A discrimination between the two kinds of chains is possible by thermoelectric power (TEP) measurements, as the sign of the TEP coefficient is positive for hole-like (TTF) and negative for electron-like conduction (TCNQ) [18]. From the sign of the change of the TEP at T_H it follows that this transition takes place mainly on theTCNQ chains, whereas the T_L transition is driven by the TTF chains. Essentially the same conclusions can be drawn from an analysis of EPR data [19], though recent results show that at T_H a small gap ($<k_B\ T_H$) also develops slowly on the TTF chains [20].

At T_I no maximum of $\partial \log \rho/\partial T$ is observed, though small anomalies exist [21] indicating possibly that T_I is the temperature where the gap on the TTF chains begins to rise slowly above k_BT. Further information about the phase transitions in TTF-TCNQ can be obtained by applying hydrostatic pressure. The resulting phase diagram is shown in Fig.1. Below about 15 Kbar, transitions at T_H and T_L are observed as at ambient pressure [22]. However, the sign of the TEP anomaly at T_H shows that above P $\approx$ 8 Kbar the Peierls order develops mainly on the TTF-chains [23], i.e. with increasing pressure the role of the TCNQ-chains is taken over by the TTF-chains.

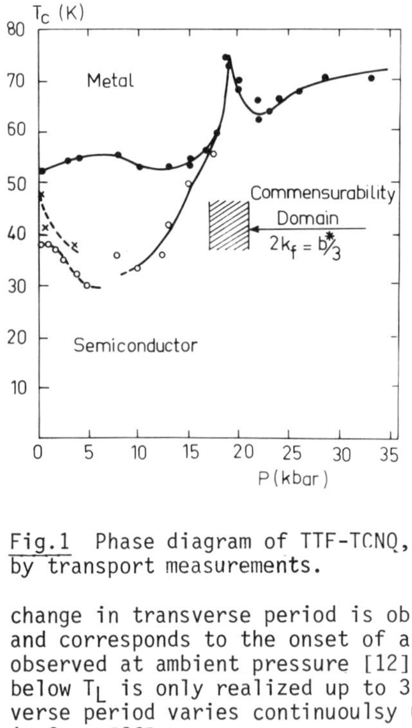

Fig.1 Phase diagram of TTF-TCNQ, derived by transport measurements.

The structure of the low temperature phases has been investigated under pressure up to 5 Kbar by elastic neutron scattering. Apart from T_H - and T_L - transitions additional features arise : (i) the T_I transition is split into two transitions at T_{I1} and T_{I2} [24]. The upper transition temperature (T_{I1}) grows with increasing pressure and corresponds to the condensation of a longitudinal deformation of wave number $2k_F$ (in addition to the transverse mode at T_H [25]). No change in transverse period is observed at T_{I1}. T_{I2} decreases with pressure and corresponds to the onset of a variable transverse modulation as already observed at ambient pressure [12]. (ii) The transverse 4a - modulation below T_L is only realized up to 3 kbar [24]. At higher pressure the transverse period varies continuoulsy up to about 4 kbar and above this pressure is 2 a. [26].

The splitting of the T_I transition into two transitions indicatesthat the origin of structural changes may be more complicated than assumed in current theories [27,28]. It is tempting to associate the T_{I2}-transition with a condensation of the $4k_F$ -mode, for which there are some indications at ambient pressure [25], however no detailed theory is currently available.

Above about 15 kbar the T_H- and T_L-transitions join into a single transition [22]. The transition temperature has a sharp maximum at 19 Kbar (T_C = 74K, Fig.1) and hysteresis shows that the transition is first order near this pressure. This effect can be understood assuming an increase of charge transfer from 0.59 at ambient pressure to 0.66 at 19 kbar. This assumption has been confirmed by the neutron scattering experiments showing a direct proof of x3 commensurability at low temperature above 15 Kbar [29]. At charge transfer 0.66 the wavelength π/k_F of the Peierls distortion is three times the lattice constant. Then third-order unklapp terms are allowed in a free-energy functional. These lead to a firstorder transition with an increased transition temperature [30], as the CDW can gain energy by locking into the lattice. At small deviations from exact commensurability the low temperature state is still commensurate, however T_C decreases due to a loss in elastic energy. The behaviour of TTF-TCNQ around 19 Kbar fits exactly into this picture with an increase in transition temperature of about 10 K and a 4 kbar wide commensurability region [22].

The isostructural selenium analog compound TSF-TCNQ shows several marked differences from the behaviour of TTF-TCNQ : (i) there is only one phase transition (T_P = 29K), the Peierls order being established simultanously on both chains [14] with constant transverse modulation 2a [31] (ii) at ambient pressure the charge transfer is 0.63 [31], so that a pressure of only 6.25 kbar is required to achieve commensurability [33]. The increase of T_P at commensurability is considerably smaller than in TTF-TCNQ. Moreover, contrasting with TTF-TCNQ no hysteresis of the phase transition is observed at commensurability. Consequently the pinning potential must be significantly smaller in TSF-TCNQ than in TTF-TCNQ.

3. Fluctuation effects above the Peierls transition

In strictly one-dimensional systems strong fluctuations suppress phase transitions at finite temperatures. Though in quasi-1D systems ordered phases at low temperature are stabilized by inter-chain coupling, above the transition temperature remnants of the strong 1D fluctuations can be expected to exist. Fluctuations above a Peierls transition are associated with short-wavelength ($\approx \pi/K_F$) modulations of the charge density and can therefore be observed directly by X-ray scattering. In TTF-TCNQ, CDW fluctuations have been observed up to 200 K [25]. Between 54 K and 60 K fluctuations on adjacent chains are coupled, i.e. the fluctuations are three-dimensional [33] whereas above 60 K they are one-dimensional with a correlation length decreasing monotonously from 100 Å at 60 K to 5 Å at 150 K. Somewhat different behaviour is found in TSF-TCNQ [34,35], where fluctuations on neighbouring chains have short range two-dimensional correlations up to about 200 K. If the wavelength of the CDW is incommensurate with the lattice constant, i.e. if their ratio is not a small fraction, the phase of the fluctuations is not fixed relative to the lattice and can therefore be freely translated through the crystal. Such a collective notion would lead to a positive contribution to the dc conductivity [36]. The existence of this mechanism in real systems has longtime been controversial [37,38]. Experiments on TTF-TCNQ near the commensurability pressure [39] show that the conductivity decreases considerably at 19 kbar (Fig.2). This decrease is very sharp near the phase transition and can be observed up to about 250 K. It is tempting to associate this effect with a lock-in of the commensurate CDW fluctuations into the lattice, so that the fluctuation conductivity mechanism is suppressed. An alternative explanation would be changes in the longitudinal single particle scattering time $\tau_{//}$ or in the density of states $N(E_F)$, although these effects are inde-

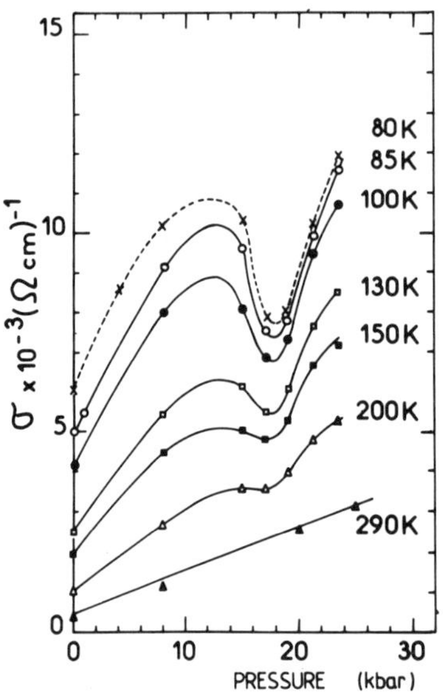

Fig.2 Pressure dependence of $\sigma_{//}$ in TTF-TCNQ.

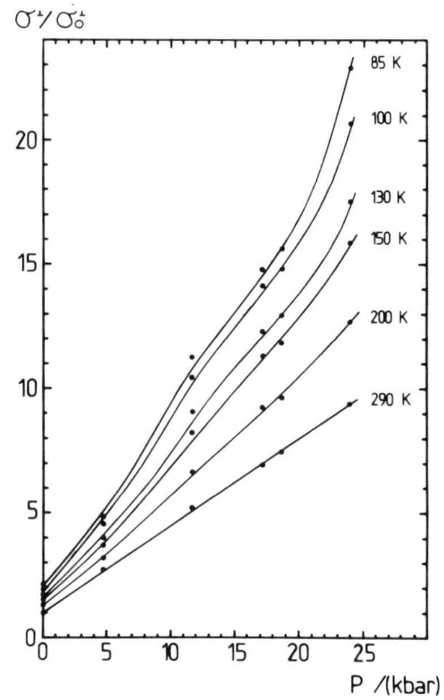

Fig.3 Pressure dependence of $\sigma_\perp$ in TTF-TCNQ along the a-direction.

pendent of the phase dynamics of the fluctuations and therefore much less influenced by commensurability [40]. One can, however, discard the existence of drastic changes in the band structure at 19 Kbar, as optical reflectance shows only a smooth and monotonous pressure dependence [41]. A direct measure of $\tau_{//}$ [and also of $N(E_F)$] is given by the transverse conductivity, which is clearly a single particle process and has been shown to be proportional to $\tau_{//}$ [42]. Therefore, if the drop in longitudinal conductivity were due to changes in the single particle properties, a comparable effect in the transverse conductivity should be seen, in contrast to the experimental result [43], where only a small effect, if any, is found (Fig.3). Thus, pinning of a CDW fluctuation contribution provides a natural explanation of the conductivity drop at commensurability. Additional evidence for an important contribution of CDW fluctuations to the longitudinal conductivity comes from experiments on irradiated samples [44] showing a suppression of the 19kbar conduction dip. Defects, as introduced by irradiation should very effectively pin the friction-less motion of CDW fluctuations, thus suppressing the CDW fluctuation conductivity even in the incommensurate situation. Correspondingly, no significant drop of conductivity has been observed in the irradiated samples near commensurability. Were the drop of conductivity due to changes in the single particle scattering one would expect (from Matthiessen's rule) a similar drop to appear in the irradiated samples. This is not observed. We thus conclude that an important part of the metallic conductivity of (incommensurate, unirradiated) TTF-TCNQ is due to the CDW fluctuation mechanism. Of course, near the transition temperature, where the fluctuations become more and more three-dimensional this mechanism is very easily suppressed by impurity pinning. Therefore in the region of three-dimensional fluctuations just above 54K, the main effect of the fluctuations is single particle scattering, leading to a decrease of conductivity [45].

In TSF-TCNQ, a drop in longitudinal conductivity around commensurability has also been observed [37]. However, this effect is largest at about 100 K and does not increase upon further decrease in temperature, contrary to TTF-TCNQ. At 100K, the effect is of the same magnitude as in TTF-TCNQ at the same relative temperature (in units of T_p). At lower temperatures interchain coupling between fluctuations is known to be important [34,35]. Such a coupling is expected to suppress fluctuation conductivity and to enhance impurity pinning. This presumably explains the smallness of the conductivity dip in TSF-TCNQ.

It is also interesting to notice that the conduction dip at commensurability is independent of the presence or absence of $4k_F$ scattering (no $4k_F$ scattering in TSF-TCNQ [31].

This rules out the explanation of the dip in terms of $4k_F$ scattering increasing the single particle rate, $1/\tau_{//}$, since when $2k_F = b^*/3$, $2k_F$ and $4k_F$ scatterings differ only by a reciprocal lattice vector.

4. Superconducting transition

Recently, the phase transition into the second kind of ordered state, namely superconductivity, has been found in the compound $(TMTSF)_2\ PF_6$ under hydrostatic pressure of 12kbar [46]. Contrary to TTF-TCNQ and related compounds, only the TMTSF-chains are conducting here, the PF_6-group serving only as electron acceptor [3]. Therefore there is no possibility to have a three-dimensional semimetal-like Fermi surface, such as for two chain conductors like HMTSF-TCNQ [47]. As each PF_6-group takes one electron, the conduction band is one-quarter filled by holes. Moreover a weak dimerization [3] of the intermolecular distance along the high conductivity axis suggests the existence of a small additional gap in the first Brillouin zone and therefore a half-filled band situation. At ambient pressure a transition into an insulating state has been observed at $T_c \approx 15K$ [3], preceded by a conductivity maximum of nearly $10^5 (\Omega cm)^{-1}$. Though by analogy with the TTF-TCNQ family it is tempting to identify this as a Peierls transition in the absence of structural investigations no definitive statement can be made. It is, however, interesting to note that preliminary X-ray experiments down to 25K do not show any precursor effects of a CDW [48]. This is in contrast to usual systems with a Peierls transition where precursor effects are observed up to four times the transition temperature or even higher [49].

Upon applying hydrostatic pressure the metal-insulator transition temperature decreases and vanishes at about 9kbar (Fig.4) [46]. Instead, at a pressure of 12kbar the resistance is observed to vanish completly at $T_c = 0.9K$, indicating a transition into a superconducting state (Fig.5a). Zero resistance is not a proof of bulk superconductivity but may be due to surface effects or superconducting metal inclusions. Therefore, the low temperature state has been further investigated measuring the low-frequency ($\nu = 68Hz$) ac susceptibility [50]. The result (Fig.5b) shows a large diamagnetism below T_c, suggesting nearly complete flux expulsion, i.e. the superconductivity is a bulk effect. This is corroborated by the fact that currents as large as 1mA do not influence significantly T_c, whereas in a typical filamentary superconductor like $(SN)_x$ in samples of comparable cross section already currents of about $10\mu A$

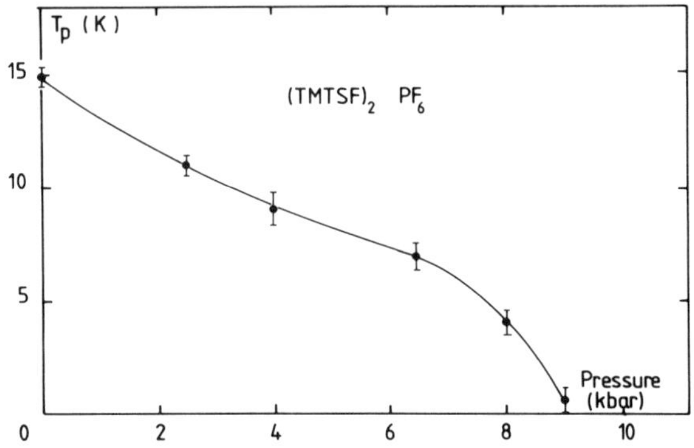

Fig.4 Phase boundary between the insulating (low T) and metallic (high T) phases of $(TMTSF)_2\ PF_6$.

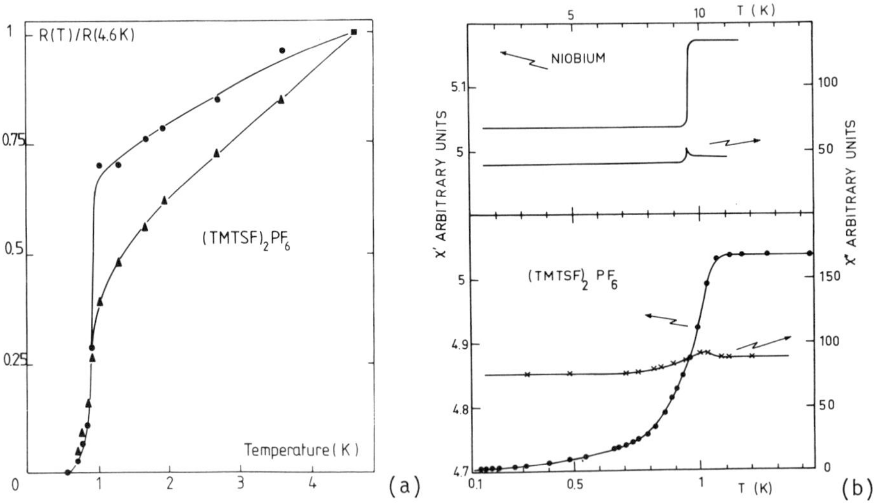

Fig.5 The superconducting transition of $(TMTSF)_2\ PF_6$ observed by (a) resistivity on sample I (triangles) and II (dots), and (b) magnetic susceptibility, together with the same effect as observed in niobium.

change T_c [51]. However, for a determination of critical current densities in $(TMTSF)_2\ PF_6$ a better control of sample heating coming from contact-resistances is required.

The critical magnetic field perpendicular to the main conducting axis has been determined by resistivity measurements [46], as shown in Fig.6 for the same two samples as in Fig.5a. Sample II shows a temperature dependence typical for a type II superconductor. The curve in Fig.6 has been taken at a current of 100μA and gives a low temperature value $H^{\perp}_{c2}$ = 140 Oe. At lower currents H_{c2} increases to 200 Oe at 10μA and 223 Oe at 1μA. All these values are considerably higher than thermodynamic critical fields of type I superconductors of comparable T_c, for example in zinc (T_c = 0.88K) one has H_c = 53 Oe [52]. Apart from a considerably higher value of $H^{\perp}_{c2}$ in sample I it is also interesting to note that a upward curvature of the $H^{\perp}_{c2}$ (T) curve is clearly observed.

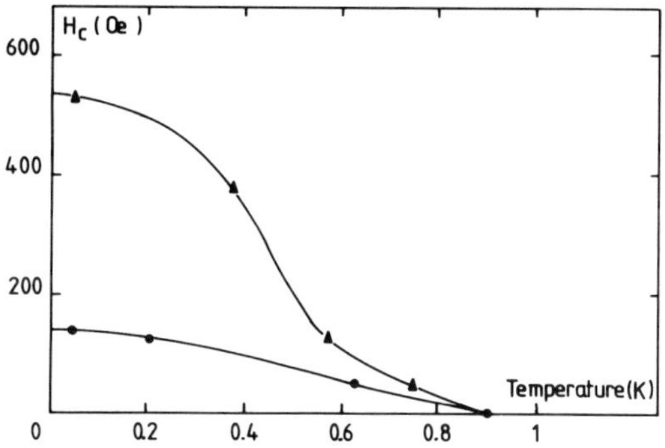

Fig.6 Perpendicular upper critical fields of the same samples as in Fig. 5 a.

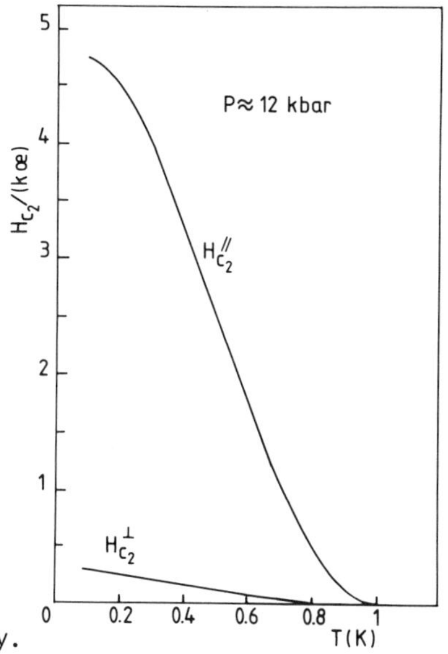

Fig.7 Parallel and perpendicular upper critical fields obtained from susceptibility.

The perpendicular critical fields have also been determined from AC susceptibility measurements [50]. From the value of magnetic field where the suceptibility deviates first from its low field value the lower critical field is determined to be $H^{\perp}_{c1}$ = 30 Oe, and $H^{\perp}_{c2}$ is found to be 330 Oe, in rough agreement with the results obtained from resistivity. These values lead to a Ginzburg-Landau parameter $\kappa^{\perp} = (H^{\perp}_{c2}/2H^{\perp}_{c1})^{1/2} = 2.35$

It should, however, be noted that the susceptibility measurements have been performed on a collection of many crystals, so that due to the strong sample dependence of H_{c2} (see Fig.6) only an average value can be deduced. Also demagnetization effects have been neglected [53].

Applying the magnetic field along the chain direction the parallel critical field $H^{\parallel}_{c2}$ has been obtained as shown in Fig.7 [54]. As to be expected in a quasi-one-dimensional conductor there is a large anisotropy $H^{\parallel}_{c2}$ / $H^{\perp}_{c2}$ = 15. As for $H^{\perp}_{c2}$, these data have been obtained on a collection

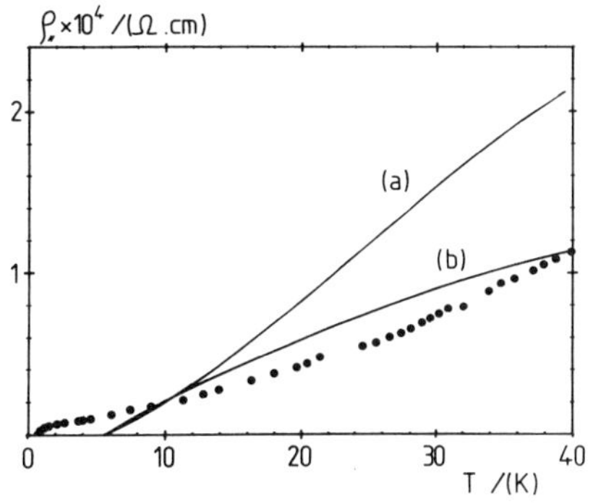

Fig.8 The longitudinal resistivity of $(TMTSF)_2 PF_6$ (dots) and the theoretical results without (curve a) and with (curve b) inclusion of residual normal resistance.

of many crystals with possible misalignment with respect to the magnetic field. Theoretical calculations show that for an anisotropy of this order a misalignment of only 10° may easily reduce $H_{c2}^{//}$ by a factor larger than two [55]. Therefore the result $H_{c2}^{//}/H_{c2}^{\perp} = 15$ should be considered as a lower limit on the real value, which may be larger by a factor two or more. Using Ginzburg-Landau theory, one can use the critical fields to calculate the coherence length $\xi_{//}$ and $\xi_{\perp}$ parallel and perpendicular to the chain direction. From

$$\frac{\phi_o}{2\pi H_{c2}^{\perp}(0)} = \xi_{//}(0)\ \xi_{\perp}(0) \qquad (1)$$

(ϕ_o is the flux quantuum) one obtains $[\xi_{//}(0)\ \xi_{\perp}(0)]^{1/2} \approx 10^3$ Å, a value much larger than a reasonable electron mean free path. Thus, $(TMTSF)_2 PF_6$ is to be considered as a "dirty limit" superconductor.

The anisotropy of the coherence length is equal to the anisotropy of the critical field [55], i.e. one deduces $\xi_{//}/\xi_{\perp} = 15$. On the other hand, a theoretical calculation [56] shows that $\xi_{//}/\xi_{\perp}$ is equal to the anisotropy of the band structure. At room temperature the transverse conductivity is diffusive [3] with an anisotropy $\sigma_{//}/\sigma_{\perp} \approx 10^3$ [3,56]. As for diffusive transverse conductivity $\sigma_{//}/\sigma_{\perp}$ is proportional to the <u>square</u> of the band structure anisotropy [42] one would expect $\xi_{//}/\xi_{\perp} \approx 30$. In view of the above mentioned effect of crystal misalignement this compares favorably with experiment.

A last interesting point is the upward curvature of $H_{c2}^{\perp}(T)$ observed in one sample (see Fig.6). From Ginzburg-Landau theory such an effect is expected to occur when $\xi_{\perp}(T)$ decreases below the interchain distance [55], as in this situation the circular Meissner currents between adjacent chains are strongly suppressed. This explanation implies that the coherence length in sample I is smaller than in sample II, in agreement with the larger transition region observed in the resistivity of sample I (Fig.5a). On the other hand, the interchain distance in $(TMTSF)_2 PF_6$ is smaller than 10 Å, much smaller than the above deduced value for $\xi_{\perp}$. However, this discussion is based on Ginzburg-Landau theory, which may be inadequate in an extremely anisotropic system. It would therefore be interesting to clarify the relation between the critical fields and the coherence length using a microscopic model.

5. Superconducting fluctuation effects

Apart from the transition into the superconducting state at $T_c = 0.9$K there are several unusual features in the temperature dependence of the conductivity of $(TMTSF)_2 PF_6$ under 12kbar pressure [46,56] :

(i) At helium temperatures the longitudinal conductivity $\sigma_{//}$ exceeds $10^5(\Omega cm)^{-1}$, an order of magnitude larger than the figure obtained in other organic conductors.

(ii) The longitudinal resistivity decreases strongly with temperature in the whole measured temperature range down to T_c (Figs.5a,8) in many samples below 50K a downward curvature of the $\rho(T)$ curves is observed. This behaviour is in striking contrast to usual metals where below about 10K the resistivity takes its temperature independent residual value.

(iii) Below 50K an important decrease of the conductivity anisotropy is found (Fig.9). At helium temperature the anisotropy is by a factor 3 smaller than above 50K. Let us try to explain these findings in terms of single electron conduction. At room temperature the transverse conductivity is $\sigma_\perp \approx 0.25(\Omega cm)^{-1}$. This implies, formally, a transverse mean free path much smaller than the interchain distance, i.e. the transverse conduction is diffusive. In this case the conductivity anisotropy is equal to the square of the ratio of the electron transfer integrals along ($t_{//}$) and perpendicular ($t_\perp$) to the chain direction [42] leading to $t_{//}/t_\perp \approx 30$. With decreasing temperature $\sigma_\perp$ increases and reaches $\approx 125(\Omega cm)^{-1}$ at 1.2K. Such a value is consistent with coherent band conduction. In this case one expects $\frac{\sigma_{//}}{\sigma_\perp} = \frac{t_{//}}{t_\perp}$ so that there is a sharp decrease in anisotropy when the transverse conductivity changes from diffusive to coherent. This effect is observed (Fig.9), though $\sigma_{//}/\sigma_\perp$ drops only to about 300 and not to 30, which would be expected from the above discussion. In spite of this qualitative agreement between a single particle conduction model and the experimentally found anisotropy, there arise very serious difficulties from this explanation :
(i) Given the carrier density of about $10^{22}cm^{-3}$, the conductivity of 5×10^5 $(\Omega cm)^{-1}$ at 1.2K leads to a mean free path of 2800 Å, i.e. 700 intermolecular spacings. Such a large mean free path requires a degree of purity and perfection unlikely to be met in organic conductors.
(ii) The strong temperature dependence of $\sigma_{//}$ even at helium temperatures is hard to explain, as usual lattice scattering mechanisms are expected to be frozen out.

In view of these difficulties and the notoriously large fluctuation effects in quasi-one-dimensional conductors, it is tempting to associate the unusual behaviour of the conductivity with superconducting fluctuation effects. The influence of such fluctuations on conductivity and susceptibility of a quasi-one-dimensional conductor has been calculated using Ginzburg-Landau theory [56,57]. Though this approach suffers from serious limitations in quasi-one-dimensional conductors, it provides a convenient starting point for the understanding of fluctuation effects. One obtains

$$\sigma'_{//} = \left(\frac{v_F}{2t_\perp a} \right)^2 \frac{e^2 m\alpha'}{16} \xi_{//}(T) \left[1 - \frac{a}{(a^2+4\pi\xi_\perp{}^2)^{1/2}} \right] \qquad (2a)$$

$$\sigma'_\perp = \frac{e^2 m\alpha'}{16} \xi_{//}(T) \left[1 - \frac{3a}{2(a^2+4\pi\xi_\perp{}^2)^{1/2}} + \frac{a^3}{2(a^2+4\pi\xi_\perp{}^2)^{3/2}} \right] \qquad (2b)$$

for the longitudinal and transverse paraconductivity, respectively. Here v_F is the Fermi velocity, a the interchain distance, m the carrier mass, and α' a constant depending on the microscopic properties of the system, especially on the single particle scattering time.

We now consider two limiting cases . (i) Near the transition temperature, $\xi_\perp$ is much larger than a, i.e. the fluctuations are three-dimensional (3D). Then, one finds the typical square root divergence of a three-dimensional superconductor, however $\sigma'_{//}$ is enhanced over its value in an isotropic superconductor by a factor $(v_F/2t_\perp a)^2$, which may be very large in organic conductors due to the small interchain tunneling integral. This enhancement is due to the small transverse coherence length compared to $\xi_{//}$ which leads to a large mean square amplitude of the superconducting order parameter. The anisotropy of the paraconductivity is $(v_F/2t_\perp a)^2$.

(ii) Sufficiently above T_c one has $\xi_\perp < a$, i.e. one-dimensional fluctuations (1D). Then eq.(2a) reduces to the well-known form of a 1 D superconductor. However, a transverse supercurrent between adjacent chains requires a well developed order parameter on the two chains. As the probability for this to happen decreases with increasing temperature, for $\xi_\perp < a$ $\sigma^{1D}_\perp$ decreases much faster than $\sigma^{1D}_{//}$. Correspondingly, the anisotropy of the fluctuation conductivity increases with increasing temperature.

Another quantity of interest is the fluctuation induced diamagnetism, which for a field applied perpendicularly to the chain direction is [56,57] :

$$\chi_\perp = - \frac{kT}{6\pi} \frac{e^2}{c^2} \xi_{//}(T) \frac{\xi_\perp(T)}{(\xi^2_\perp(T)+a^2/4)^{1/2}} \qquad (3)$$

This shows that the fluctuation diamagnetism is not enhanced over its value in an isotropic superconductor. This result may be understood noting that the circular Meissner currents which are at the origin of the diamagnetism have to pass between adjacent chains. Due to the weak coupling the currents between chains are strongly suppressed. This effect compensates (for perpendicular field and $\xi_\perp >> a$) the large amplitude of the fluctuations, so that for $\xi_\perp >> a$ $\chi_\perp$ is the same as in an isotropic superconductor. For $\xi_\perp \lesssim a$, the interchain currents are further suppressed, so that $\chi_\perp$ is reduced. In a parallel field, the Meissner currents pass only between the chains and never along the chains. Therefore $\chi_{//}$ is always much smaller than $\chi_\perp$.

Up to now, we have considered the fluctuation effects only in lowest order. However, it is well known that in quasi-one-dimensional conductors fluctuation effects are strong, and therefore interactions between fluctuations have to be considered. By comparison with exact solutions of strictly one-dimensional models it can be shown that over a wide temperature range (even below the mean-field transition temperature T°_c) these interactions are adequatly treated in a Hartree-Fock-like procedure [57,58]. This results in replacing T°_c by some effective temperature $T^\circ_c + \Gamma(T)$. In the 3D fluctuation region $\Gamma(T)$ varies only very slowly, so that in this temperature range higher order fluctuation effects may be approximately accounted for by replacing T°_c by the real T_c in eqs.(2) and (3). In the transition region between 3D and 1D fluctuations $\Gamma(T)$ decreases quite rapidly, so that no simple picture can be obtained. However, in the 1D region Γ is again nearly constant, so that T°_c can be replaced by some effective temperature T_0 (of course larger than T_c), and in a simple model one has $T_0 \approx T^\circ_c/3$ [56]. The Ginzburg-Landau results presented here gives the contribution of the Azlamazov-Larkin diagram, however, it cannot account for the Maki-Thompson diagrams. Inclusion of these diagrams and of the effect of the coupling between the BCS and the Peierls instability in one-dimensional conductors [1] requires a microscopic calculation. However we expect that the most interesting results, namely the strong enhancement of the fluctuation conductivity due to the large anisotropy (eq.2a) and the reduction of the fluctuation diamagnetism with respect to an isotropic supraconductor (eq.3) should be unchanged.

We shall now attempt to see whether the large value and the temperature dependence of the conductivity of $(TMTSF)_2$ can be related to the theory of paraconductivity in a very anisotropic conductor proposed above. It is then convenient to deal with resistivity, which follows from eq.(2a) as

$$\rho^{3D}_{//} = 5.10^{-4} \left(\frac{T}{T_F}\frac{\theta}{T_F}\right)^{1/2} \frac{B}{T} \left(\ln \frac{T}{T_c}\right)^{1/2} (\Omega\text{cm}) \qquad (4a)$$

$$\rho^{1D}_{//} = 10^{-3} \left(\frac{T}{T_F}\frac{\theta}{T_F}\right)^{1/2} \left(\ln \frac{T}{T_0}\right)^{3/2} (\Omega\text{cm}) \qquad (4b)$$

for the 3D and 1D regions, respectively. We have set a = 4 Å, $\theta = 1/(k_B\tau)$ is the lifetime broadening of the Fermi surface and B determines crossover between 3 D and 1 D fluctuation regions. The theoretical value of B is quite sensitive to approximations made in the calculations, and therefore in the following we consider B as an adjustable parameter.

First, we pay attention to the T-domain very close to the superconducting transition at $T_c = 0.9$ K. Near T_c the transverse coherence lenght becomes larger than the interchain lattice spacing and the 3D form of the resistivity, eq. (4a) is expected to fit the data of Fig. 8. The experimental value $\rho(1K) = 2.10^{-6}$ Ωcm substituted into eq.(4a), with $\theta/T_F = 0.2$, leads to B = 0.85K and hence to a cross-over temperature $T^* = 1.75$K. This gives an excellent quantitative agreement with the measured resistivity up to 2K.

Secondly, the 1D form of $\rho_{//}$ is expected to explain the temperature dependence of the resistivity as long as the paraconductivity of superconducting origin dominates over all other conductivity channels. We may reasonably believe that such a condition is fulfilled at helium temperature if the paraconductivity is important. A value of $T_0 = 5.6$K is necessary in eq.(4b) if we want $\rho^{1D}_{//}(10K)$ to be equal to the experimental value of $2\ 10^{-5}$ Ωcm. At higher temperatures the $\rho^{1D}_{//}(T)$ theoretical curve (with $\theta/T_F = 0.2$, $T_0 = 5.6$K) deviates significantly from the observed data. For example at 40K, the theory leads to twice too much resistivity, (Fig.8). The agreement between the theory and experimental data can be improved, realizing that the paraconductivity at 40K calculated from eq.(4b) amounts to 4166 $(\Omega\text{cm})^{-1}$. This value is no longer large compare to other mechanisms of conduction, in particular the single particle channel limited by impurities and (or) lattice imperfections. If we take a T-independent single particle contribution below 40K equal to 4000 $(\Omega\text{cm})^{-1}$ we obtain a theoretical resistivity $\rho = (\sigma^{1D}_{//} + \sigma_0)^{-1}$ which follows more closely the experimental data, see Fig.8. Moreover, the observed shape of the temperature dependence of $\rho_{//}$ between 40K and 4.2 K can vary slightly from sample to sample. In some samples, we have noticed a steep descent of the resistivity below 40K, resembling closely the theoretical behavior.

The decrease of the conductivity anisotropy with decreasing temperature in the 1D region is also in agreement with the paraconductivity picture, as discussed below eq.(2b). In this picture $\sigma_{//}/\sigma_{\perp}$ begins to decrease when paraconductivity becomes the dominant conduction mechanism. From Fig.9 this should be the case below about 40K, in agreement with the above discussion of the longitudinal resistivity.

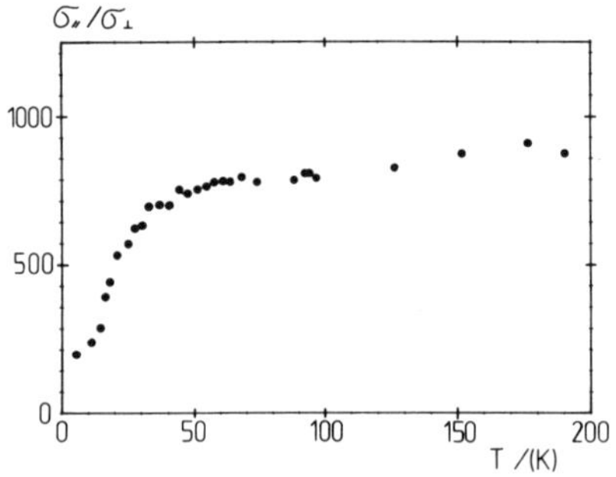

Fig.9 Conductivity anisotropy of a single sample of $(TMTSF)_2\ PF_6$.

Attempts to detect strong fluctuation diamagnetism above T_c in $(TMTSF)_2\ PF_6$ have not given significant results [50]. From the sensitivity of the experimental set-up, we find that the fluctuation diamagnetism is smaller than, say 10^{-4} times the diamagnetism corresponding to a Meissner effect, namely $1/4\pi$. However these results are in agreement with the result for $\chi_{\perp}$ (eq.(3)) which has clearly established that the fluctuation diamagnetism above T_c is not enhanced in a quasi-one-dimensional conductors by dimensionality effects.

In conclusion, the assumption of paraconductivity dominated conduction below 40K accounts well for the experimental observations :

(i) With reasonable values of the theoretical parameters quantitative agreement between experiment and theory is achieved for the conductivity.
(ii) The nonsaturating temperature dependence of the conductivity is well accounted for.
(iii) The temperature dependence of $\sigma_{//}/\sigma_{\perp}$ is well explained in this picture as well as
(iv) the absence of a large fluctuation effect on the magnetic susceptibility. In view of the fact that points (i) and (ii) are hard to reconcile with a single particle conduction picture, the paraconductivity picture gives a very convincing (even quantitative) explanation of the experimental results.

Acknowledgements - We wish to acknowledge the help of our colleagues at Orsay and Copenhagen who participated in the experimental efforts. We have also benefited from discussion with J. Friedel on several theoretical aspects of this work.

References

1. J. Solyom: Adv. Phys. 28, 201 (1979)
2. M.J. Cohen, L.B. Coleman, A.F. Garito, A.J. Heeger: Phys. Rev. B10, 1298 (1974)
3. K. Bechgaard, C.S. Jacobsen, K. Mortensen, H.J. Pedersen, N. Thorup: Solid State Commun. 33, 1119 (1980)
4. A.A. Bright, A.F. Garito, A.J. Heeger: Phys. Rev. B10, 1328 (1974)
5. P.M. Grant, R.L. Greene, G.C. Wrighton, G. Castro: Phys. Rev. Lett. 31, 1311 (1973)
6. L.B. Coleman, M.J. Cohen, M.J. Sandmann, D.J. Yamagishi, A.F. Garito, A.J. Heeger, Solid State Commun. 12, 1125 (1973)
7. J.P. Ferraris, D.O. Cowan, V.V. Walatka, J.H. Perlstein: J. Amer. Phys. Soc. 95, 948 (1973)
8. D. Jérome, W. Müller, M. Weger: J. Phys. (Paris) Lett. 35, L77 (1974)
9. J.R. Cooper, D. Jérome, M. Weger, S. Etemad: J. Phys. (Paris) Lett. 36, L219 (1975)
10. F. Denoyer, R. Comès, A.F. Garito, A.J. Heeger: Phys. Rev. Lett. 35, 445 (1975)
11. S. Kagoshima, H. Anzai, K. Kajimura, T. Ishiguro: J. Phys. Soc. Jap. 39, 1143 (1975).
12. R. Comès, S.M. Shapiro, G. Shirane, A.F. Garito, A.J. Heeger: Phys. Rev. Lett. 35, 1512 (1975)
13. R.E. Peierls: Quantum Theory of Solids (Clarendon, Oxford 1955), p. 108
14. S. Etemad: Phys. Rev. B13, 2254 (1976)
15. J.P. Pouget, S.K. Khanna, F. Denoyer, R. Comès, A.F. Garito, A.J. Heeger: Phys. Rev. Lett. 37, 437 (1976)
16. J.B. Torrance: In Chemistry and Physics of One-dimensional Metals, ed. by H.J. Keller (Plenum, New York 1977), p. 137
17. D. Jérome, L.G. Caron: "Magnetic Properties of TTF-TCNQ", in *Quasi-One-Dimensional Conductors I*, ed. by S. Barišic, A. Bjeliš, J.R. Cooper, B. Leontič, Lecture Notes in Physics, Vol. 95 (Springer, Berlin, Heidelberg, New York 1979) p. 91
18. P.M. Chaikin, R.L. Greene, S. Etemad, E.M. Engler: Phys. Rev. B13, 1627 (1976)
19. Y. Tomkiewicz, A.R. Taranko, J.B. Torrance: Phys. Rev. Lett. 36, 751 (1976)
20. J.R. Cooper, J. Lukatela: In ref.17, p. 174.
21. Y. Tomkiewicz: In "Proceedings of the NATO ASI on Low-Dimensional Solids, ed. by L. Alcacer (Reidel, to be published)
22. R.H. Friend, M. Miljak, D. Jérome: Phys. Rev. Lett. 40, 1048 (1978)
23. C. Weyl, D. Jérome, P.M. Chaikin: to be published
24. S. Megtert, R. Comès, C. Vettier, R. Pynn, A.F. Garito: In ref.21
25. S.K. Khanna, J.P. Pouget, R. Comès, A.F. Garito, A.J. Heeger: Phys. Rev. B16, 1468 (1977)
26. C.R. Fincher et al.: Phys. Rev. B, to be published
27. P. Bak, V.J. Emery: Phys. Rev. Lett. 36, 978 (1976)
28. A. Bjelis, S. Barisic: Phys. Rev. Lett. 37, 1517 (1976)
29. S. Megtert, private communication
30. T. Ohmi, H. Yamamoto: Progr. Theor. Phys. 58, 743 (1977)
31. C. Weyl, E.M. Engler, K. Bechgaard, G. Jehanno, S. Etemad: Solid State Commun. 19, 925 (1976)
32. J.F. Thomas, D. Jérome: Solid State Commun., to be published
33. R. Comès: In ref. 16, p. 315
34. S. Kagoshima, T. Ishiguro, T.D. Schultz, Y. Tomkiewicz: Solid State Commun. 28, 485 (1978)
35. S. Megtert, J.P. Pouget, R. Comès: In Molecular Metals, ed. by W.E. Hatfield (Plenum, New York 1979), p. 87

36. J. Bardeen: Solid State Commun. 13, 357 (1973)
37. G.A. Thomas et al.: Phys. Rev. B13, 5105 (1976)
38. M.J. Cohen, L.B. Coleman, A.F. Garito, A.J. Heeger: Phys. Rev. B13, 5111 (1976)
39. A. Andrieux, H.J. Schulz, D. Jérome, K. Bechgaard: Phys. Rev. Lett. 43, 227 (1979)
40. H.J. Schulz: Solid State Commun. 34, 455 (1980), and to be published
41. B. Welber, P.E. Seiden, P.M. Grant: Phys. Rev. B18, 2692 (1978)
42. M. Weger: J. Phys. (Paris) Colloque 39, C6-1456 (1978)
43. A. Andrieux, H.J. Schulz, D. Jérome, K. Bechgaard: J. Phys. (Paris) Lett. 40, L385 (1979)
44. S. Bouffard, to be published
45. P.M. Horn, D. Guidotti, Phys. Rev. B16, 491 (1977)
46. D. Jérome, A. Mazaud, M. Ribault, K. Bechgaard: J. Phys. (Paris) Lett. 41, L95 (1980)
47. M. Weger: Solid State Commun. 19, 1149 (1976)
48. J.P. Pouget, private communication
49. In the sulfur analog compound $(TMTTF)_2PF_6$ weak structural precursor effects are observed at 50K; J.P. Pouget, private communication
50. M. Ribault, G. Benedek, D. Jérome, K. Bechgaard: J. Phys. (Paris) Lett., 15 August 1980
51. R.I. Civiak, C. Elbaum, L.F. Nichols, H.I. Kao, M.M. Labes: Phys. Rev. B14, 5413 (1976)
52. P.G. de Gennes: Superconductivity of Metals and Alloys (Benjamin, New York 1966)
53. D. Saint James, E.J. Thomas, G. Sarma: Type II Superconductivity (Pergamon, Oxford 1969)
54. D. Mailly, M. Ribault, to be published
55. L.A. Turkevich, R.A. Klemm: Phys. Rev. B19, 2520 (1979)
56. H.J. Schulz, D. Jérome, A. Mazaud, M. Ribault, K. Bechgaard, to be published
57. H.J. Schulz, to be published
58. D. Jérome, H.J. Schulz: In Extended Linear Chain Compounds , ed. by J.S. Miller (Plenum, New York, in press)

Non-Linear Transport in the Fröhlich Mode Conductor, $NbSe_3$

R.M. Fleming

Bell Laboratories
Murray Hill, NJ 07974, USA

ABSTRACT

$NbSe_3$ is an anisotropic metal with two charge-density wave (CDW) phase transitions at 144K and 59K. At temperatures just below each of the two CDW phase transitions the conductivity can be dramatically enhanced by the application of small dc fields or by relatively low frequency rf fields. It is now believed that the non-linear conductivity is a result of a CDW which is depinned from lattice and moves in an applied dc field. The experimental evidence supporting this conclusion will be reviewed.

INTRODUCTION

Although the possibility of electron transport via a moving charge-density wave (CDW) was recognized as early as 1954 [1], until now none of the many materials which have a CDW at low temperatures have shown an enhancement of dc conductivity from this mechanism. As anticipated by Lee, Rice and Anderson [2], in all CDW materials found thus far the phase of the CDW is pinned to the host lattice by either impurities or by commensurability. For a one dimensional material the result is the well known Peierls-Fröhlich insulating state, which was reviewed for us earlier in this conference by T. M. Rice. My talk today concerns a remarkable material, $NbSe_3$, where the CDW does play an active role in electron transport. Although the CDW in $NbSe_3$ is pinned to the lattice by impurities at zero electric field, the CDW may be depinned with the application of extremely small applied fields (~5 mV/cm). The depinned CDW provides an additional contribution to the electrical conductivity in a manner very similar to the original Fröhlich model. I will review today the experimental evidence for Fröhlich mode conductivity in $NbSe_3$ as well as some novel effects associated with a moving CDW.

The temperature dependence of the resistance of $NbSe_3$ is shown in Fig. 1. $NbSe_3$ is not highly one-dimensional and consequently it does not have a complete Peierls transition to an insulating state. Prominent increases in the resistance at $T_1 = 144K$ and $T_2 = 59K$ are associated with the onset of CDW energy gaps in the band structure and subsequent loss of carriers. Similar behavior is seen in other CDW materials such as layered compounds [3]. The feature which makes $NbSe_3$ unique is the non-ohmicity which is seen at temperatures just below T_1 and T_2 [4,5]. The application of a very modest electric field (<1 V/cm) is sufficient to completely remove the two resistive anomalies. At high electric fields the temperature dependence of the resistance below T_1 and T_2 is very close to the value one would obtain by extrapolation from high temperatures. This increased conductivity at finite electric fields is now believed to be a result of a moving or "sliding-mode" CDW and is the focus of my talk today. At temperatures below about 20 K, $NbSe_3$ shows little non-linear behavior, a feature which is not presently understood in the context of a depinned CDW.

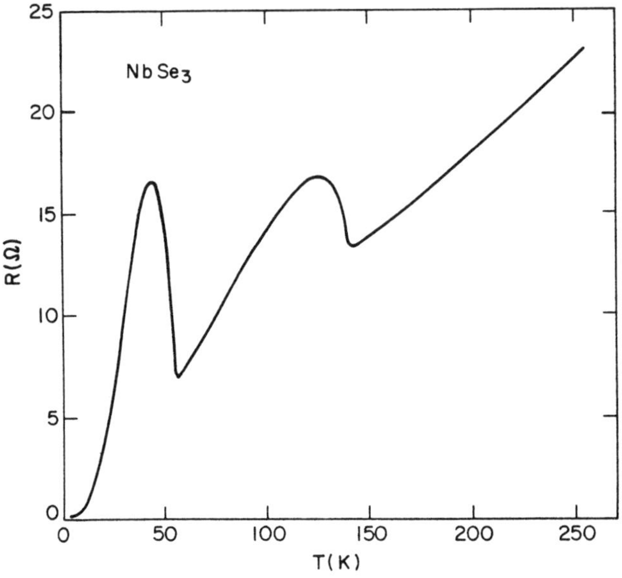

Fig. 1 Temperature dependence of the resistance in $NbSe_3$. The large anomalies at 144K and 59K are the result of the formation of two separate CDWs.

$NbSe_3$ has a monoclinic crystal structure [6]. The niobium atoms are located at the center of trigonal prisms which are stacked to form chains along the b-axis. An examination of the crystal structure shows that despite the chain-like morphology, $NbSe_3$ is not highly one-dimensional. This assumption is supported by Fermi surface measurements [7,8] which show anisotropic three dimensional Fermi surface cross sections and by diffuse X-ray scattering measurements [9] which show anisotropic three dimensional critical scattering above the two CDW phase transitions.

CHARGE-DENSITY WAVES

The two phase transitions at T_1 and T_2 are each associated with the formation of separate and nearly independent CDWs [9,10]. Both CDWs are incommensurate with wave vectors which are temperature independent. The high temperature CDW has a reduced wave vector which is exactly along b*, $\vec{q}_1 = (0,0.243,0)$. The low temperature CDW is also incommensurate along b* but also has commensurate components along a* and c*, $\vec{q}_2 = (0.5,0.263,0.5)$. Like the layered compounds second harmonic scattering is also seen at $2\vec{q}_1$ indicating that the CDW is slightly non-sinsoidal. This has important implications in the discussion of CDW domains [11]. A curious feature of the diffraction geometry is that the sum of the two CDW wave vectors is nearly commensurate; within the resolution of the x-ray scattering experiments $\vec{q}_1 + \vec{q}_2 = (\frac{1}{2}\,\frac{1}{2}\,\frac{1}{2})$. Preliminary results at 8K from a high resolution scattering experiment using synchrotron radiation support this conclusion [12]. Despite this evidence for an interaction between the CDWs, however, the depinning of the two CDWs is apparently independent.

In establishing the presence of Fröhlich mode conductivity, it is important to show that the non-linear behavior does not somehow result from a field-induced destruction of the CDW state. In such a case the conductivity might be expected to revert to a value extrapolated from temperatures above the phase transition in agreement with experiment [4,5]. This hypothesis was checked by making a simultaneous measurement of the sample resistance and the CDW amplitude [9]. The temperature dependence of the CDW amplitude, which is proportional to the intensity of a CDW satellite observed by x-ray diffraction, is shown in Fig. 2. Note that at dc currents sufficient to suppress approximately 60% of the low temperature resistive anomaly, no change in the intensity of the CDW satellite reflection is seen. This indicates that both the CDW amplitude and wave vector are field independent and leads one the obvious conclusion that the CDW itself is responsible for the enhanced conductivity at high electric fields. The extremely small magnitude of the fields necessary to see non-linear effects plus the fact that $NbSe_3$ is metallic argue strongly against other field-dependent mechanisms which can increase the carrier density such as impact ionization, hot electrons, etc.

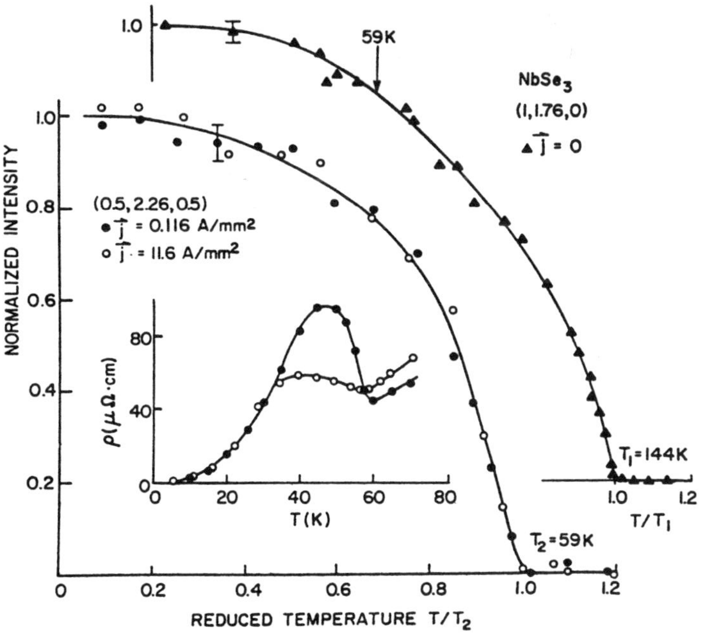

Fig. 2 Temperature dependence of the integrated X-ray scattering intensity for both CDW superlattices. The inset is a simultaneous measurement of the sample resistance.

NON-LINEAR CONDUCTIVITY

The field dependence of the conductivity in $NbSe_3$ was originally studied by MONCEAU et al [4,5] using dc techniques. Near each of the resistive anomalies they fit the conductivity to an empirical relation

$$\sigma(E,T) = \sigma_a(T) + \sigma_b(T)e^{-E_0(T)/E}. \tag{1}$$

E_0 is a characteristic field which is a measure of the width of the transition from the low field conductivity, σ_a, to the high field conductivity, $\sigma_a + \sigma_b$. ONG and MONCEAU [5] find that E_0 is strongly temperature dependent. E_o has minima near the resistance maxima and diverges both at low temperatures and near the two phase transitions. The exponential form of (1) originally lead to the speculation that the non-ohmic conductivity resulted from Zener tunneling across CDW energy gaps. It is now felt that a single practice tunneling model is not correct, principally because of the extremely small gaps (10^{-3} kT) needed to account for the observed values of E_o. BARDEEN [13] however, has been able to obtain (1) with reasonable values of E_o but within a 1D commensurate model involving coherent tunneling of the CDW.

A model which is closer in concept to the original ideas of Fröhlich is that of Lee and Rice [14]. They consider in detail the competition between impurity pinning and the electric field energy and balance the energy gained from the impurity against the elastic energy cost of deforming the CDW. LEE and RICE identify two classes of impurities, strong and weak. In the strong pinning case the CDW coherence is shorter than the mean impurity spacing and the CDW phase is determined by a single impurity. This is expected for charged impurities. In the weak pinning case the CDW phase is determined not by a single impurity but by an average over a collection of impurities. This behavior is expected for isoelectronic impurities. In both cases a key feature of the Lee and Rice model is that the depinning field should scale with the number of impurities. They find that $E_c \propto n_i$ for strong pinning and $E_c \propto n_i^2$ for weak pinning. Furthermore the depinning of the CDW is expected to be associated with a sudden onset of extra conductivity. Although the rise of E_o near the transition temperature is a feature of the Lee and Rice model, they offer no explanation for the functional form of (1) or for the increase of E_o at low temperatures.

An important feature of a CDW which is depinned by electric field is the observation of a sudden onset of extra conductivity at a well-defined threshold field, E_T [14,15]. In a detailed measurement of the field-dependent resistance using ac current modulation techniques, Fleming and Grimes [16,17] showed that a threshold field behavior is present near both of the CDW phase transitions. Figure 3 shows the field dependence of the resistance and a dc measurement of I-V at 59K. The data clearly show ohmic behavior for $E > E_T$ and a sudden onset of non-linear conductivity at $E = E_T$. The field dependence of the differential resistance at 125K also shows a threshold field, but E_o and E_T are about a factor of ten larger [17]. The points in Fig. 3 are data

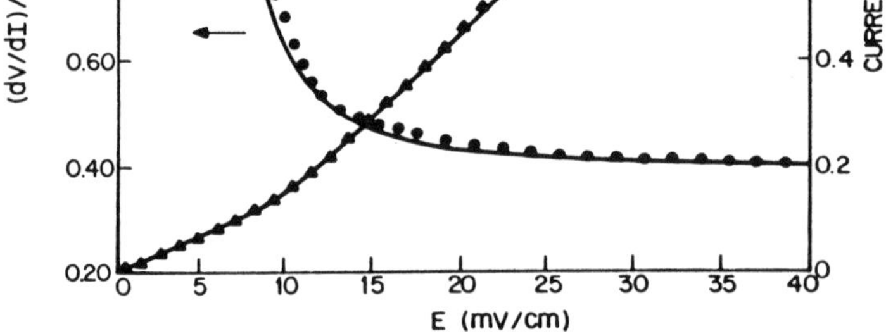

Fig. 3 Normalized differential resistance and dc current as a function of electric field at 51K.

taken from analog chart recorder traces. The solid lines are fits to the data using the empirical relations

$$J(E) + \sigma_a E \quad ; (E<E_T)$$
$$J(E) = \sigma_a E + \sigma_b (E-E_T) e^{-E_o/E-E_T} \quad ; (E>E_T) \tag{2}$$
$$\frac{dV}{dI} = \left[\frac{dJ}{dE}\right]^{-1}$$

Equation (2) leads to essentially the same field-dependent conductivity $\sigma(E) = \frac{1}{E} J(E)$ as (1) except for the inclusion of a threshold electric field. As pointed out earlier, the extremely small values of the threshold field (~5 mV/cm at 59K) argue strongly for a CDW depinning mechanism.

NOISE

Perhaps the most unusual aspect of the non-linear conductivity of $NbSe_3$ is that of noise. FLEMING and GRIMES [16,17] showed that for dc biased samples with $E > E_T$, quasi-periodic noise is observed near both CDW transitions. The spectral content of the noise at 48K is shown in Fig. 4 at several values of E. The noise is characterized by discrete frequencies superimposed upon a broad-band background. At fields just above threshold the periodic portion of the signal contains a single frequency together with numerous harmonics. At higher fields additional frequencies can be identified. All periodic components initially appear near zero frequency and at low fields the frequency increases approximately linearly with field. At higher electric fields WEGER et al [18] report that the frequency increases as E^2. The phase coherence of the narrow-band noise was analyzed by repetitively summing the signal using a digital transient recorder. A trigger was provided by the periodic noise signal itself in a manner similar to an oscilloscope. The result shown in Fig. 5 contains only the coherent portion of the noise. One can see that the fundamental frequency maintains phase coherence over many cycles. The qualitative behavior of the noise near the high temperature CDW transition is similar, but the coherence and amplitude of the periodic portion of the signal are not nearly so well defined.

Noise was not anticipated by theoretical models but its presence is consistent with CDW motion. In the most simple picture the CDW has a preferred local phase determined by the pinning centers. Each time the phase of the moving CDW attains the preferred value the CDW tends to lock to the lattice. If the CDW or domains of CDW have little coherence the motion would result in a broadband noise. If the CDW maintains phase coherence the noise might have a periodic character with a frequency given by $F = V/\lambda$ where V is the CDW velocity and λ is the CDW wavelength. The problem is that if one uses reasonable estimates for V and λ the calculated frequencies are about 10^2 larger than the observed frequencies. MONCEAU et al [19] point out that one should really only use that fraction of the current contributed by the moving CDW,

$$J_{CDW} = J\left[1 - \frac{R(E)}{R(O)}\right]. \tag{3}$$

The frequencies are then given by $f = J_{CDW}/\rho_c n_o e\lambda$, where ρ_c and n_o are the fraction of condensed electrons and the normal carrier density respectively. This approach gives reasonable frequencies, but only if one uses calculated values of n_o rather than experimental values from Hall effect measurements [7,20] WEGER et al [18] have proposed a phenomenological model for the narrow band noise where they assume a distribution of CDW domain lengths and the mechanism of coupled relaxation oscillators. Their model predicts a frequency dependence $f \propto E^2$ in agreement with their observations.

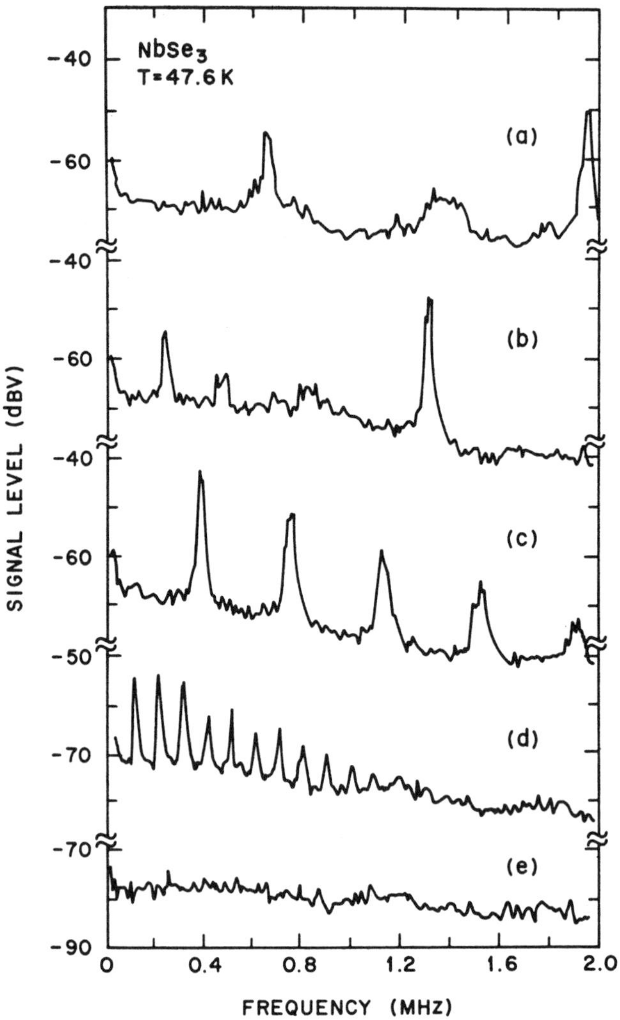

Fig. 4 Noise spectrum of $NbSe_3$ at a number of dc currents: (a) 15.7 mV/cm, (b) 13.6 mV/cm, (c) 11.0 mV/cm, (d) 9.2 mV/cm and (e) E = 0.

The onset of broadband noise provides a convenient way to measure the threshold field for depinning the CDW [17]. Near the maxima of the resistive anomalies the noise onset and the onset of non-linear conductivity coincide closely. At lower temperatures E_0 increases and the onset of non-linear conductivity becomes more difficult to determine. The noise onset, however, remains sharp over a wide temperature range and below T_2 one can identify separate depinning for each of the two CDWs. Figure 6 shows the rectified output of an amplifier with a 75 kHz bandpass filter. For temperatures above 59K there is a single onset of noise from the high temperature CDW. At temperatures below 59K one can identify two separate onsets of noise corresponding to depinning each of the two CDWs. In contrast to E_0 which diverges at about 20K,

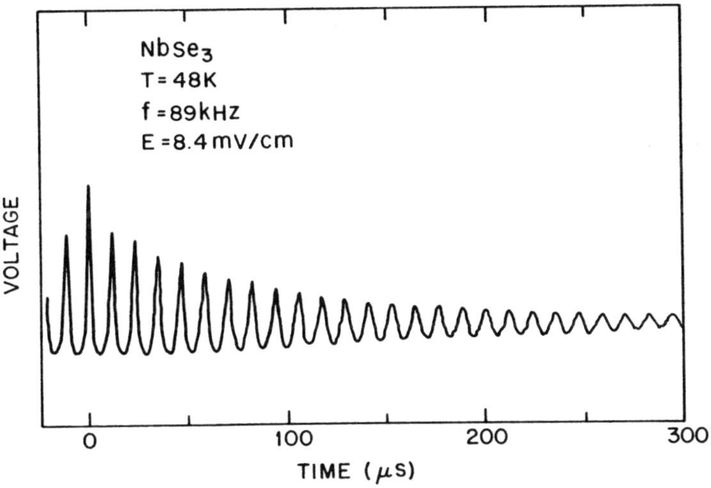

Fig. 5 Coherent portion of noise voltage as a function of time obtained by repetitively summing noise signal.

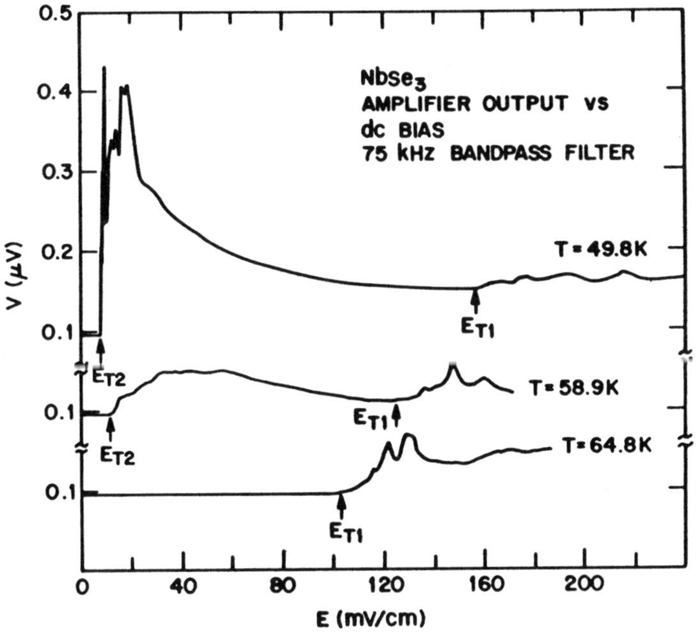

Fig. 6 Rectified amplifier output as a function of dc electric field. At temperatures below T_2, two noise onsets can be seen corresponding to the separate depinning of each of the two CDWs.

E_{T2} rises only moderately at low temperatures. At 17K one can still observe excess noise at 35 mV/cm but there is no non-ohmic behavior. As pointed out earlier, this behavior is not presently understood.

IMPURITY PINNING

If the non-linear conductivity in $NbSe_3$ is a result of a depinned CDW, the depinning field should scale with the number of pinning sites. ONG et al [21,22] have studied the non-linear properties as a function of the concentration of Ta (isoelectronic) or Ti (charged) impurities. They use the residual resistivity ratio $R_{273K}/R_{4.2K}$ as a measure of the number of pinning sites. They find that for Ta impurities E_o or E_T scales as n_i^2 in agreement with the Lee and Rice model for weak pinning. Ti impurities pin the CDW much more strongly than Ta, as expected, but no smooth variation of E_o with the impurity concentration is seen. Neutron irradiated samples also show an increase of E_o and E_T with radiation dose [23], but in this case E_T scales linearly with the radiation time. Undoubtably radiation damage and chemical doping result in different types of defects. In all cases however, the conclusion can be made that the depinning field increases as the number of pinning sites is increased.

FREQUENCY DEPENDENT CONDUCTIVITY

A final test of Fröhlich conductivity in $NbSe_3$ is the frequency dependence of the conductivity. LEE, RICE and ANDERSON [2] showed that for a pinned CDW in one dimension the oscillator strength at dc in the Fröhlich model is shifted to finite frequency. ONG et al [5,21,22] showed that 9.8 GHz the resistive anomalies are removed in $NbSe_3$. Furthermore, if one dopes the

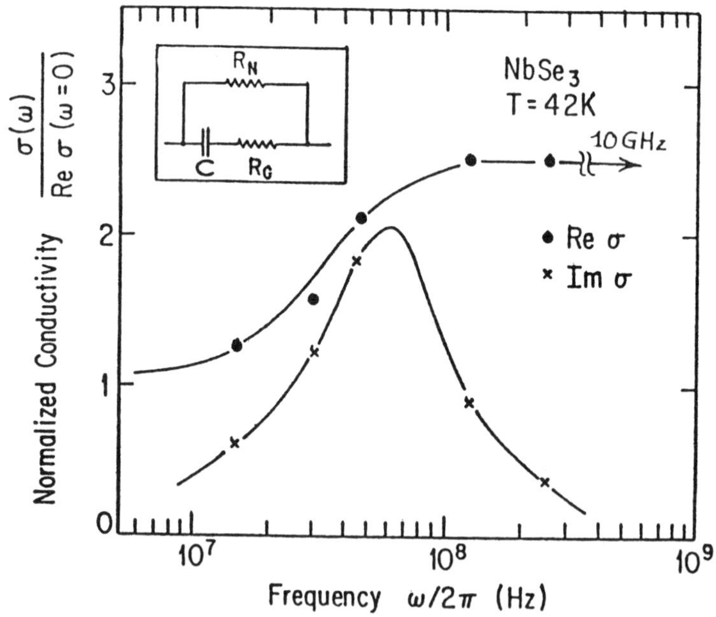

Fig. 7 Frequency dependence of the conductivity from [25]. A broad increase in Re $\sigma(\omega)$ indicates a distribution of pinning energies.

samples with Ta the resistive anomalies are again seen, indicating stronger pinning. GRÜNER et al [24] have made measurements of the conductivity over an extended frequency range. Their results, which are shown in Fig. 7, can be most easily interpreted if one thinks of the response of a CDW in terms of a phenomenological resistor-capacitor circuit shown in the insert of Fig. 7. For a one dimensional material all of the conduction electrons condense into a CDW and the circuit would consist of only a capacitor. In this case one would find a resonance at the pinning energy. $NbSe_3$ however is not highly one dimensional and a substantial fraction of the electrons do not participate in the CDW. This acts to add a resistance, R_N, to the circuit in parallel with the capacitance. Now one has an overdamped oscillator and rather than a peak at the pinning energy one expects a sharp increase in the conductivity. Figure 7 shows that the observed increase is actually very broad indicating a wide distribution of pinning energies. Finally if one were to apply a dc field and measure $\sigma(\omega)$ the CDW would be depinned giving $C = 0$ and a finite R_C from the moving CDW. GRÜNER et al [25] have performed this experiment and, as expected, they see only a constant, frequency independent conductivity.

SUMMARY

It now seems well established that $NbSe_3$ exhibits non-linear conductivity as a result of a new form of electron transport, a moving CDW. This is essentially the mechanism of Fröhlich conductivity with the added features that the material is three dimensional and that interactions of the CDW with the host lattice are important. Chief among the observations which support Fröhlich made in conductivity in $NbSe_3$ are:

1. Non-linear conductivity at extremely small electric fields.
2. A field independent CDW amplitude.
3. A sudden onset of non-linear conductivity at a threshold field.
4. The occurrence of quasi-periodic noise.
5. A depinning field which scales with the number of pinning sites.
6. A frequency dependent conductivity at low dc fields and a frequency independent conductivity at high dc fields.

Several aspects of $NbSe_3$ are still not completely understood. There is no explanation for the rise of the depinning at low temperature or a microscopic mechanism for the functional form of (1). The narrow band noise is likely a result of coherent motion of the CDW but estimates of the frequencies from models are larger than the observed frequencies. There is presently no convincing explanation for the high-field conductivity which is equal to an extrapolation of the high temperature conductivity. Perhaps the most puzzling aspect of Fröhlich mode conductivity is its apparent absence in other materials. Although many materials with CDW are now known, $NbSe_3$ has thus far been the only one to show convincing evidence of non-linear conductivity by this mechanism.

ACKNOWLEDGEMENTS

In much of the work reported here I have benefited greatly from collaborations with C. C. Grimes, D. E. Moncton and D. B. McWhan.

REFERENCES

1. H. Fröhlich, Proc. R. Soc. London A 223, 296 (1954).
2. P. A. Lee, T. M. Rice, and P. W. Anderson, Solid State Commun. 14, 703 (1974).
3. For a review see F. J. DiSalvo, in Electron-Phonon Interactions and Phase Transitions, edited by Tormed Riste (Pelnum, New York, 1977), p. 107.
4. P. Monceau, N. P. Ong, A. M. Portis, A. Meerschaut, and J. Rouxel, Phys. Rev. Lett. 37, 602 (1976).
5. N. P. Ong and P. Monceau Phys. Rev. B 16, 3443 (1977).
6. J. L. Hodeau, M. Marezio, C. Roucau, R. Ayroles, A. Meerschaut, J. Rouxel and P. Monceau, J. Phys. C 11, 4117 (1978).
7. R. M. Fleming, J. A. Polo, Jr. and R. V. Coleman, Phys. Rev. B 17, 1634 (1978).
8. P. Monceau, Solid State Commun. 24, 331 (1977).
9. R. M. Fleming, D. C. Moncton and D. B. McWhan, Phys. Rev. B18, 5560 (1978).
10. K. Tsutsumi, T. Takagaki, M. Yamamato, Y. Shiozaki, M. Ido, T. Sambongi, K. Yamaya, and Y. Abe, Phys. Rev. Lett. 39, 1675 (1975).
11. See for example R. M. Fleming, D. E. Moncton, D. B. McWhan and F. J. DiSalvo, Phys. Lett. 45, 576 (1980).
12. R. M. Fleming, D. E. Moncton and J. D. Axe, unpublished data.
13. John Bardeen, Phys. Rev. Lett. 42, 1498 (1979).
14. P. A. Lee and T. M. Rice, Phys. Rev. B 19, 3970 (1979).
15. J. B. Sokoloff, preprint.
16. R. M. Fleming and C. C. Grimes, Phys. Rev. Lett 42, 1423 (1979).
17. R. M. Fleming, Phys. Rev. B, to be published.
18. M. Weger, C. Grüner and W. G. Clark, Solid State Commun. 35, 243 (1980).
19. P. Monceau, J. Richard and M. Renard, Phys. Rev. Lett. 45, 43 (1980).
20. N. P. Ong and P. Monceau, Solid State Commun. 26, 487 (1978).
21. N. P. Ong, J. W. Brill, J. C. Eckert, J. W. Savage, S. K. Khanna and R. B. Somoano, Phys. Rev. Lett. 42, 811 (1979).
22. J. W. Brill, N. P. Ong, J. C. Eckert and J. W. Savage, preprint.
23. P. M. Chaikin, private communication.
24. G. Grüner, L. C. Tippie, J. Sanny and W. G. Clark, Phys. Rev. Lett. 45, 935 (1980).

Part VI

Disorder and Localization

Lattice Dynamics and Spectral Properties of Disordered Chains

M. Goda

Faculty of Engineering, Niigata University
950-21 Niigata, Japan

Some rigorous treatments on special frequencies and localization of eigenstates in one-dimensional systems are outlined. Special frequencies are understood as generalized spectral gaps which are predicted by the general Saxon-Hutner-type theorem. Localization of eigenstates is understood by the exponential growth property of particular solution. Furstenberg's convergent theorem is generalized for this purpose.

1. Introduction

The studies of disordered systems in one-dimension have been drawing the increasing attention of mainly theoretical physicists for more than the last twenty years. Concerning linear phenomena, these studies in this period were motivated, roughly speaking, by two facts.

One of them was that we had an expectation that some characteristic features might be found in these systems. The expectation was much enhanced by the numerical work with negative factor counting method on the state density of phonons [1] and also by a conjecture of MOTT and TWOSE [2]. Since then enormous studies have been made on these systems [3-5] and many exact results have been obtained from theoretical point of view.

On the other hand, there was another viewpoint in studying one-dimensional systems, that these systems were simple examples to check the validity of the approximate methods which were developed for studying two- or three-dimensional real systems. Meanwhile, the Green function method has been developed and contributed to the study of disordered systems [6,7].

Recent developments on producing pseudo-one-dimensional systems [8,9] have yielded much experimental information about real one-dimensional systems, and we have now one other motivation.

The main purpose of this talk is to review, keeping the above-mentioned three movements in mind, some rigorous treatments on special frequencies and localization of eigenstates in one-dimensional disordered systems. Though systems are limited in one dimension and hence no topological disorder is considered, we can find some essential and characteristic features of disordered systems in these rigorous studies.

2. Eigenvalue Equation, Transfer Matrix

Many typical systems of phonons, tightly binding electrons, Frenkeltype excitons and magnons can be described (for simplicity in the cases where only nearest-neighbour interaction exists) by the following type of eigenvalue equation

$$\alpha_n a_n = t_{n,n+1} \cdot a_{n+1} + t_{n,n-1} \cdot a_{n-1} \tag{1}$$

where, $\alpha_n = (E - \varepsilon_n)$ (ε_n; atomic energy at the site n and $t_{n,n\pm1}$; the transfer energies) in the case of electrons. In the case of excitons the correspondence is analogous to the above. In the case of phonons we put suitably

$$\alpha_n = (K_{n,n+1} + K_{n,n-1}) - M_n \cdot \omega^2 , \quad t_{n,n\pm1} = K_{n,n\pm1} , \tag{2}$$

where M_n is the mass at the site n and $K_{n,n\pm1}$ are the force constants, and in the case of Ferromagnetic magnons

$$\alpha_n = \frac{E}{S_n} - 2\left(\frac{S_{n+1}}{S_n} J_{n,n+1} + \frac{S_{n-1}}{S_n} J_{n,n-1}\right) , \quad t_{n,n\pm1} = -2J_{n,n\pm1} , \tag{3}$$

where S_n is the magnitude of spin at site n and $J_{n,n\pm1}$ are the exchange interactions. In the latter case, excitation amplitude $\{a_n\}$ has been modified by $\{S_n\}$ as $\tilde{a} = \tilde{S}^{\frac{1}{2}}\phi$. Kronig-penny models and an electron in continuous random potential can be considered in an analogous way [5], but we do not touch on them here.

Transfer matrix $\tilde{T}_n$ is then defined as

$$\begin{pmatrix} a_{n+1} \\ a_n \end{pmatrix} = \begin{pmatrix} \alpha_n/t_{n,n+1} , & -t_{n,n-1}/t_{n,n+1} \\ 1 , & 0 \end{pmatrix} \begin{pmatrix} a_n \\ a_{n-1} \end{pmatrix} = \tilde{T}_n \begin{pmatrix} a_n \\ a_{n-1} \end{pmatrix} . \tag{4}$$

When we put

$$\tilde{a}_n = \begin{pmatrix} a_n \\ a_{n-1} \end{pmatrix} \quad , \quad \tilde{T}_{(n+m,n)} = \tilde{T}_{n+m-1} \cdot \;\cdots\; \cdot \tilde{T}_{n+1} \cdot \tilde{T}_n \quad .$$

the equation is simply written as

$$\tilde{a}_{n+1} = \tilde{T}_n \cdot \tilde{a}_n \quad , \quad \tilde{a}_{n+m} = \tilde{T}(n+m,n) \cdot \tilde{a}_n \quad . \tag{6}$$

The assembly $\{\tilde{T}_i ; i = n, \ldots , n+m-1\}$ contains all the information about the relation between $\tilde{a}_n$ and $\tilde{a}_{n+m}$; it transfers correctly $\tilde{a}_n$ to $\tilde{a}_{n+m}$. The attractive feature of the representation (6) comes from the tridiagonal form of the Hamiltonian in our simple cases. Thus the study of the assembly $\{\tilde{T}_n\}$ provides us with a necessary and sufficient knowledge about the eigenvalues and the eigenstates.

3. Phase Character in Wave Functions

The method of transfer matrix described by (6) is useful to study the eigenvalues and eigenstates with some boundary conditions (10), and has been widely accepted.

We should mention here a quite general relation between the numbers of the eigenvalues and the nodes of the eigenstates. We start our discussion on phonon system with fixed boundary condition $a_0 = 0$, $a_{N+1} = 0$. Then N roots of the polynomial $a_{N+1}(\omega^2) \propto [\tilde{T}(N+1,1)]_{(1,1)}$ give the eigenvalues and thus real eigenvectors

$$\lambda_1 = \omega_1^2 \leqq \lambda_2 \leqq \cdots \leqq \lambda_N \quad , \quad (\{a_n(\lambda_i)\} \quad ; \quad i = 1,2,\ldots, N) \quad . \tag{7}$$

Instead of each wave function itself, sequence of state ratios

$$\{\zeta_n\} \quad ; \quad \zeta_n \equiv a_n/a_{n-1} \tag{8}$$

or a sequence of phases

$$\{\theta'_n\} \quad ; \quad \tan \tfrac{1}{2}\theta'_n = a_n/a_{n-1} \tag{8'}$$

is meaningful for discussing the nature of the wave function as the absolute value of each function has no physical meaning. The recurrence relation is obtained in the form of a continued fraction

$$\zeta_{n+1} = \alpha_n/t_{n,n+1} - t_{n,n-1}/\zeta_n \quad . \tag{9}$$

As seen in Fig.1, we have known that for each wave function the phase θ'_n increases monotonically with n along the chain, and hence we can understand each wave function well with the concept of phase in a quite analogous way like a continuous wave function.

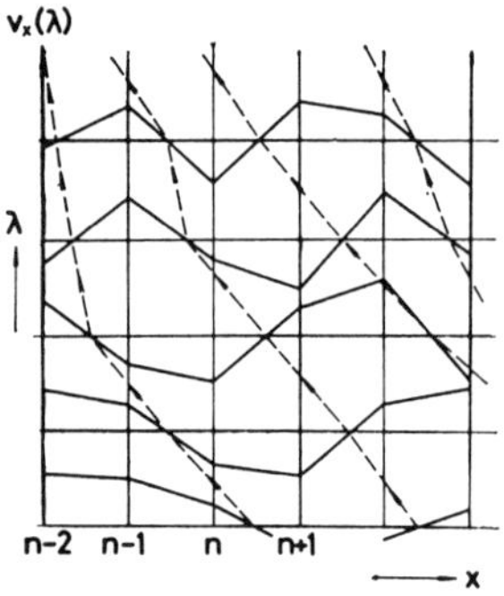

Fig.1. The shift of the phase angles and the nodes of the eigenstates $\{a_n(\lambda)\}$ with successively higher values of λ (from [3])

As a result it is also obvious that each eigenvector $\{a_n(\lambda_k)\}$ has k-1 nodes and

$$\theta'_N(\lambda_K) = 2\pi \cdot k \quad , \tag{10}$$

that is, the state density $D(\lambda)$ and the accumulated state density $M(\lambda)$ are written as

$$M(\lambda) = \int_{-\infty}^{\lambda} D(\lambda')d\lambda' = \lim_{N\to\infty} \frac{1}{2\pi N}\, \theta'_N(\lambda) \quad . \tag{11}$$

We can thus define the spectral gap as an interval of λ where

$$D(\lambda) = \lim_{N\to\infty} \frac{1}{2\pi N} \frac{d\theta'_N(\lambda)}{d\lambda} = 0 \quad . \tag{12}$$

From a theoretical point of view it is sometimes more convenient to use a representation which diagonalizes one kind of transfer matrix as

$$\tilde{Q}^{(0)} = \tilde{P}\tilde{T}^{(0)}\tilde{P}^{-1} = \begin{pmatrix} e^{2i\beta} & 0 \\ 0 & e^{-2i\beta} \end{pmatrix} \quad . \tag{13}$$

and

$$\tilde{Q}_n = \tilde{P}\tilde{T}_n\tilde{P}^{-1} = \begin{pmatrix} A_n & B_n \\ B^*_n & A^*_n \end{pmatrix} \quad , \quad \tilde{x}_n = \begin{pmatrix} X_n \\ X_{n-1} \end{pmatrix} = \tilde{P}\tilde{a}_n \quad . \tag{14}$$

Then each new state ratio of complex value

$$Z_n = X_n/X_{n-1} \tag{15}$$

has unit modulus and is a Cayley-type transform of ζ_n, which is a conformal mapping and maps the whole real axis ζ onto the unit circle of a complex plane. Accordingly, new phase angles

$$\{\theta_n \; ; \quad e^{i\theta_n} = Z_n\} \tag{16}$$

have the same properties as $\{\theta_n'\}$ has, as is seen in (10), (11) and (12). Obviously $\tilde{x}_{n+1} = \tilde{Q}_n\tilde{x}_n$ and the new state ratios have the recurrence relation

$$Z_{n+1} = (A_nZ_n + B_n)/(B_n^*Z_n + A_n^*) \quad . \tag{17}$$

which is a conformal mapping which transforms any circle to another circle and is called Möbius transform. The properties of the transform (17) are discussed in the next section.

4. Special Frequencies (Energies)

In 1960 Dean calculated $M(\lambda)$ of finite, diatomic and isotopic disordered chains with masses M_1 and M_2 by counting the negative factor of ζ_n by the successive numerical calculation of (9) as each negative ζ_n corresponded to a node. A result is shown in Fig.2. It was surprising that quite characteristic feature was seen in its fine structure. Many peaks of the state density were explained by him as narrow impurity bands of clusters. However, there was a doubt that these seriously deep valleys might be the feature of a finite system. The answer for the doubt was given first by Mastuda [11] that there are some special frequencies in infinitely long chains at which state densities were exactly zero. A sufficient condition on which the special frequencies appear have been studied extensively by Hori [3,12].

Consider for example diatomic isotopic ($t_{n,n\pm1} = 1$) disordered chains with masses $M^{(1)}$ and $M^{(2)}$ ($M^{(1)} < M^{(2)}$). We have two different transfer ma-

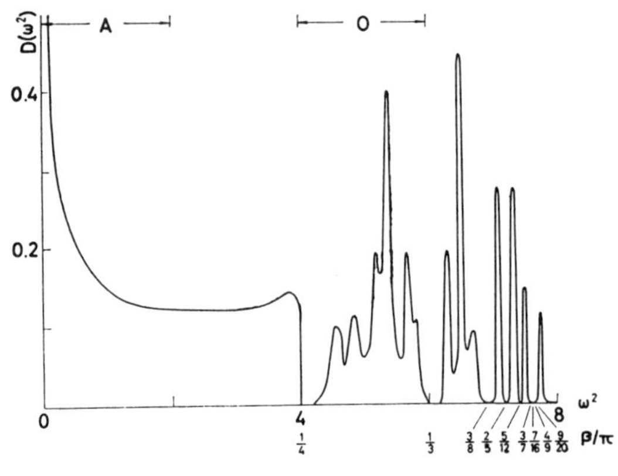

Fig.2. The squared-frequency spectrum of the isotopically disordered chain with mass ratio 2. The deep valleys correspond to particular rational values of β/π (from [1])

trices $\tilde{T}^{(1)}$ and $\tilde{T}^{(2)}$ defined in (2) and (4) corresponding to the kind of mass at the site. The system defined by a sequence of masses $\{M_n\}$ is described also by the sequence of transfer matrices $\{\tilde{T}_n\}$ or $\{\tilde{Q}_n\}$, where $\tilde{P}$ is chosen as $\tilde{Q}^{(1)} = \tilde{P}\ \tilde{T}^{(1)}\ \tilde{P}^{-1}$ is diagonalized in the form (13) for $\lambda = \omega^2$ in the band of pure $M^{(1)}$ system. Almost all sequences of the transfer matrices $\{\tilde{Q}_n\}$ can be reconstructed as sequence of the elements

$$\tilde{Q}^{(2m)} \equiv \tilde{Q}^{(2)}(\tilde{Q}^{(1)})^m \quad (m = 0, 1, 2, \dots) \quad . \tag{18}$$

Here we should mention that the transfer matrices $\tilde{Q}$ can be classified into three groups with corresponding eigenvalues of two different real values, two complex values, or two degenerated real values

$$\text{hyperbolic} \ ; \quad (\tfrac{1}{2}\,\mathrm{Tr}\tilde{Q})^2 > \det \tilde{Q} \tag{19}$$

$$\text{elliptic} \ ; \quad (\tfrac{1}{2}\,\mathrm{Tr}\tilde{Q})^2 < \det \tilde{Q} \tag{20}$$

$$\text{parabolic} \ ; \quad (\tfrac{1}{2}\,\mathrm{Tr}\tilde{Q})^2 = \det \tilde{Q} \quad . \tag{21}$$

Characteristic features of transforming Z_n with these three types of $\tilde{Q}$ are shown in Fig.3. Flow of the mapping is shown in the whole Z planes with arrows though only mapping of unit circles is physically necessary for us. If we imagine a regular chain with equal mass M the spectral gaps are obtained obviously from the condition (19) because phase angle θ_n converges with increasing n and hence no eigenstate can exist. It has also been confirmed that in the band frequency regions are obtained from the condition (20) and also the band edges just from (21).

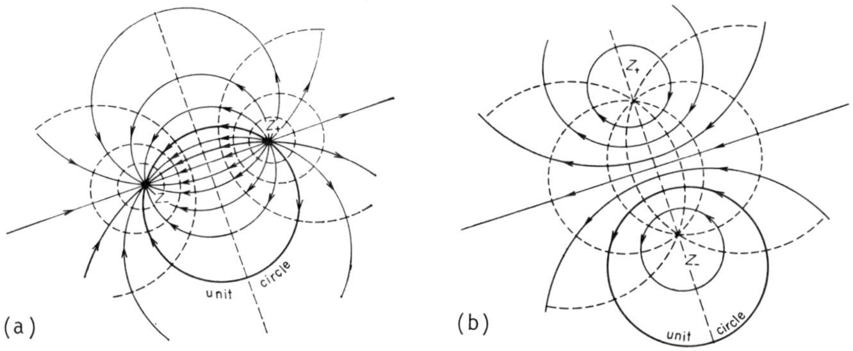

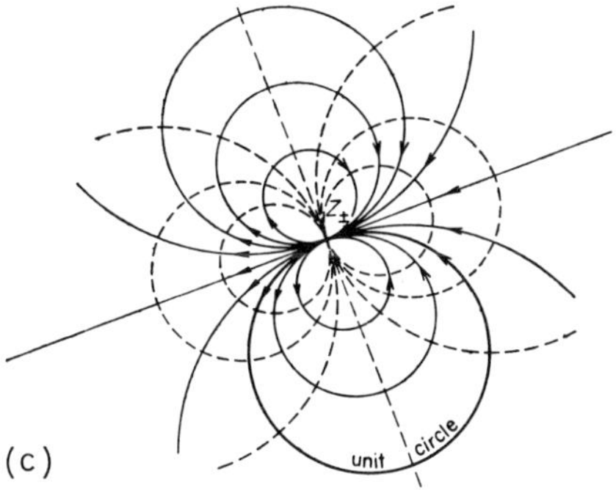

Fig.3. The flow lines and families of successively transformed circles in the hyperbolic case (a), elliptic case (b) and parabolic case (c). The unit circles are depicted by thick lines (from [12])

Now we can consider the spectral gaps of any random sequence of (18). When all $\tilde{Q}^{(2m)}$ are hyperbolic for some λ and also all sink points $Z_-^{(2m)}$ and all source points $Z_+^{(2m)}$ are grouped respectively in two separate regions on the unit circle, the phase θ_n also can not increase with increasing n after Z_n drops into the region of sink points. Hence the squared-frequency λ belongs to a spectral gap of the disordered system. When the condition is satisfied only on discrete point values of λ we call them special frequencies. We thus arrive at a generalized version of Saxon-Hutner-type theorem which is, quite intuitively stated, that the common gaps of all constituent regular systems are the gaps of the disordered systems. Constituent regular systems are in our case periodic systems with unit cells described by $\tilde{Q}^{(2m)}$ (m = 0, 1, 2, ...). In our case we have an infinite sequence of discrete frequencies determined by

$$2\beta(\lambda) = (P - q)\pi/P$$

$$q \leqq (M^{(2)}/M^{(1)} - 1) \leqq \cot^2\{\pi/2(q + 1)\} \quad , \tag{22}$$

where integers P and q are subject only to the restriction that they should be relatively prime. The sequence of special frequencies clearly explains the deep valley in the Fig.2. The special frequencies are special band gaps which have infinitesimally small ranges. It is however expected from the construction of the theory that each special frequency has Lifshitz-type band edge at least on one side and hence physically reasonable band gaps can be expected in real disordered systems. The expectation has been confirmed in infinite disordered system by Gubernatis and Taylor [13]. They used the exact method of Dyson and solved a functional equation for binary alloys.

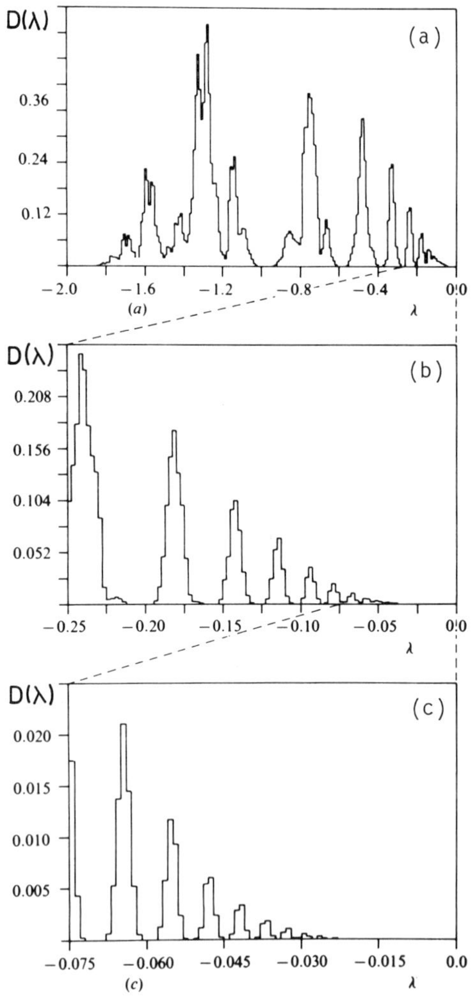

Fig.4. The spectrum of disordered binary alloy showing fine structure near the special frequencies. Note successive enlargements of the scale in (a), (b), (c) as tail is entered [13]. From [4]

Now we have come to understand the feature of special frequencies. But it is not clear how these affect physical phenomena. A related problem is discussed in the next section.

5. Localization of Eigenstates

After Anderson's pioneering work [14], MOTT and TWOSE [2] conjectured that in the case of one-dimension eigenfunctions of electrons are localized in the whole allowed energy region and absence of diffusion takes place, that is, when $a_n(0) = \delta_{n,N}$ at $t = 0$ for any fixed N then

$$\lim_{t\to\infty} |a_N(t)|^2 > 0 \quad . \tag{23}$$

We have known that some ensemble-averaging procedures often destroy the essential character of the localization problem. The probability distribution itself should be argued to avoid the difficulty.

After Borland's probabilistic argument [15] on the exponential growth property of particular solution, Matsuda and Ishii [5,16] made an extensive study on the property using Furstenberg's convergent theorem [17] on the product of random matrices. One of the essential parts of the study is to prove the exponential growth property of particular solution, that is,

$$P\left\{\lim_{n\to\infty} \frac{1}{n} \log \| [\prod_{\ell=1}^{n} \tilde{T}_\ell(\lambda)\ \tilde{a}_1 \| = 2\gamma(\lambda) > 0\right\} = 1 \tag{24}$$

on the ensemble of systems, where $\| \tilde{a}_{n+1} \|$ means the vector norm $[a^2_{n+1}(\lambda) + a^2_n(\lambda)]^{\frac{1}{2}}$.

The property (24) means for any $\tilde{a}_1 \neq 0, \| \tilde{a}_1 \| = 1$ and given λ that the wave function $|a_n(\lambda)|$ increases like $e^{\gamma n}$ for sufficiently large n with probability 1 on the ensemble, and hence almost all wave functions cannot be eigenfunctions. The rate of exponential growth $\gamma(\lambda)$ is given as an average value of $\log(\| \tilde{T}\tilde{a} \| / \| \tilde{a} \|)$,

$$\gamma(\lambda) = \iint \log \frac{\| \tilde{T}\tilde{a} \|}{\| \tilde{a} \|} d\mu(\tilde{T}) d\nu(\bar{\tilde{a}}) \tag{25}$$

where μ is a probability distribution of the transfer matrices, that is, a measure on SL(2,R). A stationary measure $\nu(\bar{\tilde{a}})$ for μ on P^1 is given by the following self-consistent equation which is equivalent to the functional equation of DYSON [18] and SCHMIDT [19],

$$\nu(\bar{\tilde{a}}) = \int \nu(\overline{\tilde{T}^{-1}\tilde{a}})d\mu(\tilde{T}) \tag{26}$$

where $\bar{\tilde{a}}$ (and $\overline{\tilde{T}^{-1}\tilde{a}}$) denotes the projected $\tilde{a}$ (and $\tilde{T}^{-1}\tilde{a}$) on P^1.

The property (24) is proved, on some additional condition, if the product of transfer matrices corresponding to a fraction of the chain is hyperbolic. For example, when $\tilde{T}_{(1)}(m)(\lambda) = [\tilde{T}^{(1)}]^m\tilde{T}^{(2)}$ is hyperbolic on our isotopically disordered systems considered in the last section, we have

$$\mathrm{Tr}\ \tilde{T}_{(1)}(m)(\lambda) > 2 \tag{27}$$

as det $\tilde{T}_{(1)}(m)(\lambda) = 1$ in this case. We have then two different real eigenvalues and corresponding eigenvectors

$$e^{+\delta(\lambda)} > e^{-\delta(\lambda)}[\delta(\lambda) > 0] \quad \text{and} \quad \tilde{x}_+ \ , \quad \tilde{x}_- \quad . \tag{28}$$

As arbitrally vector $\tilde{x}$ can be represented as a linear combination of these eigenvectors the operation of $\tilde{T}_{(1)}(m)$ on $\tilde{x} = C_+\tilde{x}_+ + C_-\tilde{x}_-$ gives

$$\tilde{x}' = \tilde{T}_{(1)}(m)\tilde{x} = C_+e^{+\delta}\tilde{x}_+ + C_-e^{-\delta}\tilde{x}_- \quad . \tag{29}$$

For sufficiently large n we have thus

$$[\tilde{T}_{(1)}(m)]^n\tilde{x} \cong C_+e^{+n\delta}\tilde{x}_+ \quad . \tag{30}$$

The expressions (29) and (30) explain the feature in the Fig.3a. If we imagine a regular system with a unit cell corresponding to $\tilde{T}_{(1)}(m)(\lambda)$ we have obviously the property (24) with $\gamma(\lambda) = \delta(\lambda)$ and this is the condition that the λ belongs to a spectral gap of the constituent regular system. We thus reconfirm that in the case of regular system (19) is the condition on which the parameter λ belongs to a spectral gap. It is just the condition on which the property (24) holds in the disordered systems.

In the case of disordered systems it was proved that a product of some finite sequences of transfer matrices [like $\tilde{T}_{(1)}(m)(\lambda)$] can always be hyperbolic when any two transfer matrices such as $\tilde{T}^{(1)}$, $\tilde{T}^{(2)}$ have no common eigenvector, even if neither $\tilde{T}^{(1)}$ nor $\tilde{T}^{(2)}$ are hyperbolic. This fact guarantees [5,16] the property (24) for all frequencies and energies for various one-dimensional systems, including models in Sect.2 with diagonal randomness only and Kronig-Penny models. As a result it has been proved, when the property (24) is guaranteed for the systems, that we have for the Green function

$$|G_{n,m}(\lambda - i0)| < O(\exp\{-\gamma(\lambda)|n - m|\}) \tag{31}$$

in the limit $|n - m| \to \infty$ and that weak absence of diffusion [5] occurs, that is, when $a_n(0) = \delta_{n,N}$ at $t = 0$ then

$$\int_0^\infty dt|a_N(t)|^2 = \infty \quad . \tag{32}$$

A generalization of Furstenberg's theorem was carried out by GODA [20-22]. By using it we can treat the convergence property of a Markov-chain of the transfer matrices which corresponds to systems with both diagonal and off-diagonal randomnesses. As a result almost the same properties as (31) and (32) have been obtained in a wide variety of one-dimensional systems mentioned in Sect.2.

ABOU-CHACRA et al. [23] presented a self-consistent theory of localization and studied Bethe lattice. ABRAHAM et al. [24] presented a scaling theory of electronic systems and stated that almost all eigenstates were localized in one and also two dimensions. Golidsheid et al. [25] proved that a Schrödinger operator with random potential has purely point spectrum in the case of one dimension, and MOLCHANOV [26] proved that the eigenfunctions corresponding to the point spectrum decrease exponentially. The last two facts guarantee the absence of diffsuion (23). Thus we can conclude that the Mott and Twose's conjecture is almost justified in these idealized one-dimensional disordered systems.

6. Concluding Remarks

We have seen two characteristic features of one-dimensional disordered systems. Both of them can be understood by the hyperbolic nature of the transfer matrices or some product of transfer matrices.

Concerning the relation between the spectral density and localization of eigenstates, we have known Molchanov's work and also HERBERT and JONES's [27] and THOULESS's [28] relation. Localization of eigenstates is related further to the conductivity [29]. About the direct relationship between the special frequencies and experimentally observable physical phenomena, however, we have very little at present.

Concerning one-dimensional linear systems we have seen that Mott and Twose's conjecture have been justified, the pseudo-one-dimensional conductors, however, bring us a new type of problem [8,9,30]. How can one-dimensional systems be conductive? Interaction between particles, nonrigidity of the lattice and residual interaction in three dimensions are some candidates for the mechanism of the conductivity. The problem stimulates us though we have not discussed it here.

Acknowledgements. The author would like to express his sincere thanks to Professor J. Hori for introducing him to the theory of disordered systems. It is to be regretted that he cannot attend this conference.

References

1 P. Dean: Proc. R. Soc. London A*254*, 507 (1960)
2 N.F. Mott, W.D. Twose: Adv. Phys. *10*, 107 (1961)
3 J. Hori: *Spectral Properties of Disordered Chaines and Lattices* (Pergamon, New York 1968)
4 J.M. Ziman: *Models of Disorder* (Cambridge Univ. Press, Cambridge 1979)
5 K. Ishii: Prog. Theor. Phys. Suppl. *53*, 77 (1973)
6 F. Yonezawa, K. Morigaki: Prog. Theor. Phys. Suppl. *53*, 1 (1973)
7 R.J. Elliott, J.A. Krumhansl, P.L. Leath: Rev. Mod. Phys. *46*, 465 (1974)
8 A.J. Epstein, J.S. Miller: Sci. Am. *241*, 48 (1979)
9 R.J. Birgeneau, G. Shirance: Phys. Today (1978), 32
10 H. Matsuda, K. Okada, T. Takase, T. Yamamoto: J. Chem. Phys. *41*, 1527 (1964)
11 H. Matsuda: Prog. Theor. Phys. *31*, 161 (1964)
12 J. Hori: Prog. Theor. Phys. Suppl. *36*, 3 (1966)
13 J.E. Gubernatis, P.L. Taylor: J. Phys. C*6*, 1889 (1973)
14 P.W. Anderson: Phys. Rev. *109*, 1492 (1958)
15 R.E. Borland: Proc. R. Soc. London A*274*, 529 (1963)
16 H. Matsuda, K. Ishii: Prog. Theor. Phys. Suppl. *45*, 56 (1970)
17 H. Furstenberg: Trans. Am. Math. Soc. *108*, 377 (1963)
18 F. Dyson: Phys. Rev. *92*, 1331 (1953)
19 H. Schmidt: Phys. Rev. *105*, 425 (1957)
20 M. Goda: Prog. Theor. Phys. *62*, 608 (1979)
21 M. Goda: submitted to Prog. Theor. Phys. Suppl.
22 M. Goda: this conference.
23 R. Abou-Chacra, P.W. Anderson, D.J. Thouless: J. Phys. C*6*, 1734 (1973)
24 E. Abrahams, P.W. Anderson, D.C. Lecciardello, T.V. Ramakrishnam: Phys. Rev. Lett. *42*, 673 (1979)
25 I.Ya. Golidsheid, S.A. Molchanov, S.A. Pastur: Funct. Anal. Appl. *11*, 1 (1977)
26 S.A. Molchanov: Izv. AHCCCP Ser. Mat. *42*, 70 (1978)
27 D.C. Herbert, R. Jones: J. Phys. C*4*, 1145 (1971)
28 D.J. Thouless: J. Phys. C*5*, 77 (1972)
29 B.I. Halperin: Adv. Chem. Phys. *13*, 123 (1967)
30 J.T. Devreese, R.P. Evrard, V.E. van Doren: *Highly Conducting One-Dimensional Solids* (Plenum, New York 1979)

Excitation Dynamics in Random One-Dimensional Systems

S. Alexander
Racah Institute of Physics, The Hebrew University, Jerusalem, Israel

J. Bernasconi and W.R. Schneider
Brown Boveri Research Center, CH-5405 Baden-Dättwil, Switzerland

R. Orbach
Department of Physics, University of California
Los Angeles, CA 90024, USA

1. Introduction

In a number of recent publications, [1] - [5], we have discussed the asymptotic form of the dynamics of a general type of random one-dimensional chains. The equations we discuss are of the form

$$C_n \frac{dV_n}{dt} = W_{n,n+1}(V_{n+1}-V_n) + W_{n,n-1}(V_{n-1}-V_n) , \qquad W_{n,n+1} = W_{n+1,n} , \tag{1}$$

where the C_n or the $W_{n,n+1}$ are independent positive random variables. Equations of this type arise in a variety of physical contexts. They can represent a master equation for hopping-type transport over random barriers (the $C_n=1$, the $W_{n,n+1}$ random hopping rates); a master equation for excitation transfer along a one-dimensional array of traps of random depth (the C_n random Boltzmann factors, the $W_{n,n+1}=1$); an electric transmission line (the C_n random capacitors or the $W_{n,n+1}$ random conductances); a random Heisenberg ferromagnetic chain at low temperatures (the $C_n=1$, V_n representing spin wave destruction operators, and the $W_{n,n+1}$ random near-neighbor exchange integrals). Replacing dV_n/dt in (1) by d^2V_n/dt^2, they represent a harmonic chain with random masses (the C_n) or random force constants (the $W_{n,n+1}$).

This mathematical problem has a long history. The first successful attack was by DYSON [6]. We also mention the classical contributions of SCHMIDT [7] and DOMB [8]. We note that (1) implies a conservation law

$$\textstyle\sum_i Q_i = \text{const.} , \qquad Q_i = C_i V_i , \tag{2}$$

which is not violated by the randomness. In this, and the positiveness of the W and C, it differs from the one-dimensional Schroedinger equation or from DYSON's case I.

Our work differs from that of most investigators in that we have focused on the asymptotic properties of the solutions for long times or low frequencies. In this limit, the solutions exhibit some universal scaling properties and one can determine certain quantities exactly.

The properties in which we are interested are determined by the small W and large C limits of their respective random distributions, $\rho(w)$ and $r(c)$. We classify them accordingly among three representative types of probability distributions.

Class (a): $\rho(x)$ such that

$$\langle x^{-1}\rangle = \int_0^\infty dx\ x^{-1}\ \rho(x) < \infty \tag{3a}$$

or equivalently

$$\langle y\rangle = \int_0^\infty dy\ y\ r(y) < \infty\ , \qquad y = 1/x\ . \tag{4a}$$

Class (b):

$$\rho(x) \to \text{constant, as } x \to 0\ . \tag{3b}$$

$$r(y) \to y^{-2}\ , \text{ as } y \to \infty\ . \tag{4b}$$

Class (c):

$$\rho(x) \to x^{-\alpha} \text{ as } x \to 0\ , \qquad 0 < \alpha < 1\ . \tag{3c}$$

$$r(y) \propto y^{-(2-\alpha)} \text{ as } y \to \infty\ , \qquad 0 < \alpha < 1\ . \tag{4c}$$

The interesting situations are those where $x = W$ (or $1/C$), or alternatively, $y = C$ (or $1/W$). These criteria are sufficient to determine the asymptotic behavior.

There are two other distributions which are of some interest. The first is the interrupted chain. This corresponds to a situation where $\rho(x)$ has the form

$$\rho(x) = a\ \delta(x) + \rho_1(x)\ . \tag{5a}$$

We have discussed such situations elsewhere [9]. The other represents the limit of case c when $\alpha \to 1$. Then,

$$\rho(x) \to (x|\ell nx|^{\mu})^{-1} \qquad \text{as} \qquad x \to 0\ ; \qquad \mu > 1\ . \tag{5b}$$

We have not yet examined this possibility.

Class (a) is the normal (or regular) case. It implies a frequency independent diffusion constant, a nonvanishing dc conductivity, and for example a constant velocity of sound. It is convenient to emphasize the analogy to critical phenomena. We write

$$\langle n^2\rangle = \xi^2 \propto t^{2\nu} \simeq \omega^{-2\nu}\ , \tag{6}$$

where $\xi(t)$ is, for example, the distance traveled by a diffusing particle in time t. On the one hand, in the normal or ordered case, one finds the classical index

$$\nu = 1/2\ . \tag{7}$$

For class (c), on the other hand, one finds

$$\nu = (1-\alpha)/(2-\alpha)\ . \tag{8}$$

This leads to anomalous behavior. Class (b) is marginal with logarithmic corrections.

Anomalous frequency dependent transport properties are observed almost universally in anisotropic quasi-one-dimensional materials [10], [11]. We believe this reflects the extreme sensitivity to randomness in one dimension. A particle confined to a one-dimensional channel can be trapped even by very rare high barriers (e.g., due to defects). They are always present, and they tend to dominate the low frequency transport properties. For two- or three-dimensional systems, barriers have no effect on the asymptotic transport properties [12]. Such situations are treated naturally in our formalism. To compare with experiment, it is essential to perform the mapping of the physical system onto the mathematical model carefully. One also must treat crossover effects properly because there are no experiments in truly one-dimensional systems, with infinite barriers, and at zero electric field. This will be the main purpose of Section 4.

To avoid confusion, we add a comment on the relationship of the problem we discuss with the electronic localization and metallic conductivity problem [15] - [17]. The two problems differ in fundamental ways. The equations are different because of the conservation law (2) which has no analogy in the Schroedinger equation for random potentials. The result is that our equations (1) have a (trivial) localization edge at the lowest eigenvalue ($\varepsilon = 0$) for V_n = constant. Near the edge, the eigenfunctions are (we believe) localized to a length ξ [$= \xi(\varepsilon)$, Eq. (6)]. The physical interpretation of the quantities (and therefore their relation to experiments) are of course different. As a result, one is not really interested in the fact that the eigenfunctions are localized when interpreting experiments. [They are probably also localized in the normal case (a).] The transport properties are dominated by ν, i.e., by the manner in which ξ changes as one approaches the edge. We are also emphasizing a different type of randomness dominated by large localized fluctuations in real space, rather than fluctuations in phase along the chain [17].

The problem we are considering is, presumably, of less fundamental interest. We do, however, have the advantage that we are dealing with a phenomenon which is readily observable and, in fact, quite common experimentally.

The remainder of this paper is constructed as follows. In Section 2 we describe briefly our fundamental approach to the solution of (1) and the exact asymptotic results we were able to obtain. Most of this material has been discussed in previously published work, and in a forthcoming review article [18]. In Section 3, we describe a qualitative scaling approach ([3], [18]) which allows us to extend our discussion to more complex situations, and in particular to a (qualitative) discussion of crossover effects. We discuss in Section 4 three relevant crossover effects, namely: 1) the effect of having a highest barrier [small w cutoff in $\rho(w)$], resulting in a crossover from class (c) [or (b)] to class (a) behavior; 2) interchain transfer leading to a crossover from a one-dimensional to a (normal) three-dimensional behavior; 3) the effect of finite electric fields for which one predicts a crossover of the type discussed by SCHER, MONTROLL, and ALEXANDER [19].

Our analysis has been applied successfully to the interpretation of the conductivity in the one-dimensional superionic conductor Hollandite [4], [20]. Similar effects are frequently observed in quasi one-dimensional electronic conductors and have been studied in detail by GRUNER and collaborators [10] in $Qn(TCNQ)_2$. We discuss in Section 5 the mapping of an interrupted metallic strand model, relevant to these materials, onto a master equation (1).

2. Exact results

Our procedure is similar but not identical to DYSON's [6]. We discuss the Laplace transform of (1), with ω the conjugate variable to t. One defines two variables:

$$G_n = W_{n,n+1}(V_n - V_{n+1})/V_n \tag{9a}$$

$$\phi_n = \omega R_n = (C_n)^{-1}(I_{n-1,n} - I_{n,n+1})/I_{n-1,n} \quad , \tag{9b}$$

where the V_n are the variables in (1) and the $I_{n,n+1}$ are the currents on the relevant branches associated with the conserved quantity Q [defined in (2)]. The quantity G_n is essentially the line admittance to the right as seen from the point immediately to the right of junction n. Similarly, R_n is the line impedance to the right as seen from the left side of n. These quantities obey the recursion relations

$$G_n = \omega/[\omega/W_{n,n+1} + \phi_{n+1}] \tag{10a}$$

$$\phi_n = \omega/(\omega C_n + G_n) \quad . \tag{10b}$$

When $W_{n,n+1}$ is the random variable ($C_n \equiv C$), we eliminate the ϕ_n and absorb C into the units of ω. Similarly, for random C_n ($W_{n,n+1} = W$), we absorb W and eliminate the G_n. In both cases, one is led to the same integral equation for the distribution of the relevant quantity.

$$f_\omega(x) = \int_0^\infty dx' \, f_\omega(x') \int_0^\infty dy \, \rho(y) \, \delta\left\{x - \left[\frac{1}{y} + \frac{1}{\omega + x'}\right]^{-1}\right\} \tag{11}$$

where x is either G or ϕ, and y is w or c^{-1}, respectively. By looking at the Stieltjes transform of this equation, BERNASCONI, SCHNEIDER, and WYSS [5] were able to determine the asymptotic form (for small ω) of the solutions of this equation exactly for classes (a) and (c). Quite generally, they find a scaling form for the solutions

$$f_\omega(x) = \frac{1}{\varepsilon(\omega)} \, h\left[\frac{x}{\varepsilon(\omega)}\right] \quad . \tag{12}$$

For class (a) one finds

$$h(x) = \delta(x-1) \tag{13}$$

and

$$\varepsilon(\omega) = \langle 1/W \rangle^{-1/2} \, \omega^{1/2} \quad . \tag{14a}$$

For class (c) distributions,

$$\varepsilon(\omega) = \omega^{1/(2-\alpha)} \quad . \tag{14c}$$

For class (c) distributions, h(x) depends on α explicitly, and can be expressed in terms of the H-function of FOX [21]:

$$h^{\alpha}(x) = \frac{\beta\gamma}{\Gamma(\beta)} H^{20}_{12}\left(\gamma x \middle| \begin{matrix} & (-1,1) & \\ (-\beta,\beta) & & (0,\beta) \end{matrix}\right) \quad , \tag{15}$$

where

$$\beta = 1/(2-\alpha); \qquad \gamma = [\beta^2\Gamma(\alpha)]^{\beta} \quad . \tag{16}$$

An effective medium approach [1], [18] which assumes the form (13) for $h(x)$, and the scaling approach we describe below, give the correct scaling behavior for $\varepsilon(\omega)$, Eqs. (14a) and (14c), but not Eq. (15). These approaches are thus correct in a qualitative, scaling sense, but not quantitatively. For case (b), these latter approaches lead to

$$\varepsilon(\omega) = [2/\rho(0)]^{1/2} [\omega/(-\ln\omega)]^{1/2} \quad . \tag{14b}$$

There are arguments to believe [18], [22] that this [and (13)] are indeed correct for the (limiting) case (b).

Once $f_{\omega}(x)$ is known, one can calculate the average diagonal part of the Green's function [for (1)] exactly, that is $\langle V_0(\omega)\rangle$. This also gives the averaged density of states and sets a lower limit on the $\langle V_n(\omega)\rangle$ versus n decay length.

We now assume that for $\omega \to 0$ the n dependence of $\langle V_n(\omega)\rangle$ can be described by a single correlation length $\xi(\omega)$. This can be expressed by the general scaling hypothesis [5], [18]

$$\langle V_n(\omega)\rangle = \langle V_0(\omega)\rangle F^{\alpha}[n/\xi(\omega)] \tag{17}$$

which immediately leads to

$$\xi(\omega) = \omega^{-\nu} \tag{18}$$

where ν is given by (7) and (8). One thus confirms the corresponding effective medium and scaling results. Together with standard fluctuation dissipation arguments [4], [5], [8] equation (17) also relates the frequency dependent conductivity $\sigma(\omega)$ to $\langle V_0(\omega)\rangle$

$$\sigma(\omega) \propto [-i\omega\langle V_0(-i\omega)\rangle^2]^{-1} \quad . \tag{19}$$

Our exact results for $\langle V_0(\omega)\rangle$ then lead to the following expressions for $\sigma(\omega)$ (in the random W case):

$$\sigma(\omega) = \frac{n_0 e^2}{k_B T} \langle 1/W\rangle^{-1} \quad , \qquad \text{class (a)} \; ; \tag{20a}$$

$$\sigma(\omega) = \frac{n_0 e^2}{k_B T} \frac{2}{\rho(0)} [-\ln(i\omega)]^{-1} \quad , \qquad \text{class (b)} \; ; \tag{20b}$$

$$\sigma(\omega) = \frac{n_0 e^2}{k_B T} D_0^{(\alpha)} [C_0^{(\alpha)}]^{-2} (-i\omega)^{\alpha/(2-\alpha)} \quad , \qquad \text{class (c)} \; ; \tag{20c}$$

where $D_0^{(\alpha)}$ and $C_0^{(\alpha)}$ depend on the detailed form of $F^{\alpha}(x)$ and $h^{\alpha}(x)$ respectively. Equation (20b) is an effective medium result. For brevity we shall not discuss the somewhat more complex random C case [18].

We would like to add one further comment here. All the expressions we give are the leading asymptotic terms. It is not obvious a priori how small ω must become to make the correction terms negligible. In the effective medium or scaling approximation one finds correction to $\varepsilon(\omega)$ which can be computed in a straightforward way [23]. They become increasingly important as $\alpha \to 0$. There are, however, additional problems which are only partly resolved. An exact asymptotic expansion of $f_\omega(x)$ to higher orders can be performed [22] and the results differ from the effective medium results. We also note that while $f_\omega(x)$ determines $\langle V_0(\omega)\rangle$ and the density of states, it is not sufficient to determine $\langle V_n(\omega)\rangle$ and therefore $\sigma(\omega)$ [without the additional assumption (17)]. It does not even determine the variance $\langle V_0{}^2(\omega)\rangle$.

3. Scaling arguments

The basic ideas of the arguments we wish to present here were described by ALEXANDER and BERNASCONI [3]. It is actually possible to make the argument somewhat more rigorous in the sense that one can show that the average behavior we calculate does indeed give the leading asymptotic power laws correctly. We have done this in [18]. Because we are mainly interested in the effect of barriers on transport properties, we shall discuss the random W case and use the language of a diffusion problem.

The basic idea is a consistency argument. We are interested in the effect of rare high barriers on the motion of a particle. During a time t, or at frequency $\omega \sim 1/t$, a particle will only travel some finite distance $\xi(t)$ on the average. If the high barriers are very rare, most segments of length ξ do not contain any of them. Thus, they cannot seriously affect particle motion on this time (length) scale. More precisely, one should be able to compute quantities like the mean square displacement $\langle x^2\rangle$ as though these high barriers did not exist. This is sufficient to determine the diffusion constant and conductivity. Of course, if one were to look in detail at the probability distribution, one would find that some particles are indeed always trapped by the rare high barriers so that they do influence the shape of the distribution but not the long time averages. We do require that fluctuations in the time taken to get across the distance ξ (assuming there are no very high barriers on the segment) should not be too large. We shall try to make this argument self consistent.

Consider first a class (a) distribution. We divide it into segments of length ℓ (in units of the lattice constant). The W are transfer rates (or conductances) linked in series. One therefore has to add their inverses. Thus, at low frequencies, the coarse grained chain of segments will have conductances

$$W_\ell = \left(\sum_{i=1}^{\ell} \frac{1}{W_i} \right)^{-1} \quad . \tag{21}$$

On the average, one has

$$\overline{W_\ell} = \frac{1}{\ell} \langle 1/W\rangle^{-1} \quad . \tag{22}$$

Thus, the average conductance per step on the original chain is

$$W_{av} = \langle 1/W\rangle^{-1} \quad . \tag{23}$$

However, if we have chosen a small ℓ there are still large fluctuations in the coarse grained conductances W_ℓ. They disappear when we choose a large enough ℓ, say L:

$$L > \frac{\langle 1/W^2 \rangle}{\langle 1/W \rangle^2} = \xi \quad . \tag{24}$$

This defines a length scale ξ for the class (a) distribution. When we look at the chain on a scale larger than ξ the randomness is averaged out and the chain looks as though we had an ordered chain with the average conductances (23).

Consider now a class (c) distribution for which $\langle 1/W \rangle$ is not defined. We choose some small cutoff $W_{c.o.}$. The overall density of W smaller than $W_{c.o.}$ on the chain is small

$$p_<(W_{c.o.}) = \int_0^{W_{c.o.}} \rho(w)\, dw \propto W_{c.o.}^{1-\alpha} \quad . \tag{25}$$

Thus, these small W are not important when one considers distances ξ,

$$\xi \leq [p_<(W_{c.o.})]^{-1} \propto W_{c.o.}^{-(1-\alpha)} \quad . \tag{26}$$

Application of (24) to the cutoff distribution leads to the reverse condition,

$$\xi \geq W_{c.o.}^{-(1-\alpha)} \quad . \tag{27}$$

Thus, both conditions are roughly obeyed if we associate a unique $W_{c.o.}(\xi)$ with the length ξ,

$$W_{c.o.}(\xi) \propto \xi^{-1/(1-\alpha)} \quad . \tag{28}$$

For distances ξ we can also calculate W_{av} with this cutoff.

$$W_{av}(\xi) \simeq \langle 1/W \rangle^{-1}_{W_{c.o.}(\xi)} \simeq \xi^{-\alpha/(1-\alpha)} \quad . \tag{29}$$

Finally, since randomness is not important for the truncated distribution on a scale ξ, one has

$$\xi^2 \simeq W_{av}(\xi)\, t \simeq W_{av}(\xi)/\omega \quad , \tag{30}$$

and using (29),

$$\xi(\omega) \simeq t^{\nu} \simeq \omega^{-\nu} \; ; \qquad \nu = (1-\alpha)/(2-\alpha) \quad . \tag{31}$$

The conductivity and diffusion constant on this length scale will be proportional to $W_{av}(\xi)$. Thus, combining (29) and (31), one finds

$$\sigma(\omega) \propto \omega^{\alpha/(2-\alpha)} \tag{32}$$

in agreement with the results obtained previously, (20c), [4], [5].

4. Crossover effects

It is straightforward to apply the above considerations to situations which are only approximately singular.

4.1 Low W cutoff

Obviously the assumption that the barriers can be infinitely high (W arbitrarily small) cannot be taken literally. There must be a smallest W (= W_{min}). Strictly speaking, this leads to a class (a) situation. There may be a considerable range, however, in which the system behaves as though it were class (c). To be specific, assume

$$\rho(w) \propto w^{-\alpha} \qquad W_{min} < W < 1$$
$$= 0 \qquad \text{otherwise} \tag{33}$$

at high frequencies,

$$W_{c.o.}(\omega) \simeq \omega^{1/(2-\alpha)} >> W_{min} \tag{34}$$

and the cutoff has no influence. In particular, the conductivity has the class (c) frequency dependence (32). Crossover occurs when $W_{c.o.}(\omega)$ becomes comparable to W_{min}, i.e., for a frequency

$$\omega_{c.o.} \simeq W_{min}^{2-\alpha} \quad . \tag{35}$$

Then, at frequencies less than this value,

$$\sigma(\omega) \propto W_{min}^{\alpha} \propto (\omega_{c.o.})^{\alpha/(2-\alpha)} \qquad \omega < \omega_{c.o.} \quad , \tag{36}$$

independent of frequency.

4.2 Interdimensional crossover

There are no truly one-dimensional systems in nature. Interchain transfer must take place at sufficiently long times, and lead to three-dimensional behavior. We assume that this can be described by a mean one-dimensional time

$$\tau_{1d} \simeq \Omega^{-1} \tag{37}$$

where Ω is the interchain transition rate. We use an argument similar to that of SCHER, MONTROLL, and ALEXANDER [19]. For large ω,

$$\omega > \Omega \tag{38}$$

the system is one dimensional and, for example,

$$\langle x^2 \rangle_{\omega} = \xi_{\parallel}^2(\omega) \propto \omega^{-2\nu} \quad . \tag{39}$$

At lower frequencies

$$\omega < \Omega \tag{40}$$

one crosses to an anisotropic three-dimensional system. The jump rate is Ω and one has proper random walk with anisotropic steps. The step length perpendicular to the chains is the interchain distance (defined as unity). Thus,

$$\langle x_{\perp}^2 \rangle \propto \Omega/\omega \quad , \qquad \omega << \Omega \quad . \tag{41}$$

The particle travels along the chain a distance $\xi(\Omega)$ at each step. Thus,

$$\langle x_{\parallel}^2\rangle \propto \xi_{\parallel}^2(\Omega)\Omega/\omega \simeq \Omega^{\alpha/(2-\alpha)}/\omega \quad . \tag{42}$$

Thus, the motion is highly anisotropic

$$\langle x_{\parallel}^2\rangle/\langle x_{\perp}^2\rangle = \sigma_{\parallel}/\sigma_{\perp} \propto \xi_{\parallel}^2(\Omega) >> 1 \quad . \tag{43}$$

We note that the anisotropy depends explicitly on α and not only on Ω. To avoid confusion, we mention that we have used the average rate on the chain ($\langle W\rangle$) or the maximum W_{max} as our frequency scale, so that $\Omega << 1$.

4.3 Field dependence of the conductivity

We have seen that the frequency dependence of the barrier dominated conductivity reflects the distance a particle travels on the chain in a time $1/\omega$. This is reflected in the relationship between ξ and the effective cutoff (28)

$$W_{c.o.}(\xi) \propto \xi^{-1/(1-\alpha)} \quad . \tag{28}$$

This causes a peculiar field dependence of the conductivity. The mechanism is similar to the enhanced trapping probability discussed elsewhere [19]. In the presence of a field, a drift velocity is superimposed on the random motion. The drift distance is

$$x_d = \langle x_d\rangle_\omega \simeq \tilde{W}\,\tilde{E}\,t \simeq \tilde{W}\,\tilde{E}/\omega \quad , \tag{44}$$

where

$$\tilde{E} = eE\ell/k_BT \quad , \tag{45}$$

and ℓ is the length of the segment between adjacent barriers. Distances x are measured in units of ℓ. $\tilde{W}$ is some appropriate effective conductance to be determined self-consistently in the spirit of Section 3. Using (29) in (44), with $\tilde{W} = W_{av}$ and $x_d = \xi$ one finds:

$$x_d(\omega,\tilde{E}) \simeq (\tilde{E}/\omega)^{1-\alpha} \quad . \tag{46}$$

This should be small compared to the zero field result

$$x(\omega,0) = \xi(\omega) \simeq \omega^{-(1-\alpha)/(2-\alpha)} \quad . \tag{47}$$

For low fields, $\xi(\omega)$ is the larger distance and will determine $\tilde{W}_{av}$. The reverse is true for large $\tilde{E}$. Crossover occurs when $x_d(\omega,\tilde{E}) \simeq \xi(\omega)$. Thus,

$$\tilde{E}_c(\omega) \propto \omega^{(1-\alpha)/(2-\alpha)}$$

or

$$\tilde{E}_c(\omega)\,\xi(\omega) \simeq 1 \quad , \tag{48}$$

as should be expected. Thus, using (32),

$$\sigma(\omega,\tilde{E}) \simeq \sigma(\omega) \simeq \omega^{\alpha/(2-\alpha)} \quad , \qquad \tilde{E}\xi(\omega) < 1 \quad . \tag{49a}$$

For large fields using (46) one obtains

$$\sigma(\omega,\tilde{E}) \propto W_{av}(x_d) \simeq (\omega/\tilde{E})^{\alpha} \quad , \qquad \tilde{E}\xi(\omega) > 1 \quad . \tag{49b}$$

The two results coincide at E_c. A different way of writing this is

$$\sigma(\omega,\tilde{E})/\sigma(\omega) \simeq [\tilde{E}\xi(\omega)]^{-\alpha} \simeq [\tilde{E}_c(\omega)/\tilde{E}]^{\alpha} \quad ; \qquad \tilde{E}_c(\omega)/\tilde{E} < 1 \quad . \tag{50}$$

Thus, one has the curious effect of a conductivity which decreases at high fields, contrary to most nonlinear effects. Obviously, this can only be observed when other nonlinear effects can be neglected.

5. Application to linear chain transport

The most obvious interpretation of (1) is in terms of single particle motion. This is also the language we have used in Sections 3 and 4. It is, however, obvious that this is not a reasonable model for the charge transfer metals or for superionic conductors. One therefore has to resort to the usual linearization assumptions in setting up the linear master equation. Since the conditions are somewhat unusual, we do this explicitly. We consider a chain of relatively high conductivity interrupted by barriers. Thus, there is a chain segment of length ℓ_n between two barriers having "conductances" $W_{n-1,n}$ and $W_{n,n+1}$, respectively. We follow the ideas of S. STRASSLER [4], [24] and assume:

a) One can define a local electrochemical potential μ_n which can be assumed constant along each segment. This is equivalent to the assumption that the segment conductance is large compared to the W.

b) Density changes at the sites are sufficiently small so that the μ_n are linear in the excess site occupation probabilities, Q_n. We shall discuss this in more detail below.

c) Electric field effects on the μ_n can be assumed linear. There is a potential drop $E\ell_n$ across the segment which will lead to a polarization.

This is in fact essential if μ_n is to be constant along the segment. We assume that the resulting corrections to μ_n, which are quadratic in E, can be neglected. The resulting polarization itself contributes to the dielectric constant. Since it is frequency independent, it should dominate the low frequency dielectric constant.

d) Conditions are such that the current across the barriers is linear,

$$j_{n,n+1} = W_{n,n+1}(\mu_n - \mu_{n+1}) \quad . \tag{51}$$

e) Electrostatic interactions between the excess charges, Q_n, on different segments of the same chain can be neglected. For the situation envisaged where one has a tight bundle of chains of atomic cross-section, this seems reasonable. The effect of the excess charge on other chains can be described by an effective dielectric constant. Excess charges on a given chain are correlated with the disorder and must therefore be included explicitly. We discuss this in more detail below.

Consider now the metallic chain case. For the nth segment, one has

$$\mu_n = - eEx_n + E_F + (Q_n E_F/\ell_n) + eV_n^{e.s.} \tag{52}$$

where x_n is the position of the center of the segment. We have written ℓ_n/E_F for the density of states. We note that this chemical potential contribution to the electrochemical potential will remain linear in Q_n as long as the corrections to μ_n are small compared to E_F. Thus, the scale for nonlinearity is E_F (and not k_BT), and one can accommodate large changes in μ_n and still remain in the linear regime. The $V_n^{e.s.}$ are the electrostatic potentials on the segments due to the excess charges.

Equations (51) and (52) should be combined with the continuity equation,

$$dQ_n/dt = - j_{n,n+1} - j_{n,n-1} \quad . \tag{53}$$

If one can neglect the $V_n^{e.s.}$, one obtains a master equation of the form (1) with

$$C_n = \ell_n/E_F \quad . \tag{54}$$

Consider now the effect of including the $V_n^{e.s.}$. One has

$$Q_n = A_n V_n^{e.s.} + \sum_{n'} B_{n,n'}(V_n - V_{n'}) \tag{55}$$

where the A_n and $B_{n,n'}$ are capacitances determined by the geometry and dielectric constant. If the B are all small, one can combine the true capacitance A_n with the chemical potential term (54). This does not change the structure of (1) but now

$$C_n = [(\ell_n/E_F)^{-1} + (e/A_n)]^{-1} \quad . \tag{56}$$

We note that the two capacitive terms are added in series.

If A_n/e is small so that E_F/ℓ_n can be neglected, one would obtain a proper electric line with the B as shunting capacitors across the barriers.

When neither of these approximations is valid, the equations become complicated even when one only includes nearest neighbor B, and have a different structure from (1). This does not allow us to take advantage of our exact solution of (1) for this problem.

6. Acknowledgments

The authors wish to thank Dr. S. Strassler and Professors G. Gruner and T. Holstein for many fruitful discussions. Two of us [S.A. and R.O.] acknowledge support from the U.S. National Science Foundation and the U.S. Office of Naval Research.

References

1. J. Bernasconi, S. Alexander, and R. Orbach, Phys. Rev. Lett. 41, 185 (1978).
2. S. Alexander, J. Bernasconi, and R. Orbach, J. de Physique C6-706 (1978).
3. S. Alexander and J. Bernasconi, J. Phys. C12, L1 (1979).
4. J. Bernasconi, H. U. Beyeler, S. Strassler, and S. Alexander, Phys. Rev. Lett. 42, 819 (1979).

5. J. Bernasconi, W. R. Schneider, and W. Wyss, Z. Physik B37, 175 (1980).
6. Freeman J. Dyson, Phys. Rev. 92, 1331 (1953).
7. H. Schmidt, Phys. Rev. 105, 425 (1957).
8. C. Domb, Proc. Roy. Soc. A276, 418 (1963).
9. S. Alexander, J. Bernasconi, and R. Orbach, Phys. Rev. B17, 4311 (1978).
10. G. Gruner, Bull. Am. Phys. Soc. 25, 255 (1980).
11. J. Kommandeur, private communication. See also H. A. Pohl and M. Pollak, J. Chem. Phys. 66, 4031 (1977).
12. S. Alexander, to be published. Contrary to claims repeatedly made in the literature {e.g., Refs. [13] and [14]} it is possible to show that barriers can have no effect on the critical exponent ν when the (bond) percolation density is smaller than unity. The situation for traps is of course quite different.
13. H. Scher and M. Lax, Phys. Rev. B7, 4491 (1973); ibid. 7, 4502 (1973); H. Scher and E. W. Montroll, Phys. Rev. B12, 2455 (1975).
14. J. Klafter and R. Silbey, Phys. Rev. Lett. 44, 55 (1980); J. Chem. Phys. 72, 843 (1980).
15. R. Landauer, Phil. Mag. 21, 863 (1970).
16. A. A. Abrikosov and I. A. Ryzhkin, Sov. Phys. JETP 44, 630 (1976); Adv. Phys. 27, 147 (1978).
17. P. W. Anderson, D. J. Thouless, E. Abrahams, and D. S. Fisher, to be published, Phys. Rev. 1980.
18. S. Alexander, J. Bernasconi, W. R. Schneider, and R. Orbach, to be published, Rev. Mod. Phys. 1980.
19. H. Scher, S. Alexander, and E. W. Montroll, to be published, Proc. Nat. Acad. Sci. 1980.
20. H. U. Beyeler, L. Pietronero, and S. Strassler, to be published, Phys. Rev. 1980.
21. C. Fox, Trans. Amer. Math. Soc. 98, 395 (1961); K. C. Gupta and U. C. Jain, Proc. Natl. Inst. Sci. (India) A36, 594 (1966).
22. J. Bernasconi and W. R. Schneider, to be submitted for publication.
23. P. M. Richards and R. L. Renken, Phys. Rev. B21, 3740 (1980); J. Bernasconi and H. U. Beyeler, Phys. Rev. B21, 3745 (1980).
24. S. Strassler, private communication.

Random Exchange Spin Chains

W.G. Clark

Department of Physics, University of California at Los Angeles
Los Angeles, CA 90024, USA

1. Introduction

The subject of random exchange spin chains started in earnest in 1972, when BULAEVSKII, SHCHEGOLEV, and co-workers [1] published their measurements of the low temperature electron spin susceptibility (χ) and magnetization (M) of the TCNQ charge transfer salts $Qn(TCNQ)_2$, $Ad(TCNQ)_2$, and NMP-TCNQ, and interpreted the observed physical properties as those of a random exchange Heisenberg antiferromagnetic chain (REHAC) with spin S = 1/2. Their theoretical model, which is discussed in more detail below, employed the Hamiltonian

$$\mathcal{H} = 2 \sum_i J_i \vec{S}_i \cdot \vec{S}_{i+1} - g\mu_B H \sum_i S_i^z \qquad (1)$$

where i is the site index on a one-dimensional (1-d) chain, J_i is the random, antiferromagnetic ($J_i > 0$), nearest neighbor exchange with random probability distribution P(J), $\vec{S}_i$ is the spin operator for the electron at site i, g is the electronic g-value, μ_B is the Bohr magneton, and H the externally applied magnetic field. With their model they were able to fit the observed low temperature (T) behavior

$$\chi(T) = AT^{-\alpha} \qquad (2)$$

at low H ($g\mu_B H << k_B T$) and

$$M(H,T) = BH^{1-\alpha} \qquad (3)$$

at intermediate H ($k_B T << g\mu_B H <<$ maximum J), as well as predict for the specific heat at constant H (C_H)

$$C_H(0,T) = CT^{1-\alpha} \qquad (4)$$

$$C_H(H,T) = DTH^{-\alpha} \qquad (5)$$

for low and intermediate H respectively. In (2)-(5), A, B, C, and D are constants, and the exponent α has the same value, usually found experimentally to fall in the range 0.7-0.9. Equations (2)-(5) are now widely accepted as important signatures of REHAC behavior.

Most experimental work reported so far on random exchange spin chains has been on systems exhibiting the characteristics mentioned above. The main objectives of this paper are to review briefly the body of experimental and theoretical knowledge of the magnetic properties of S = 1/2 REHAC's.

In order, the topics to be covered are (a) examples demonstrating the "standard" thermodynamic properties of a REHAC, (b) a brief discussion of the materials in which they are seen and the microscopic origins of the disorder, (c) "early" theoretical models, (d) recent theoretical developments, (e) some recent experimental results demonstrating quasi-universal behavior and the low T crossover to three-dimensional (3-d) properties, and (f) a discussion of numerous miscellaneous points regarding REHAC's.

2. Examples of Thermodynamic Properties

The easiest and most widely investigated property of a REHAC is $\chi(T)$. Both static and ESR methods have been used successfully. An example of the latter reported by SANNY et al. [2] for the most completely studied material, $Qn(TCNQ)_2$ is shown in Fig. 1. There, it is seen that the disordered, 1-d behavior of (2) is exhibited over the wide temperature range 40 mK $<$ T $<$ 20 K, with $\alpha = 0.79 \pm 0.02$ in the as grown material. For reasons of interpretation to be covered later in this paper, it is important that the experimentally determined value of α be as accurate as possible. This means that the measurements should be extended to very low T. The accuracy of $\pm$ 2-3% for α shown in Fig. 1 is typical of what can be obtained.

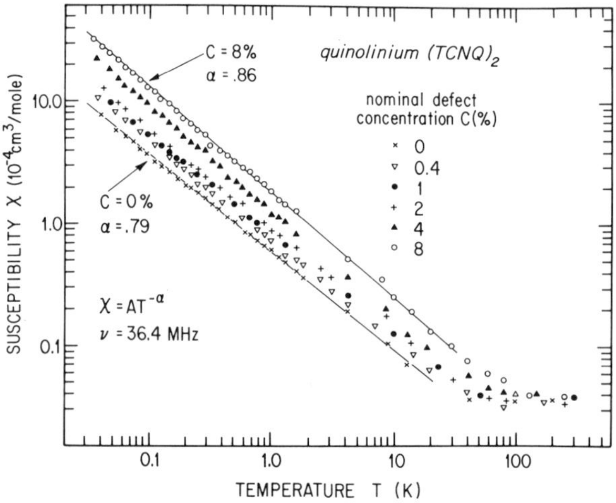

Fig. 1 Susceptibility of $Qn(TCNQ)_2$ (SANNY et al. [2]). The curve labeled c = 0% is typical of χ for as grown material, and shows the low T behavior of a REHAC with α = 0.79. The other curves show changes caused by further disorder introduced by neutron irradiation, and are interpreted in terms of the quasi-universality (see Fig. 5).

In contrast to $\chi(T)$, very few measurements of C_H have been reported for temperatures low enough to exhibit the properties associated with (4) and (5) [3]-[5]. The clearest examples are found in the work on $Qn(TCNQ)_2$ by AZEVEDO and CLARK [3] shown in Fig. 2 and Fig. 3, where $C_H(T)$ for H = 0 and $C_H(H,T)$ for constant T are shown. After subtraction of the T^3 phonon term, it is seen in Fig. 2 that the zero field T dependence of (4) is followed closely using the value $\alpha = 0.82$ obtained from measurements of $\chi(T)$ on the same sample. Similar results have been reported for $Ad(TCNQ)_2$ [4]

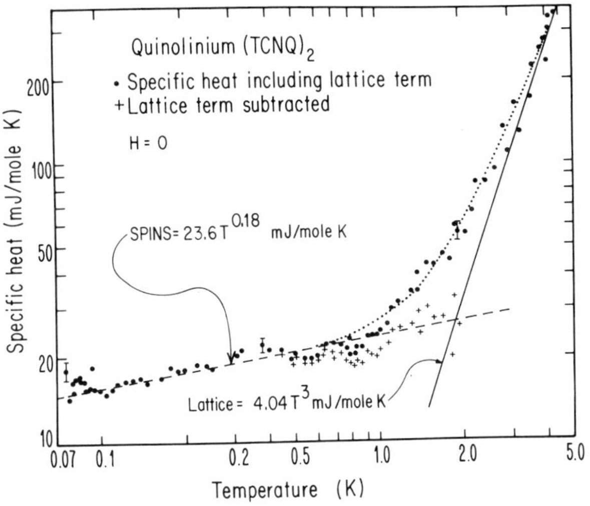

Fig. 2 Low T specific heat of $Qn(TCNQ)_2$. After subtraction of the cubic lattice term (solid line) the remaining REHAC term (4) is indicated by the dashed line for α = 0.82, as determined from χ on this sample. From AZEVEDO and CLARK [3].

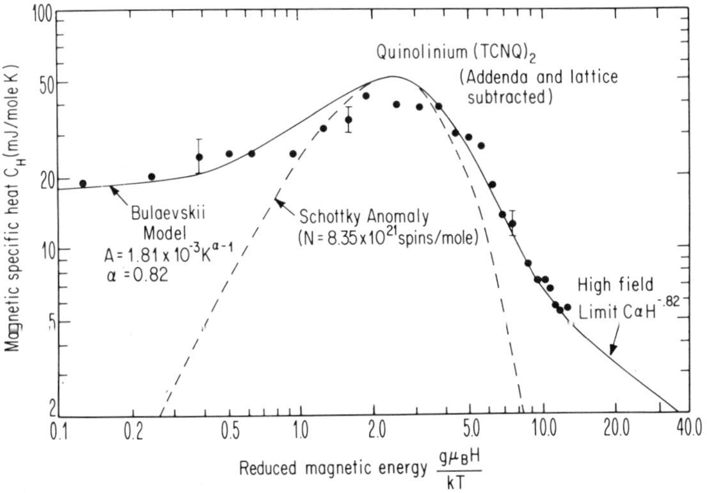

Fig. 3 Magnetic specific heat of $Qn(TCNQ)_2$ as a function of H at T = 213 mK. The solid line shows the fit given by the Bulaevskii model [1]. Equally good fits are obtained using the ECP model and the double δ function renormalization calculation [26]. A comparison with the Schottky anomaly of free spins is shown by the dashed line. From AZEVEDO and CLARK [3].

[5]. Figure 3 shows that the intermediate field case (5) is rather difficult to explore, as its onset occurs at about five times the value of H which corresponds to the maximum in C_H. The theoretical curve, which is seen to give a good fit to the data through the transition from low to intermediate H is discussed in Section 4.

The other REHAC property that has been demonstrated is (3) for $Qn(TCNQ)_2$, $Ad(TCNQ)_2$, and NMP-TCNQ in static measurements at 0.1 K up to H = 70 kOe [1].

3. Materials and the Origins of Disorder

The two main requirements for REHAC behavior are random (disordered) exchange and one dimensionality. In this section we indicate which materials have been so identified and briefly discuss several of the microscopic features which can be responsible for these disordered 1-d characteristics.

All of the materials reported so far to be REHAC's are highly conducting organic charge transfer salts, most of them based on the TCNQ molecule as an acceptor. They include $Qn(TCNQ)_2$ [1]-[3] [6]-[10], $Ad(TCNQ)_2$ [1] [4] [5], NMP-TCNQ [1], $NMP_{0.63}Phen_{0.37}$-TCNQ [11], $NMeAd(TCNQ)_2$ [8] [10] [12], 2-2' $BiP(TCNQ)_2$ [8] [10] [12], $HMTTF_{0.89}HMTSF_{0.11}$-TCNQ [10], 2-3 $BAd(TCNQ)_2$ [8] [10], HMTSF-TNAP [10], TTT_2I_3[10], irradiated TTF-TCNQ [9] [10], and irradiated N-propyl $Qn(TCNQ)_2$ [9]. (We have conserved space by using common abbreviations for these materials. The correct chemical names and molecular structures are indicated in the references.) One feature in common for all of the above substances is that their electronic conduction is associated with flat molecules which stack into 1-d chains with strong overlap of the electronic wave functions along the chain axis. This leads to highly 1-d exchange interactions along the stacks and very weak magnetic interactions between chains.

There are several different kinds of disorder which are believed capable of generating the randomness in J. One of them occurs in materials such as $Qn(TCNQ)_2$, which have asymmetric donors, uniform stacks, and complete (or nearly so) charge transfer to the TCNQ chain, which is responsible for the conduction [7] [13] [14]. The asymmetry of the donor molecule and its electric dipole moment are small enough to favor crystal growth with some degree of randomness in the orientation of the donor dipole moment [14]. This in turn leads to a random electrostatic potential on the adjacent conducting TCNQ chains, and hence to some degree of localization of the electrons [15]. Since this is driven by a random potential, the electron spacing and the exchange determined by it is random [16].

The mechanism just described is intrinsic to the material--it will be present in carefully grown crystals made from perfectly pure ingredients. Several other ways to produce random exchange have also been investigated which involve the deliberate introduction of non-intrinsic imperfections into the material. They represent some of the newest and most interesting developments in this field. One example is disorder introduced chemically into a material which is normally ordered, and in its pure phase does not show REHAC behavior. This has been demonstrated by MILJAK et al. [10] in their work on HMTTF TCNQ which has had 11% of the symmetric HMTTF molecules replaced at random by the slightly different, symmetric HMTSF molecule. The pure material shows the usual Peierls gap behavior for χ plus a small Curie tail which behaves as $\chi = CT^{-1}$ and, presumably, is due to a small concentration of isolated impurities. The alloyed sample exhibits a smearing of the transition, a shift of it to lower T, and the appearance of $\chi = AT^{-\alpha}$ with $\alpha = 0.8$ below about 20 K. This is interpreted as a crossover to REHAC behavior when the material is alloyed.

Another example which they report [9] [10] is that of pure TTF-TCNQ into which disorder has been introduced by neutron irradiation. The pure material behaves as an ordered system with a few isolated magnetic impurities. This is evident from a large drop in χ between 60 K and 20 K associated with the opening of the Peierls gap followed by a curie tail χ of the form $\chi = CT^{-1}$ at lower T. After an amount of irradiation estimated to introduce a 0.6% defect concentration the high T χ remains unchanged, but the drop in χ is shifted to lower T, is greatly reduced, and a low T term emerges which is nearly 10 times larger and varies below 6 K as $\chi = CT^{-\alpha}$, with $\alpha \simeq 0.74$. This is also interpreted as a crossover to REHAC behavior caused by disorder.

The lowest T covered in the two examples just discussed did not go below 2.5 K. Therefore, the conclusion of REHAC behavior is still somewhat tentative and needs to be investigated further by similar work down to much lower T.

4. Early Theoretical Models

All of the early theoretical models for a REHAC have used (1) with a random J as a starting point, but have employed approximations which obscure the relationship between the macroscopic thermodynamics and the microscopic P(J). These relationships are made much more evident in Section 5, where recent developments which better account for the many body aspects of the problem are described. The early models, however, are of more than historical interest, as they have provided the basic starting point for work on REHAC's, have generated useful insights regarding their properties, and in some respects still form the most complete and simplest description of the behavior at large H. The main features of three of them are indicated in the sections which follow. For complete presentations the reader should consult the original references.

4.1 Bulaevskii Model

When BULAEVSKII et al. [1] introduced their model (1) for the low T magnetic properties of several TCNQ salts, they recognized the difficulty of solving the full many-body problem, and therefore introduced an approximation strategy which is now generally called the Bulaevskii model. There are several key steps, as follows.

First, (1) is separated into a main part (H_0) and a perturbation term (H') given by

$$H_0 = 2 \sum_i J_i (S_i^x S_{i+1}^x + S_i^y S_{i+1}^y) - g\mu_B H \sum_i S_i^z \qquad (6)$$

$$H' = 2 \sum J_i S_i^z S_{i+1}^z \quad ; \qquad (7)$$

Eq. (6) is the x-y model which, even for random J, is known to have a spinless fermion representation for the thermodynamic behavior with a Fermi energy given by $-g\mu_B H$. The problem is that for random J the energy eigenvalues (ε), and hence the density of states $\rho(\varepsilon)$, are not known. It is then postulated that the Fermi liquid nature of the problem is retained when the (Ising) perturbation term (7) is added to complete (1), even though H' is not small compared to H_0. The consequent modification of the energy eigenvalues is accounted for by introducing a modified, phenomenological density of states $\rho'(\varepsilon)$ having the form

$$\rho'(\varepsilon) = a|\varepsilon|^{-\alpha} \tag{8}$$

where the constant a and exponent α are fitted to experimental results. This model predicts the thermodynamic behavior (2)-(5). It provides an excellent fit to the low T magnetic properties of several REHAC materials over a wide range of T and H [1]-[3] [6] [7] [11] [12] [17]. In general, it is a very useful model as it provides an explicit formula for the free energy, and therefore a complete thermodynamic description of a REHAC. Its shortcomings are that there is no connection between the thermodynamics and P(J), and no insight is given regarding the nature of the low energy elementary excitations of a REHAC.

4.2 Theodorou and Cohen's Model

In a series of papers THEODOROU and COHEN [16] [18] developed a microscopic, theoretical model for the properties of a REHAC. Although their work emphasizes the interpretation of χ for NMP-TCNQ, much of it is expected to be more widely applicable.

There are two major aspects to their treatment of the problem: (a) a microscopic calculation of P(J), and (b) elucidation of the corresponding REHAC thermodynamics. For the first task, they start with the Hubbard model with a random on-site energy, which is attributed to disorder in the orientation of the electric dipole moments associated with asymmetric donor molecules. In this picture, individual TCNQ molecules are occupied by zero, one, or two electrons according to the value of the on-site potential and the amount by which the band is filled. Singly occupied sites are magnetic, whereas unoccupied or doubly occupied ones (spins paired) are not. The magnetic sites form a 1-d Heisenberg antiferromagnetic with a nearest neighbor J which falls approximately exponentially as a function of separation. Since the random on-site energy locates the magnetic sites randomly, J is also random. The calculation yields the form $P(J) \propto J^{-\alpha}$ for small values of J along with a peak about a large value which corresponds to the smallest separation of magnetic sites. Although $P(J) \propto J^{-\alpha}$ is obtained for small J starting from the Hubbard model with disorder, they also point out that it occurs for nearest neighbor interactions on the basis of two plausible conditions: (a) J falls off exponentially with spin separation, and (b) equally spaced sites are singly occupied at random by a substantially smaller number of spins [19]. Their work constitutes a strong argument that $P(J) \propto J^{-\alpha}$ should occur easily in nature.

A cluster model is then used to calculate the temperature dependence of χ. It is assumed that the infinite chain can be decoupled into finite, independent segments at each link for which $J < k_BT$. Segments with an even number of spins have a nonmagnetic singlet ground state and those with an odd number behave at low T as an isolated S = 1/2 spin with a Curie χ. A simple statistical calculation based on counting the number of odd spin segments as a function of T then gives (2). One result of this approach is that the exponent α is the same for both χ and P(J), with the consequence that the singular χ requires a singular P(J). (These two points are now known to be wrong on the basis of the renormalization work described in

Section 5, where it is pointed out why the thermal decoupling $J < k_BT$ severely underestimates the number of weak links in a random chain). A good fit to the measured χ of NMP-TCNQ is obtained using their P(J) and the classical Heisenberg model. This analysis provides an evaluation of their three free Hubbard model parameters, the transfer t, Coulomb repulsion U, and width of the disorder parameter σ [16] [18]. Because of a substantial number of approximations and their assumption of a half filled band for NMP-TCNQ (it is only about 1/3 filled [20]), it is difficult to evaluate the relevance of their values assigned to t, U, and σ.

THEODOROU and COHEN also used the cluster model to obtain the same low T behavior of C_H indicated in (4). Their approach does not, however, give a complete thermodynamic description, and is difficult to apply for large H.

4.3 The ECP Model

A third model which has been used successfully to describe the properties of a REHAC is the exchange coupled pair (ECP) model introduced by CLARK and co-workers [7] [10] [21] [22]. Detailed accounts and formulas of the ECP model appear in the literature. Here, we point out the main features and results.

The central point of the ECP model is to avoid solving the difficult many body problem (1) by decoupling it into a spectrum of easily solvable systems in much the same spirit as the independent electron model of a metal. The process is shown in Fig. 4a,b. Figure 4a is a schematic representation of a REHAC in which the circles represent spin sites (not necessarily equidistant in real space) and the height of the arches connecting them represents the magnitude of the exchange. This many body problem is decoupled by simply setting the odd numbered J's equal to zero, as in Fig. 4b. The result is a system of independent ECP's which have the well-known energy levels and quantum numbers indicated in Fig. 4d. The complete thermodynamics of a single ECP is trivial. That of the ECP model is simply the

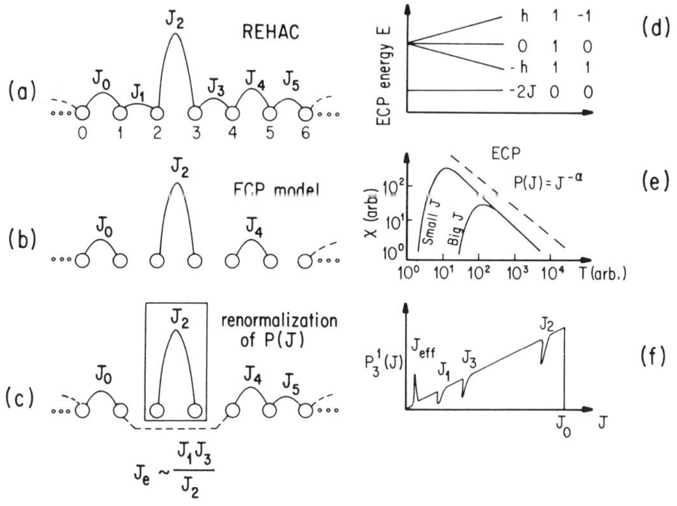

Fig. 4 Schematic representation of a REHAC (a), its ECP approximation (b) [22], the decimation procedure (c) [24], ECP energy levels (d), ECP χ due to superposition (e), and the renormalization of $P_3(J)$ after the first decimation.

superposition of the properties of all the pairs suitably weighted by P(J). This is shown schematically in Fig. 4e for χ. The solid lines indicate χ for two different pairs; one with a large value of J and the other with a small value of J. In each case for $k_BT >> J$, a Curie law $\chi \propto T^{-\alpha}$ is obtained, whereas for $k_BT << J$, χ varies as $\exp\{-J/k_BT\}$. As T is decreased, fewer pairs contribute. By using $P(J) \propto J^{-\alpha}$, $\chi = AT^{-\alpha}$ is obtained as well as (3)-(5).

There are several important features of the ECP model. It gives closed expressions for all thermodynamic quantities at all values of T and H. In those cases where its results have been compared with those of the Bulaevskii model, the two are found to be nearly indistinguishable [3] [22]. It also provides a very good fit to the available low T experimental properties in the sense that they follow the limiting cases (2)-(5). Because of the similarity of the ECP and Bulaevskii model results, the fit shown in Fig. 3 for C_H over a wide range of H also applies for the ECP model. A shortcoming of the ECP model is that it assumes that the P(J) used to calculate the thermodynamic quantities is the same as the microscopic density of couplings. In analogy to work which shows this to be incorrect for a random ferromagnet [23], it was argued [22] that there should be a renormalization of α on going from $P(J) \propto J^{-\alpha}$ for a REHAC to the P(J) used with the ECP model. The very recent renormalization work described in the next section confirms this point, but shows that the detailed renormalization of α for the antiferromagnetic case is different from that of the ferromagnetic one. Another limitation is that in eliminating the many body character of a REHAC, the ECP model omits most of its dynamic properties [22]. Some speculation about why the ECP model works as well as it does has already appeared [22]. An additional reason may be that the elementary excitations of a REHAC actually resemble ECP's, as suggested by the decimation procedure described below.

5. Recent Theoretical Developments

During the last year several important advances in the theoretical understanding of a REHAC have appeared. All of them use various renormalization schemes to handle the many body character of the system. They represent substantial progress over the models described in Section 4 in that they allow many different forms for the microscopic P(J) to be follows through to the calculation of the thermodynamic properties. In this section we briefly indicate some of the main points of the decimation calculation of MA et al. [24]. The renormalization group work of HIRSCH and JOSE [25], and the numerical calculation of SOOS and BONDESON [26].

The most specular result of the recent theoretical work is that it shows a quasi universal thermodynamic behavior for a REHAC almost independent of the detailed nature of P(J). For example, four very different functional forms for P(J) all lead to χ close to (2) and, where investigated, lead to C_H close to (4) and (5). They are [$P_i(J) = 0$ for unspecified values of J] $P_1(J) = (1-\alpha)\, J_0^{\alpha-1} J^{-\alpha}$, $0 < J < J_0$; $P_2(J) = J_0^{-1}$, $0 < J < J_0$; $P_3(J) = 2J_0^{-1}J$, $0 < J_0 < J$; and $P_4(J) = 0.1\ \delta(J-0.3J_0) + 0.9\ \delta(J-J_0)$. It is not surprising that the singular $P_1(J)$ leads to a singular $\chi(T) \propto T^{-\alpha}$, but nearly the same kind of divergence for $T \to 0$ is also obtained using $P_2(J)$, which is constant as $J \to 0$, $P_3(J)$, which does to zero as $J \to 0$, and even $P_4(J)$, which is zero for all $J < 0.3J_0$.

It is instructive to consider the ECP model [22] and the decimation calculation [24] to see intuitively how a random exchange system leads to (2). On the basis of the ECP model, it is seen from Fig. 4e that a χ which is large at low T is associated with small values of J. The question then is, how can a REHAC generate a substantial number of effective exchange couplings which are very weak, even though only a few [or none, as for $P_4(J)$] of the J's are small? This can be answered by considering the sequence J_1, J_2, J_3 in Fig. 4a. Because $J_2 >> J_1, J_3$, the energy levels of spins 2 and 3 are determined mainly by J_2, with J_1 and J_3 acting as small perturbations; i.e., spins 2 and 3 act like an ECP. To the lowest approximation, they will occupy their nonmagnetic ground state (Fig. 4d) at low T, thereby cutting the magnetic chain. But in fact, J_1 will mix in some of the triplet state, which according to standard perturbation theory couples spin 1 across through J_3 to spin 4 with an effective exchange constant $J_e \sim J_1 J_3 / J_2$. At low T, this segment of the REHAC can be approximated by an isolated ECP (spins 2 and 3) with its thermodynamics plus a modified chain with two fewer spins and a new $J_e << J_1, J_2, J_3$, as indicated in Fig. 4c. Using $P_3(J)$ as an example, after one such step the modified chain has a new distribution $P_3^1(J)$ in which weight has been subtracted from large values of J and added to a small one (Fig. 4f). This can be repeated many times, further shifting weight to small values of J for the modified chains. The process just described is the main content of the decimation calculation [24], and shows how small values of J are generated in a random system. The detailed calculations [24]-[26] show that it leads approximately to (2).

There are several points to be made regarding these new developments. First, since many different forms of P(J) give about the same χ, etc., it is not possible to know P(J) on the basis of a few measurements of magnetic properties. Instead many different kinds of experimental and theoretical knowledge will be needed to ascertain P(J) for a given material. Two of the calculations [24] [25] indicate some deviations from (2)-(5) in that α in (2) depends weakly on T, and the exponent for χ is slightly different from that for C_H in (4). These effects are small, however, and have not been seen experimentally (see Section 2). They may turn out to be artifacts of the forms used for P(J) which do not have a prominent peak in P(J) at a large value of J, as calculated using the Hubbard model [2] [18]. The importance of such a peak is suggested by the calculations of SOOS and BONDESON [2], which provide an excellent fit over the entire range of T for data of the kind shown in Figs. 1-3 using $P_4(J)$, which does have most of its weight at J_0. Their numerical method, unlike the other renormalization work [24] [25] is also well adapted to including a wide range of U.

There are two further points which have been emphasized by HIRSCH and JOSE [25]. The first is that quantum fluctuations play a key role in the properties of a REHAC. The classical REHAC, as studied by THEODOROU [18], required a singular P(J) to have a divergent χ, whereas the renormalization group approach shows that the S = 1/2 case does not. In addition, scaling arguments have been developed which show that disorder in the exchange interaction leads to localization of the spin states in a REHAC [25].

6. Recent Experimental Results

In this section we mention two recent experimental studies which have added new understanding of the properties of REHAC's. The first is a set of measurements of $\chi(T)$ for neutron irradiated $Qn(TCNQ)_2$ by SANNY et al. [2]. The object of this investigation was to study the effect of systematically increasing the amount of structural disorder on the magnetic properties. Figure 1 shows $\chi(T)$ for samples from the same preparation batch, but with neutron induced defects in various contrations up to 8%. The behavior of A and α for these data is shown in Fig. 5, where it is seen that α changes by only about 10%, but that A increases linearly by about a factor of three. This insensitivity of α to substantial changes in the sample disorder is interpreted as evidence of the quasi universality property explained by the renormalization calculations [24]-[26]. Furthermore, one of them [26] explicitly predicts the linear variation of A. Similar experimental results over a much more limited range have also been reported by MILJAK et al. [9]. The notion of quasi universal behavior for a REHAC is also supported by the fact that various ways of introducing disorder (see Section 3) lead to $\chi \propto T^{-\alpha}$ at low T, even though the microscopic P(J)'s could be expected to differ substantially.

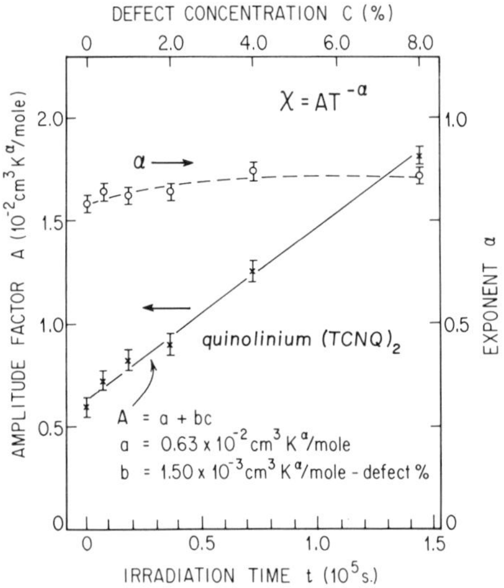

Fig. 5 Exponent α and amplitude factor A in neutron irradiated Qn(TCNQ)$_2$. The near independence of α and linear dependence of A on the defect concentration is evidence of quasi-universality. From SANNY et al. [2].

Another recent experiment was prompted by the questions how low in T does $\chi \propto T^{-\alpha}$ persist, and what properties emerge at still lower T? These questions have been answered for Qn(TCNQ)$_2$ by BOZLER et al. [6] in a set of experiments down to 0.45 mK. One of their results is shown in Fig. 6, where it is seen that (1) is followed down to about 5 mK. At lower T, a broad maximum in χ is observed. They also report that over about the same temperature range, both an internal field and substantial broadening of the ESR line appear. Several features similar to those seen in spin glasses become evident, and one interpretation is that at low enough T the REHAC goes over to a 3-d dipolar spin glass. Much more work is needed to clarify these points.

7. Miscellaneous Points

In this section we conclude with a few miscellaneous observations regarding the REHAC's discussed in this paper. The first point is that the exchange is very isotropic (Heisenberg) in REHAC materials, such as Qn(TCNQ)$_2$. This is indicated by the fact that extremely narrow ESR lines are seen in

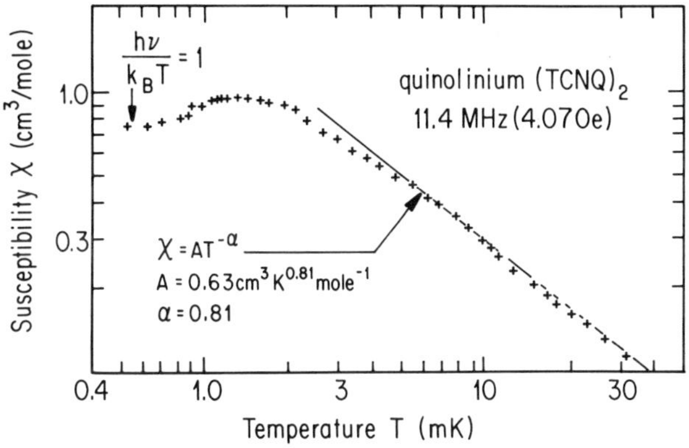

Fig. 6. Ultralow T behavior of χ in Qn(TCNQ)$_2$. REHAC behavior is seen down to ∿ 5 mK. The deviation at lower T is interpreted as a crossover to 3-d behavior. From BOZLER et al. [6].

powdered samples [7] [17]. If a substantial anisotropy were present, it would lead to a very broad ESR line.

In Qn(TCNQ)$_2$, no evidence of a sharp, magnetic phase transition is seen down to below 1 mK, even though the value of χ at 300 K indicates the material is magnetically dense with about one unpaired spin per formula unit. This lack of a magnetic transition is in marked contrast to concenttrated magnets without disorder, and probably indicates that the disorder prevents the relevant coherence length from reaching the critical size for 3-d ordering [27].

It is natural to ask if the behavior $\chi \propto T^{-\alpha}$ is really 1-d. Prior to the renormalization calculations [24]-[26] the answer was a clear yes! A divergent P(J) was needed for (2), and only the 1-d case gave a divergent P(J) [19]. But now the renormalization calculations indicate that in 1-d, the observed χ (T) $\propto T^{-\alpha}$ is expected for almost any P(J), including those that would occur in a 2-d or 3-d system. Would the same quasi-universality be observed if the exchange were 2-d or 3-d? Qualitatively, the answer appears to be no. For example, the decimation procedure which generates the small J's (Fig. 4c) would be less active in 2-d and 3-d, for in those cases there are other exchange paths to circumvent the barrier imposed in 1-d by a single large J. In the case of Qn(TCNQ)$_2$ 1-d exchange is also indicated by the fact that all known 2-d and 3-d systems of comparable electron spin concentration exhibit magnetic transitions at much higher T than does Qn(TCNQ)$_2$.

I would summarize our present understanding of REHAC's as follows:

The main features of the thermodynamics are reasonably well known, both theoretically and experimentally. On the other hand, much less is known about finer details, such as interaction strengths, the microscopic P(J), etc. The dynamic spin behavior at low T is not understood theoretically. There are experimental results on nuclear and electron spin relaxation [7] [17] [28] for which there is no quantitative, or in some cases, even qualitative understanding. Finally, the nature of the elementary excitations of a REHAC has not yet been elucidated, nor is it known what is their degree of localization and how it is related to P(J).

I wish to thank P. Chaikin, G. Grüner, R. Orbach, and S. Alexander for stimulating discussions upon several of the topics covered here, as well as J. Hirsch, J. José, and Z. G. Soos for keeping me informed about their progress in the theoretical treatment of REHAC's. This work was supported by National Science Foundation Grant DMR 77-23577.

References

1. L. N. Bulaevskii, A. V. Zvarykina, Yu. S. Karimov, R. B. Lyobovskii, and I. F. Shchegolev, Zh. Eksp. Teor. Fiz. 62, 725 (1972) [Sov. Phys. JETP 35, 384 (1972)].
2. J. Sanny, G. Grüner, and W. G. Clark, Solid State Commun., to be published.
3. L. J. Azevedo and W. G. Clark, Phys. Rev. B 16, 3252 (1977).
4. P. Delhaes, G. Keryer, S. Flandrois, and J. P. Manceau, Phys. Status Solidi B 80, 125 (1977).
5. W. Duffy, Jr., F. Weinhaus, D. Strandburg, and J. Deck, Phys. Rev. B 20, 1164 (1979).
6. H. M. Bozler, C. M. Gould, T. J. Bartolac, W. G. Clark, K. Glover, and J. Sanny, Bull. Am. Phys. Soc. 25, 217 (1980). H. M. Bozler, C. M. Gould, and W. G. Clark, submitted to Phys. Rev. Lett.
7. W. G. Clark, J. Hammann, J. Sanny, and L. C. Tippie, "Disordered Magnetic Chain Behavior of Quinolinium $(TCNQ)_2$ and Related Materials at Low Temperatures", in Quasi-One-Dimensional Conductors II, Proc. Dubrovnik, 1978, ed. by S. Barišić, A. Bjeliš, J.R. Cooper and B. Leontić, Lecture Notes in Physics, Vol. 96 (Springer, Berlin, Heidelberg, New York 1979) p. 255
8. M. Miljak, J. Cooper, and G. Grüner, KFKI Report No. 78-3 (1978).
9. M. Miljak, B. Korin, J. R. Cooper, K. Holczer, G. Grüner, and A. Jánossy, J. Magn. and Magn. Mater. 15-18, 219 (1980).
10. M. Miljak, B. Korin, J. R. Cooper, K. Holczer, and A. Jánossy, J. Phys. (Paris) 41, 639 (1980).
11. J. Hammann, W.G. Clark, A.J. Epstein and J.S. Miller, "Low Temperature Magnetic Susceptibility of the Organic Alloy $NMP_{0.63}Phenazine_{0.37}$ TCNQ", in Quasi-One-Dimensional Conductors II, Proc. Dubrovnik, 1978, ed. by S. Barišić, A. Bjeliš, J.R. Cooper and B. Leontić, Lecture Notes in Physics, Vol. 96 (Springer, Berlin, Heidelberg, New York 1979) p.310
12. J. Hammann, L. C. Tippie, and W. G. Clark, ibid., p. 309.
13. J. F. Kwak, G. Beni, and P. M. Chaikin, Phys. Rev. B 13, 641 (1976).
14. K. Holczer, G. Mihály, A. Jánossy, G. Grüner, and M. Kertész, J. Phys. C 11, 4707 (1978).
15. N. M. Mott and W. D. Twose, Adv. Phys. 10, 107 (1960).
16. G. Theodorou and M. H. Cohen, Phys. Rev. Lett. 37, 1014 (1976).
17. L. C. Tippie and W. G. Clark, submitted to Phys. Rev. B.
18. G. Theodorou, Phys. Rev. B 16, 2254, 2264, 2273 (1977).
19. G. Theodorou and M. H. Cohen, Phys. Rev. B 19, 1561 (1979).
20. J. P. Pouget, S. Metgert, R. Comés, and A. J. Epstein, Phys. Rev. B 21, 486 (1980).
21. W. G. Clark, L. C. Tippie, G. Frossati, and H. Godfrin, J. Phys. (Paris) Suppl. 39, C6-1160 (1978).
22. W. G. Clark and T. C. Tippie, Phys. Rev. B 20, 2914 (1979).
23. J. Bernasconi, S. Alexander, and R. Orbach, Phys. Rev. Lett. 41, 185 (1978). S. Alexander and J. Bernasconi, J. Phys. C 12, L1 (1979).

24. S. Ma, C. Dasgupta, and C. K. Hu, Phys. Rev. Lett. 43, 1434 (1979). C. Dasgupta and S. Ma, Phys. Rev. B 22, 1305 (1980).
25. J. E. Hirsch and J. V. José, J. Phys. C Letts. 13, L53 (1980). J. E. Hirsch and J. V. José, Phys. Rev. B., to be published.
26. Z. G. Soos and S. R. Bondeson, Solid State Commun. 35, 11 (1980). S. R. Bondeson and Z. G. Soos, Phys. Rev. B, to be published.
27. Y. Imry, P. Pincus, and D. Scalapino, Phys. Rev. B 12, 1978 (1975).
28. L. J. Azevedo, W. G. Clark, E. O. McLean, and P. F. Seligmann, Solid State Commun. 16, 1267 (1975).

Universality in Quantum Random Magnetic Chains [1]

J.E. Hirsch[2]

James Franck Institute, The University of Chicago

J.V. José[2]

Physics Department, Rutgers University

1. Introduction

In recent years a great deal of interest has been devoted to the study of disordered systems. The disorder can be of two types: i) In terms of stochastic variables denoting random, masses, exchange constants, on site energies, symmetry breaking fields, etc. but conserving a one-to-one correspondence between the points or bonds of a periodic lattice and ii) amorphous systems where the disorder is also in the topology of the lattice. Here we will talk about the first type of models and will deal with magnetic systems with random exchange constants.

Most studies carried out in random magnetic systems use classical models, i.e. models for which quantum effects are neglected. The justification for studying these models comes from the well-known success of the Universality hypothesis in critical phenomena. This applies to homogeneous or periodic systems in the critical region, but there is no proof that it should also apply to strongly random problems and at sufficiently low temperatures where quantum effects become relevant.

In this short note we shall address ourselves to the study of a 1-D Heisenberg random antiferromagnetic model at low temperatures. Here we find that the quantum case ($s = 1/2$) is strikingly different from the classical ($s = \infty$) one. Of course, this is a 1-D system without a phase transition and we cannot make any statements that would apply to, for instance, the still not understood spin-glass problem. However, it does suggest the rethinking of Universality hypothesis as applied to strongly random systems at low temperatures.

The organization of this note is as follows: Section 2 defines the model and reviews the experimental motivation for studying the problem. The model is of theoretical interest in itself. Section 3 describes briefly the methods used to attack the problem and some of the results. Section 4 discusses the results in the light of the scaling theory of localization.

2. The Model

The interest in studying the random exchange Heisenberg antiferromagnet (REHAF) arose from the pioneering experiments by SHEGOLEV [1] and BULAEVSKII et al. [2]. They measured the thermodynamic properties of charge transfer

[1]This is a summary of a more complete version to appear in Phys.Rev.B [8], prepared and presented by one of us (J.V.J.).

[2]Present addresses (J.H.) I.T.P. Santa Barbara, CA 93106; (J.J.) Northeastern U., Boston, MA 02115.

complexes of TCNQ. They found that the susceptibility, χ, and the specific heat, C, behave like

$$\chi \sim T^{-\alpha_\chi} \; ; \; C \sim T^{1-\alpha_C} \tag{1}$$

respectively, where the magnetization, M, as a function of the magnetic field h goes as $M \sim h^{1-\alpha_M}$. The exponents (α_χ, α_C, α_M) are smaller than 1 and, within their experimental accuracy, seemed to be equal. In order to understand these findings Bulaevskii et al proposed a REHAF model defined by the Hamiltonian

$$H = \Sigma_i J_i \vec{S}_i \cdot \vec{S}_{i+1} \quad , \tag{2}$$

with $\{J_i\}$ a positive set of independent random variables (i.r.v.). $\{\vec{S}_i\}$ are the Pauli spin-1/2 operators. The thermodynamic properties result from the quenched average of the configurational free energy $F\{J_i\}$. The problem is defined after specifying the form of $P_o(J_i)$. Without knowing what this function was Bulaevskii et al. mapped the model to an interacting fermion problem. Assuming the validity of a Landau type fermi liquid theory, they guessed what the form of the density of states would be in order to get the experimental results given in (1). This implied assuming $\alpha_\chi = \alpha_M = \alpha_C = \alpha$=constant.

A more microscopic approach was carried out by THEODORU and COHEN (TC) [3] starting from a Hubbard Hamiltonian with random onsite energies. At low temperatures they showed that Hubbard Hamiltonian reduces to (2). From a simple statistical argument they obtain the $P_o(J)$ as

$$P_o(J) \sim 1/J^\alpha \quad , \tag{3}$$

with $J \varepsilon [0, J_o]$ and α a constant between 0 and 1.

More recently, CLARK and TIPPIE [4] studied the RHAF with (3) assuming that every other spin is decoupled. This approximation is rather crude because it replaces the N-spin problem with the 2-spin one, which can be solved explicitly even in the presence of a field. Nevertheless, their results are the same as those of Bulaevskii et al or TC.

Within the last year three different groups have investigated the model given in (2) and (3) incorporating many spin effects. Their methods of approach are different but their conclusions are similar and complementary. BONDESON and SOOS [5] have used direct diagonalization numerical methods to calculate χ, M and C for chains of up to 10 spins. This method is best for high T's. MA, DASGUPTA and HU [6] have used a method that involves successive elimination of strongly coupled spins together with a change in the cutoff of $P_o(J)$ to calculate the thermodynamic properties of the REHAF. Lastly, the present authors extended the quantum real space RG method to finite temperatures and with random exchanges to evaluate χ and C at low temperatures [7,8]. In the three approaches it was concluded, to differing degrees, that the specific form of $P_o(J)$ is irrelevant to derive a χ,M,C behavior as expressed in (1). In the remainder of this paper we will discuss the work of ref. [7,8].

3. Methodology and Results

The approach is an extension of the real space RG developed by DRELL et al [9] at T=0, to finite T's and with random J's. The RG entails thinning out the degrees of freedom in the system. The method consists of 3 steps. First the lattice is divided into cells. Here the cells are taken to be of 3 spins to allow the mapping of a spin-1/2 Hamiltonian to a spin-1/2 Hamiltonian. Next, the cell Hamiltonian is diagnonalized exactly and the lowest eigenvalues are used to construct the new, or renormalized, cell Hamiltonian with the same form as in (2) but with a new exchange constant J_1. This operation is denoted symbolically by $J_1=RJ_o$. The probability law for the set of new J's is also obtained as $P_1(J)=RP_o(J_o)$. Repeating this procedure N-times leads to the asymtotic result [7,8]

$$P_n(J_n) = \frac{(1-\alpha)^{3^n}}{(3^n-1)!} \frac{(n\, J_n)^{3^n-1}}{J_n^{\alpha}} \quad . \tag{4}$$

Here we took $P_o(J)$ as given in (3) but with arbitrary value for α (positive or negative), the 3 comes from the size of the cell and 3^n=n. This calculation is done at T=0 and the renormalizations are of J and P exclusively. However, it turns out (see below) that, for 1-D systems, the form of $P_n(J_n)$ determines the low temperature properties of the model.

To obtain $[F]_{av}$ it is realized that at low T the main contributions come from having small energy gaps. Dividing the energy eigenvalues of the cell into high lying, ΔE_1, low lying, ΔE_2, and the coupling between high and low ΔE, we take as the zeroth order approximation $\Delta E_3=0$. This approximation is equivalent to dividing the chain into thermally disconnected segments. The free energy becomes equal to the series

$$[F]_{av} = \frac{-T}{3} \sum_{n=0}^{n} \frac{1}{3^n} [P_n(\Delta E_1 \text{ or } \Delta E_2 < T)] \quad . \tag{5}$$

The Universality of the results arises from analyzing this result. First we note that it depends strongly on the renormalized $P_n(J_n)$. If we plot the sequence of $P_n(J)$ for n=0,1,2,3,... for arbitrary $P_o(J)$ we find that after a few iterations $P_n(J) \to \delta(J)$. Taking the derivatives with respect to the appropriate thermodynamic fields we find that [7,8] i) C take the forms given in (1) but with $\alpha_\chi \neq \alpha_c$, ii) α_χ and α_c are weakly dependent functions of temperature ($\sim \ln 1/T$), iii) in the limit when $T\to 0, \alpha_\chi, \alpha_c \to 1$, but with $\alpha_\chi > \alpha_c$ for all T. Some of these results are further supported by the analysis of [6,5].

The method outlined above has been used [10] to study the anisotropic Heisenberg model. It is found that the calculations are essentially the same as above for the x-y model, thus leading to the same conclusions as in the Heisenberg case, whereas the Ising model limit behaves classically. The results found for the x-y model via this approach compare satisfactory with the exact solution obtained by Dyson on the random chain of oscillators problem. This leads further support to the results obtained for the Heisenberg model.

4. Connections to Localization

Following a recent paper by Anderson et al. [11] we look into the scaling properties of the scaled probability law $P_n(J_n)$. In order for $P_n(J_n)$ to characterize a well defined probability law we ought to check the behavior of its different moments. Because of the log's in (6) it is clear that $P_n(J_n)$ has long "tails" and is strongly centered around $J_n=0$. In fact $(\langle(\nabla J_n{}^2)\rangle/\langle J_n\rangle^2) \gg 1$, when n is large. This quantity measures the amount of fluctuation associated with $P_n(J_n)$. Clearly, $P_n(J_n)$ is not a well behaved function for the statistical point of view. The suggestion of Anderson et al. is to look for the appropriate scaling variable "Y" for the problem in terms of which the probability law, T (Y), obeys the central limit theorem.

First we notice that $Y=|\ell nJ|$ is the "appropriate" scaling variables in our problem. We can immediately write the RGβ - function for $\langle Y\rangle$ giving [8]

$$\beta = \frac{d\langle Y\rangle}{d\ell nN} = \langle Y\rangle \quad . \tag{6}$$

This equation is of exactly the same form as that derived by ABRAHAMS et al. [12] in the localized region of the random electronic problem [13]. Integrating (6) leads to

$$J_{scaling} \sim e^{-N/\lambda} \quad , \tag{7}$$

that is characteristic of a localization region. From this result we suggest that the Universality of the results mentioned above can be understood in terms of the scaling theory of localization.

We turn now to the classical case. Again we want to study the scaling properties of $P_0(J_0)$ for the classical Heisenberg model. Using a decimation transformation it is found that [8]

$$P_n(J_n) = P_{n-1}(J_{n-1}) + \text{weaker corrections} \quad , \tag{8}$$

i.e. the probability law remains essentially unmodified by the renormalization procedure. Looking at the recursion relations for J_n it is found that

$$\frac{1}{J_n} = \sum_{i=1}^{N} \frac{1}{J_i} \quad , \tag{9}$$

meaning that the scaling variable for the classical case is $1/J$. Running through the same arguments leading to (11), starting with $P_0(J) \sim \exp -\lambda/J$, we get

$$J_{scaling} \sim 1/N \tag{10}$$

in contrast to (7). The result (10) is reminiscent of an extended-like behavior for $J=J(N) \sim 1/N^d$ with d=dimensionality. Thus we see that we go from the s=1/2 localized result of (7) to the extended-like $s=\infty$ case.

References

1. I.F.Shegolev: Phys. Stat. Solid (a) 12, 9 (1972)
2. L.N.Bulaevskii et al: Zh. Eksp. Teor. Fitz., 62, 725 (1972) (Sov. Phys. JETP 35, 384 (1972))
3. G.Theodoru and M. Cohen: Phys. Rev. Lett. 37, 1014 (1972)
4. G.Clark and C.Tippie: Phys. Rev. B20, 2914 (1979)
5. S.R.Bondeson and Z.G.Soos: S. St. Comm. 35, 11 (1980)
6. S.K. Ma, C.Dasgupta and C.K.Hu: Phys. Rev. Lett. 43, 1434 (1979) and C.Dasgupta and S.Ma: Phys. Rev. B (to appear)
7. J.Hirsch and J.José: J. Phys. C13, L53 (1980)
8. J.Hirsch and J.José: Phys. Rev. B (in press)
9. S.Drell et al: Phys. Rev. 16, 1969 (1977)
10. J.Hirsch (preprint)
11. P. Anderson et al: (to appear Phys. Rev. B)
12. E.Abrahams et al: Phys. Rev. Lett. 42, 673 (1979)
13. Recently Lukes has reanalyzed Anderson model in d=1 and has also obtained the Γ-distribution in his case (preprint).

Localization in One Dimension

D.J. Thouless

Department of Engineering and Applied Science, Becton Center, Yale University
New Haven, CT 06520, USA

Abstract

A brief survey of the theory of localization in one-dimensional systems and of its application to conduction in thin metallic wires is given.

Soon after ANDERSON [1] had shown that strong static disorder could localize electrons in solids, MOTT and TWOSE [2] and LANDAUER [3] argued that any disorder was sufficient to localize electrons in a one-dimensional system. For various reasons it is not a straightforward matter to study this effect in real systems. One possibility is to measure the electrical resistance of long very thin wires. The conditions necessary to observe one-dimensional behavior in thin metal wires were discussed by THOULESS [4]. Electrons are localized by static disorder in a length of the wire that corresponds to a zero-temperature resistance of about 20 kOhm, but inelastic scattering of electrons from one energy level to another produced by thermal motion destroys the localization, so rather low temperatures and high impurity concentrations are needed to observe localization. The effects of localization in such systems have proved harder to see than was expected, either because the rate of inelastic scattering in such highly disordered materials is much higher than was expected, or because many-body effects of electron-electron interactions mask the localization [5].

A second possibility for observation of localization is in long conducting molecules. Here there are two effects competing with the localization within the molecule. In the first place the thermal motion of the molecule can cause an electron to undergo a transition from one localized state to another, just as in the wire. Secondly the electron can tunnel from one molecule to a neighboring molecule. Both these effects must be relatively small for localization to be observed.

The theory of a strict one-dimensional system (essentially a chain with no breadth) was worked out by LANDAUER [3], and the theory is given here with some modifications introduced by ANDERSON et al. [6]. We suppose the random medium can be represented by a set of reflecting barriers at random positions, so that if a wave e^{ikx} is incident at the nth center, a wave $r_n e^{-ikx}$ will be reflected and a wave $t_n e^{ikx}$ transmitted, where $|r_n|^2 + |t_n|^2 = 1$. The wave $u_{n-1}e^{ikx} + v_{n-1}e^{-ikx}$ to the left of this center is related to the wave $u_n e^{ikx} + v_n e^{-ikx}$ to the right by the transfer matrix equation

$$\begin{bmatrix} u_{n-1} \\ v_{n-1} \end{bmatrix} = \begin{bmatrix} 1/t_n & r_n^*/t_n^* \\ r_n/t_n & 1/t_n^* \end{bmatrix} \begin{bmatrix} u_n \\ v_n \end{bmatrix} \quad . \tag{1}$$

The amplitude transmission coefficient t and reflection coefficient r for a system of N barriers are given by the product of transfer matrices in the form

$$\begin{bmatrix} 1 \\ r \end{bmatrix} = \prod_{n=1}^{N} \begin{bmatrix} 1/t_n & r_n^*/t_n^* \\ r_n/t_n & 1/t_n^* \end{bmatrix} \begin{bmatrix} t \\ 0 \end{bmatrix} \quad . \tag{2}$$

If we take the geometric mean of $u_{n-1} = u_n/t_n + v_n r_n^*/t_n^*$ over the random phase of the reflection coefficient r_n we get $\overline{u_{n-1}} = \overline{u_n/t_n}$, and so the geometric mean of t is $\prod_n t_n$; the transmission coefficient is a multiplicative function. This means that on the average multiple reflections cancel out, and only the transmission of the unreflected wave contributes to the total transmission through the barriers. The wave function therefore falls off exponentially with distance.

If the barriers scattered waves incoherently, (1) would be replaced by

$$\begin{bmatrix} |u_{n-1}|^2 \\ |v_{n-1}|^2 \end{bmatrix} = \begin{bmatrix} 1/|t_n|^2 & |r_n|^2/|t_n|^2 \\ |r_n|^2/|t_n|^2 & 1 - |r_n|^2/|t_n|^2 \end{bmatrix} \begin{bmatrix} |u_n|^2 \\ |v_n|^2 \end{bmatrix} \tag{3}$$

with the same boundary conditions $|u_0|^2 = 1$, $|v_N|^2 = 0$. The intensity reflection coefficient $|r|^2$ and transmission coefficient $|t|^2$ of the system are given by $|v_0|^2$ and $|u_N|^2$, and the solution of (3) gives

$$|r/t|^2 = \sum_{n=0}^{N} |r_n|^2/|t_n|^2 \quad . \tag{4}$$

LANDAUER [3] showed that the electrical resistance of the system is (for degenerate electrons with two independent spin states)

$$R = (\pi\hbar/e^2)|r/t|^2 \quad .$$

so (4) says that for incoherent scattering the resistance of a series of barriers is the sum of the resistances of the individual barriers. However, (1) and (2) imply that for static barriers, for which phase coherence in multiple scattering is not lost, it is $\ell n[1 + (e^2/\pi\hbar)R]$ which is an additive function of length. This leads to the result

$$R(L) = \frac{\pi\hbar}{e^2}\left[\exp\left(\frac{e^2}{\pi\hbar}\rho L\right) - 1\right] \approx \rho L\left[1 + \frac{e^2}{2\pi\hbar}\rho L\right] \quad , \tag{6}$$

where R(L) is the resistance of a length L and ρ is the resistance per unit length for short lengths.

In order that the effects of localization should be observable in thin metallic wires, the quantity ρL must be comparable with $2\pi\hbar/e^2$, which has a value of about 26 kOhm. This seems easy to attain, but the length L in this expression is not the total length of the wire, but is the length scale over which the multiple scattering is coherent, so that no inelastic (energy changing) scattering has occurred [4,7,8]. This length L is increased by working at low temperatures, so that there are few phonons or exited electrons available to cause inelastic scattering of the electrons. The resistance per unit length ρ is increased by using very fine wires in which the electrons have a short mean free path. It is of course essential that the length L be at least as great as the cross-sectional dimensions of the wire or else the behavior will not be one dimensional.

The length L is the distance an electron wave packet can diffuse before inelastic scattering. Because the electron motion over distances larger than the mean free path is essentially a random walk, when the time for inelastic scattering is much greater than the time for elastic (impurity) scattering this distance L is the mean free path λ multiplied by the square root of the ratio of the inelastic scattering time to the elastic scattering time:

$$L = \lambda(\tau_i/\tau_e)^{1/2} = u_F(\tau_i\tau_e)^{1/2} \quad . \qquad (7)$$

Early estimates of the inelastic scattering time suggested that it would be fairly easy to get into the regime where $e^2\rho L/2\pi\hbar$ is of the order unity [4], but experiments have shown that localization is not so easy to observe as was supposed. DOLAN and OSHEROFF [9,10] studied very thin films of Au-Pd. These films were laid down as long narrow strips, and if the estimates of τ_i had been correct these would have been narrow enough to have behaved one dimensionally, but in fact they seem to have a two-dimensional behavior. Two successfull sets of experiments have been performed. GIORDANO, GILSON, and PROBER [11] and GIORDANO [12] have made wires of Au-Pd whose cross-section is a triangle whose dimensions are a few hundred Ångstroms. These results show an increase of resistance as the temperature is lowered which is at most 2%. The increase is inversely proportional both to the cross-sectional area of the wire and to the square root of temperature. The area dependence is what should be expected from (6). The temperature dependence suggests, from (6) and (7), that the inelastic scattering time τ_i is proportional to T^{-1}, which is not in agreement with any simple mechanism for the inelastic scattering.

Similar results were obtained by CHAUDHARI and HABERMEIER [13,14] using very thin and quite narrow (down to about 800 Å) films of W-Re. The temperature range was more restricted, but the area dependence was the same, and the magnitude of the effect was very similar. Because W-Re is superconducting it is possible to make an independent measurement of τ_i. When a very narrow wire is made and a supercurrent passed through it the first thing that happens when the current exceeds a critical value is that the wire goes normal in one region whose size is determined by the distance electrons can diffuse before recombining into pairs; this distance is essentially L as given by (7). The ratio of the resistance of this single phase slip center to the resistance to higher currents, when the whole wire is normal, is the ratio of L to the length of the wire. There are some corrections due to superconducting coherence factors, but the result for L obtained by this method, of order 500 Å, is close to the figure needed to explain the localization results at this temperature [15].

GARLAND, GULLY, and TANNER [16] have tried to detect one-dimensional localization in a silver composite. It is argued that close to the percolation threshold the conductivity is determined by long paths with a high resistance. However, these paths are not simple, and the fact they are connected to many dead ends makes an important change in the inelastic scattering time. The interpretation of the null result of this experiment is therefore not straightforward. There is also an experiment by OVERCASH et al. [17] using Sb-doped Bi whiskers. In this case the low density of electrons helps to make the system more one dimensional, but the resistivity is fairly low, and I do not think that convincing estimates of the inelastic scattering time have been made. If this time is comparable with the times needed to explain the measurement on Au-Pd and W-Re, the localization effectsin the Sb-Bi whiskers would be somewhat too small for the sensitivity of the reported measurements.

The high rate of inelastic scattering and its proportionality to T, rather than the expected T^2 or T^3, are an unsatisfactory feature of the explanation of these results, and similar results for two-dimensional systems, in terms of localization. ALTSHULER, ARONOV, and LEE [5] have an alternative explanation of these results in terms of a theory based on the scattering by dilute static impurities of interacting electrons. This theory seems to be different from a localization theory, but it gives very similar results, with τ_i replaced by $\hbar/kT$. Perhaps if the electron-electron interaction for electrons in a dirty material were taken into account properly it would give a τ_i of this form, but it has not been shown that this is so.

This work was supported in part by NSF Grant No. DMR-7908329.

References

1. P. W. Anderson, Phys. Rev. 109, 1492 (1958)
2. N. F. Mott and W. D. Twose, Adv. Phys. 10, 107 (1961)
3. R. Landauer, Phil. Mag. 21, 863 (1970)
4. D. J. Thouless, Phys. Rev. Lett. 39, 1167 (1977)
5. B. L. Altshuler, A. G. Aronov, and P. A. Lee, Phys. Rev. Lett. 44, 1288 (1980)
6. P. W. Anderson, D. J. Thouless, E. Abrahams, and D. S. Fisher, Phys. Rev. B, to be published
7. D. J. Thouless, J. Non-Cryst. Solids 35-6, 3 (1980)
8. D. J. Thouless, Solid State Commun. 34, 683 (1980)
9. G. J. Dolan, D. D. Osheroff, and D. C. Tsui, Bull. Am. Phys. Soc. 24, 233 (1979)
10. G. J. Dolan and D. D. Osheroff, Phys. Rev. Lett. 43, 721 (1979)
11. N. Giordano, W. Gilson, and D. E. Prober, Phys. Rev. Lett. 43, 725 (1979)
12. N. Giordano, to be published
13. P. Chaudhari and H.-U. Habermeier, Phys. Rev. Lett. 44, 40 (1980)
14. P. Chaudhari and H.-U. Habermeier, Solid State Commun. 34, 687 (1980)
15. P. Chaudhari, private communication
16. J. C. Garland, W. J. Gully, and D. B. Tanner, Bull. Am. Phys. Soc. 24, 280 (1979)
17. D. R. Overcash, B. A. Ratnam, M. J. Skove, and E. P. Suttwell, Phys. Rev. Lett. 44, 1348 (1980)

Localization in Thin Wires

N. Giordano

Department of Physics, Purdue University
West Lafayette, IN 47907, USA

Abstract

Recent experimental studies of electrical conduction in very thin wires at low temperatures are reviewed and compared with theoretical predictions based on localization in one dimension. While the principal experimental results are in good qualitative agreement with the theory, there are several quantitative discrepancies. At least some of these discrepancies appear to be due to inadequacies in our present understanding of inelastic scattering processes in disordered metals, but recent suggestions that electron-electron interaction effects are playing an important role cannot be ruled out.

1. Introduction

Some years ago it was shown by MOTT and TWOSE [1], by LANDAUER [2], and by a number of other workers [3], that all of the electronic states of a disordered one-dimensional system would be spacially localized. This result implies that the conductance of such a system will vanish at absolute zero. This behavior is in sharp contrast to that of an ordinary "nonrandom" metal, which exhibits a finite nonzero conductance as $T \to 0$. These predictions concerning the behavior of disordered one-dimensional systems did not appear to be accessible to experimental study until recently when THOULESS [4] argued that all systems, regardless of their cross-sectional area, will behave one dimensionally insofar as localization is concerned provided only that they are sufficiently long. It was predicted [4] that all one dimensional systems (i.e., wires) with impurity resistances greater than about $2\hbar/e^2 (\simeq 8\ k\Omega)$ would exhibit thermally activated conductances at low temperatures, and be insulators at absolute zero. The temperature at which this behavior becomes observable, T_L, is predicted to increase both as the cross-sectional area of the wire, A, is decreased, and as the amount of randomness in the wire (i.e., the impurity resistivity) is increased. For macroscopic size wires T_L is predicted to be unattainably low, but for wires with $A = 2.5 \times 10^{-11}\ cm^2$ (a diameter of order 500 Å) T_L is predicted to be of order 1 K. This is why very thin wires are needed for studies of this kind.

In this paper we review and discuss the results of recent experiments aimed at observing the effects of localization in thin wires. In the next section (2) we will briefly discuss the theoretical predictions which are pertinent to the experiments [5]. In Section 3 we describe the experimental results which have been obtained to date, and compare them with the theoretical predictions. The final section (4) contains our conclusions and an outlook for future work in this field.

2. Theory

Since the work of THOULESS [4], a number of workers have considered the problem of localization both in one dimension and in thin wires using various theoretical approaches [5]. All of these theories make basically the same prediction for the behavior of the resistance, R, as a function of temperature. When the effect of localization is small, i.e. at high temperatures, they predict that $R = R_o + \Delta R$ with

$$\frac{\Delta R}{R_o} = \frac{\rho_e\sqrt{2D\ \tau_i}}{A\ R_T} \quad . \tag{1}$$

Here R_o is the impurity resistance (which is, of course, temperature independent), D is the electronic diffusion constant, τ_i is the electron inelastic scattering time, ρ_e is the "bulk" resistivity due to (elastic) impurity scattering, and R_T is the impurity resistance corresponding to the length of a localized state and which recent work has predicted to be 36 KΩ [6]. It is useful to rewrite (1) using the relations $D = v_F{}^2\tau_e/3$, where v_F is the Fermi velocity and τ_e the elastic scattering time, and $\rho_e = m/(ne^2\tau_e)$ where m is the mass, e is the charge, and n is the density of electrons [7]. We find

$$\frac{\Delta R}{R_o} = \left[\frac{2m\,\rho_e\,\tau_i}{3n}\right]^{\frac{1}{2}} \frac{v_F}{e\ A\ R_T} \quad . \tag{2}$$

Eq. (2) is the basic theoretical prediction which we will be comparing with the experimental results. It states that the effect of localization (i.e., the resistance increase) is proportional to A^{-1}, to $\sqrt{\tau_i}$, and to $\sqrt{\rho_e}$ (assuming, of course, that the various other factors in (2) are independent of these quantities). We also see that all of the temperature dependence comes from that of τ_i. For this reason, τ_i is a key quantity in the discussions which follow.

Unfortunately, it is impossible to make a reliable a priori estimate of τ_i. There are a number of different inelastic scattering mechanisms which could conceivably be important. The most obvious mechanisms are ordinary electron-phonon and electron-electron scattering, and these have been considered in detail by Thouless [4]. As we will see in the following section, the experiments are not consistent with either of these scattering mechanisms. This has led to the suggestion that some other scattering mechanism is dominant in these systems [8,9]. One such mechanism is the scattering of electrons from the "tunneling levels" which are believed to be present in all highly disordered systems [10,11]. We will have more to say about τ_i and the dominant inelastic scattering process after the experimental results have been discussed.

Very recently it has been pointed out that electron-electron interaction effects could lead to behavior very similar to that predicted by theories based on localization. The electron-electron interaction effects are predicted [12] to lead to behavior identical to (2) except that τ_i is replaced by $\hbar/kT$. We will see that because the predictions of the two theories are so similar, it is very difficult (although not impossible) for experiments to distinguish between them. We will comment further on this problem in section 4. In the following section we will often, for the sake of brevity, refer to "localization effects". Unless noted otherwise, such a reference is intended to include both localization and interaction effect theories.

3. Experimental Results

The first observation of localization effects in thin wires was made by DOLAN and OSHEROFF [13]. They studied relatively wide (typically $\sim 1\ \mu m$), very thin (~ 50 Å) strips of AuPd alloys made using optical lithography. From the initial estimates of THOULESS [4], it was expected that one-dimensional localization effects would be very pronounced in these samples, and a resistance increase was in fact observed at low temperatures. However, it was found that in all but the narrowest samples the resistance rise was a two-dimensional effect. That is, the magnitude of the resistance rise was independent of the width of the sample. Behavior of this kind in a two-dimensional system was completely unexpected, and this result has helped to stimulate a great deal of interest in the properties of disordered two-dimensional metals [14], a topic which is beyond the scope of this paper. The results of DOLAN and OSHEROFF showed that the effects of localization in one dimension are much smaller then originally predicted, and implied that the theory [4] had greatly overestimated the inelastic scattering time. The magnitude of the discrepancy was so large that, as noted above, it was suggested [8,9] that the inelastic scattering was due to a mechanism not considered in the original calculations.

GARLAND, GULLY, and TANNER [15] have studied percolative mixtures of Ag particles in KCℓ. Estimates based on THOULESS' predictions [4] implied that localization effects would be observable in the samples which were studied. However, measurements at temperatures as low as 0.5 mK showed no evidence for localization. Thus, as concluded from the results of DOLAN and OSHEROFF, the effects of localization were found to be smaller than originally predicted. We should note that the percolative mixtures are somewhat different from the continuous metallic structures which have been considered by the theory, and it is conceivable that the conditions necessary for observing localization could be different in these systems.

GIORDANO and coworkers [16,17] have studied continuous AuPd wires made using a lithographic technique described in [18]. These wires had triangular cross sections, with areas as small as 1×10^{-11} cm^2 ($\sqrt{A} \simeq 300$ Å). They were found to exhibit the effects of localization, as the resistance of the wires increased as the temperature was lowered. This increase was found to vary as A^{-1}, in good quantitative agreement with the theory (2). The temperature dependent part of the resistance was found to vary as $T^{-\frac{1}{2}}$ [17]. From (2), these results imply that the inelastic scattering time varies as T^{-1} and is independent of the cross-sectional area of the wire. To our knowledge, the only scattering process which behaves in this way is electron-tunneling level scattering [4,9,11]. The absolute magnitude of τ_i for this process is difficult to estimate a priori. BLACK, GYORFFY, and JÄCKLE [11] have used experimental data for the metallic glass PdSiCu to estimate $\tau_i = 2.3 \times 10^{-9}\ T^{-1}$ (sec-K) in this system. If we use this value of τ_i in (2), we find a value of $\Delta R/R_o$ which is about a factor of seven larger than observed experimentally. Given the fact that this value of τ_i was estimated for a different material than was used for the wires [19], and the general difficulties encountered in making absolute estimates of scattering times, this level of agreement may be reasonable. We should note, however, that the observed behavior is also consistent with the predictions of theories based on electron-electron interaction effects. These theories also predict the correct magnitude of the resistance rise to within a factor of order unity (a more precise statement cannot be made because of uncertainties in the values of parameters such as v_F and τ_e which enter in (2)).

To this point, we have considered the dependence of the resistance rise on the cross-sectional area of the wire and on temperature. Eq. (2) also predicts the dependence of the effect on ρ_e, the impurity resistivity. This dependence has been investigated in two ways. First, by depositing the AuPd in two different ways (sputtering and thermal evaporation) it was possible to make wires with two different values of ρ_e (but with the same composition) [16,17]. Second, it was possible to anneal the sputtered wires studied previously [16,17] (the sputtered AuPd had a higher resistivity than the evaporated AuPd) by gently heating the wires. We expect that this annealing, which was carried out at temperatures less than 200°C, well below the melting point of AuPd, did not change the cross-sectional areas of the samples. The advantage of annealing was that it allowed us to vary ρ_e continuously over a reasonably wide range.

Figure 1 shows the results of an annealing experiment. This wire was measured three times, and between measurements the wire was annealed. We see that as ρ_e decreased, the fractional resistance rise $\Delta R/R_o$ also decreased, in qualitative agreement with (2). The inset in Fig. 1 shows the variation of the resistance rise with R_o, and it can be seen that $\Delta R/R_o$ varies linearly with R_o and hence with ρ_e, in contrast with the $\sqrt{\rho_e}$ variation given by (2).

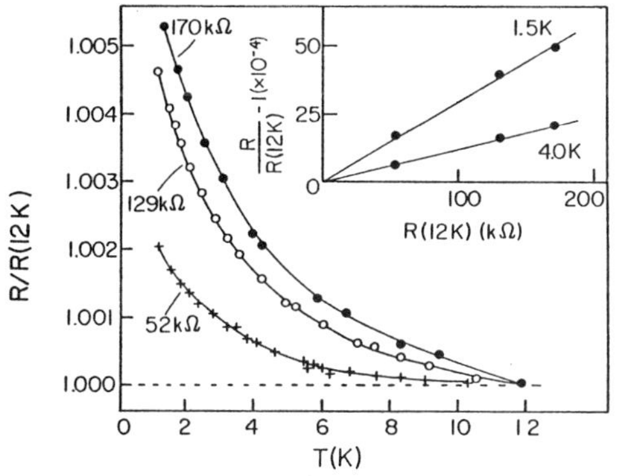

Fig.1 Resistance as a function of temperature for a AuPd wire with $\sqrt{A}$ = 570 Å. Each curve corresponds to a different resistivity, obtained by annealing as described in the text. The values of R(12 K) are indicated in the figure, and are proportional to the impurity resistivities which were 360, 270, and 110 μΩ-cm. The inset shows the fractional resistance rise at 4.0 K and 1.5 K as a function of R(12 K).

Figure 2 shows results for the resistance rise as a function of ρ_e. Here we include data for unannealed wires made from sputtered and evaporated AuPd [16, 17], together with the results of several annealing experiments of the sort shown in Fig. 1. We see again that $\Delta R/R_o$ varies approximately linearly with ρ_e, in contrast with the $\sqrt{\rho_e}$ variation predicted by (2) and shown as the dashed line in Fig. 2. It is, of course, possible that other factors in (2) vary with ρ_e. One might argue that if the inelastic scattering is due to electron-tunneling level scattering, then τ_i should change as the sample is annealed, since the number of tunneling levels would be expected to decrease with annealing. However, this would cause τ_i to increase as ρ_e decreases, and this is the wrong direction to explain the results in Fig. 2 (the data fall below the theoretical (dashed) curve, not above). It is also possible that the factor $v_F\sqrt{m/n}$ is varying as ρ_e is changed, but it seems unlikely that this quantity will change by the amount needed (50%) to account for the results at low values of ρ_e. We, therefore, have a problem in understanding the variation of $\Delta R/R_o$ with ρ_e. This variation is inconsistent with the theories based on electron-electron interaction effects, and with theories based on localization unless the inelastic scattering time exhibits an unexpected and so far unexplained dependence on ρ_e.

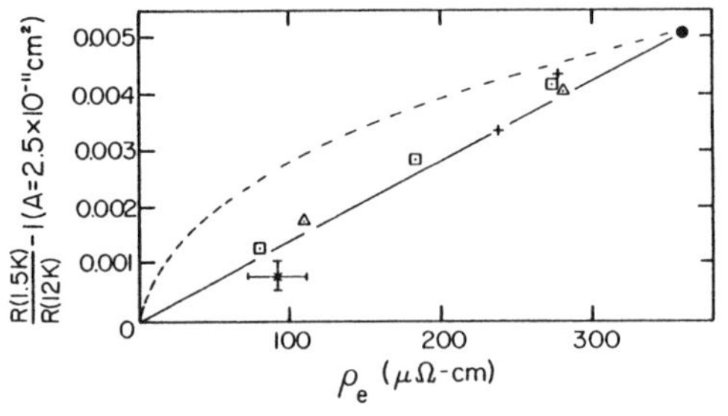

Fig.2 Magnitude of the resistance rise, as measured by the difference between the resistances at 1.5 K and 12 K normalized by the value at 12 K, for wires with $\sqrt{A}$ = 500 Å as a function of the impurity resistivity, ρ_e. The resistance rise found in thin films of the same thickness as the wires has been subtracted out [17] (this contribution corresponds roughly speaking to two-dimensional behavior; it was 17% of the total rise at the highest ρ_e and was linearly proportional to ρ_e so that it in no way affects the linearity of the curve). The (●) and (X) represent data for unannealed AuPd wires [16,17]. The other symbols correspond to sputtered wires which were annealed by various amounts. Note that the results for the annealed wires have been scaled (using the previously noted A^{-1} dependence) so as to correspond to a wire with $\sqrt{A}$ = 500 Å. The dashed curve shows the $\sqrt{\rho_e}$ dependence predicted by (2) - this curve was normalized to fit the data at the largest value of ρ_e.

CHAUDHARI and coworkers [20-22] have studied thin wires of amorphous WRe made using electron beam lithography. The wires were very thin (50 Å) strips with widths of order 1000 Å. Experiments with these wires have shown that they exhibit localization effects. The resistance rise was found to vary approximately as $T^{-\frac{1}{2}}$, and also as A^{-1} [21]. These results are in good agreement with those of GIORDANO et al. [16,17]. In fact, the resistivity ρ_e of the WRe wires is the same as for wires made from sputtered AuPd, and the two types of wires exhibit resistance increases which are the same (for wires of the same cross sectional area) to within the experimental error [21]. We should note that WRe is superconducting with $T_c \simeq 4$ K for the composition studied by CHAUDHARI et al. These authors have utilized this by using the phase slip centers [23] observable below T_c to measure the inelastic scattering time. The value found [22] is in good agreement with that expected from (2) in conjunction with the measured value of $\Delta R/R_o$. This is the first direct evidence that the inelastic scattering times are of the correct magnitude to explain, within the localization picture, the behavior of the resistance which has been observed.

All of the experiments described so far have involved wires either on substrates or imbedded in a matrix. The only work to date on freely suspended wires is that of OVERCASH et al. [24]. These authors have studied the behavior of whiskers of Bi and Bi doped with Sb, with diameters as small as 1400 Å. They find no evidence for localization even though their estimates of τ_i indicate that the effects should be readily observable. We note, however, that these estimates were made in a manner similar to those of THOULESS [4], and as we have noted, the latter are not consistent with the experiments on AuPd and WRe wires. If the results for AuPd and WRe are extrapolated using the observed dependences on A and ρ_e to parameter values appropriate to the samples of OVERCASH et al., a resistance rise smaller than their experimental resolution is found. Further work to resolve this problem is certainly called for.

Very recently, two more experiments on thin wires have been performed. WHITE, FLANDERS, and SKOCPOL [25] have studied Ni wires a few hundred Å in diameter and preliminary results indicate a resistance rise consistent with that found for AuPd and WRe wires. GOULD et al. [26] have studied heavily doped polyacetylene at very low temperatures. Polyacetylene consists of a loosely coupled bundle of fibers, with the diameter of each fiber being of order 100 Å. The resistance of these samples exhibited a logarithmic dependence on temperature below about 10 mK, indicative of two-dimensional rather than one-dimensional behavior [14]. The reason for this is not clear, but it may be due to the layered molecular structure which is believed to occur in the heavily doped material [26].

4. Summary and Conclusions

There have been a number of different experimental studies of localization in thin wires made using a variety of different techniques and composed of several different materials. From the results of these studies, a reasonably consistent picture has emerged. First, the effects are much smaller than originally expected [4]. This accounts for the null results obtained by DOLAN and OSHEROFF [13] (except for their narrowest samples), and GARLAND, GULLY, and TANNER [15]. Second, the experiments on both AuPd [16,17] and WRe [20,21] wires have shown that the effect varies as A^{-1} and as $T^{-\frac{1}{2}}$. While this suggests that localization with a scattering mechanism such as electron-tunneling scattering is dominant, it is also consistent with the interaction effect theory. The phase slip center measurements by CHAUDHARI et al. [22] are the only results to date which distinguish between the two theories. These results are direct evidence that the inelastic scattering time is as short as required by localization theories, and this means that the experiments are indeed observing the effects of localization. It seems premature, however, to rule out the possibility that the interaction effect mechanism is also contributing to the observed effects. That is, the behavior could be due to a "sum" of localization and interaction effects. A theoretical treatment which includes both of these effects would certainly be of interest.

The picture described above seems reasonable, but there are a number of unexplained problems. The dependence of the effect on ρ_e appears to be inconsistent with both theories and cannot be understood unless τ_i has a completely unexpected dependence of ρ_e. The null result obtained with the Bi whiskers as well as the unexpected behavior of polyacetylene also deserve further attention. Finally, the precise nature of the inelastic scattering mechanism is not entirely clear. While electron-tunneling level scattering is consistent with the experimental results, further study of this matter is needed.

I would like to thank the authors of [22, 25, 26] for permission to describe and discuss their results prior to publication. I also thank P. H. Keesom for the loan of some equipment, and R. Landauer for bringing some of the papers in [2] to my attention. This work was supported in part by a fellowship from the Alfred P. Sloan Foundation, by the Research Corporation, and by National Science Foundation Grants ENG77-10164, DMR78-17957, and DMR79-06716.

References

1. N. F. Mott and W. D. Twose, Adv. Phys. 10, 107 (1961).
2. R. Landauer, Philos. Mag. 21, 863 (1970) and references contained therein, including IBM J. Res. Dev. 1, 223 (1957). See also the discussion in R. Landauer, AIP Conf. Proc. 40, 2 (1978).
3. See for example, B.I. Halperin, Adv. in Chem. Phys. 13, 123 (1967); K. Ishii, Prog. Theor. Phys. Suppl. 53, 77 (1973); A. A. Abrikosov, and I. A. Ryzhkin, Adv. Phys. 27, 147 (1978); D. J. Thouless, J. Phys. C6, L49 (1973); and references contained therein.
4. D. J. Thouless, Phys. Rev. Lett. 39, 1167 (1977).
5. For a more extensive discussion of the theory, we refer the reader to the paper by D. J. Thouless at this conference.
6. P. W. Anderson, D. J. Thouless, E. Abrahams, and D. S. Fisher, to be published.
7. This expression for ρ_e is, of course, an approximation. However, for our purposes only the functional dependence of ρ_e on τ_e is important, and we would expect this to be more generally valid.
8. P. A. Lee, private communication.
9. D. J. Thouless, Solid State Comm. 34, 683 (1980).
10. See for example, P. W. Anderson, B. I. Halperin, and C. M. Varma, Philos. Mag. 25, 1 (1972); W. A. Phillips, J. Low Temp. Phys. 7, 351 (1972).
11. J. L. Black, B. L. Gyorffy, and J. Jäckle, Philos. Mag. B40, 331 (1979).
12. B. L. Altshuler, A. G. Aronov, and P. A. Lee, Phys. Rev. Lett. 44, 1288 (1980); B. L. Altshuler, D. Khmel'nitskii, A. I. Larkin, and P. A. Lee, to be published.
13. G. J. Dolan and D. D. Osheroff, Phys. Rev. Lett. 43, 721 (1979).
14. E. Abrahams, P. W. Anderson, D. C. Licciardello, and T. V. Ramakrishnan, Phys. Rev. Lett. 42, 673 (1979).
15. J. C. Garland, W. J. Gully, and D. B. Tanner, Bull. Am. Phys. Soc. 24, 280 (1979); and to be published.
16. N. Giordano, W. Gilson, and D. E. Prober, Phys. Rev. Lett. 43, 725 (1979).
17. N. Giordano, to be published.
18. D. E. Prober, M. D. Feuer, and N. Giordano, Appl. Phys. Lett. 37, 94 (1980).
19. We should note that the AuPd wires were polycrystalline, in contrast to the PdSiCu considered by [11] which is amorphous. However the AuPd wires had very high resistivities, corresponding to elastic mean free paths of order an interatomic spacing, so it is plausible that the two materials would exhibit similar properties.
20. P. Chaudhari and H.-U. Habermeier, Phys. Rev. Lett. 44, 40 (1980).
21. P. Chaudhari and H.-U. Habermeier, Solid State Commun. 34, 687 (1980).
22. P. Chaudhari, A. N. Broers, C. C. Chi, R. B. Laibowitz, E. Spiller, and J. M. Viggiano, to be published.
23. W. J. Skocpol, M. R. Beasley, and M. Tinkham, J. Low Temp. Phys. 16, 145 (1974).
24. D. R. Overcash, B. A. Ratnam, M. J. Skove, and E. P. Stillwell, Phys. Rev. Lett. 44, 1348 (1980).
25. A. E. White, D. C. Flanders, and W. J. Skocpol, to be published.
26. C. M. Gould, D. M. Bates, H. M. Bozler, A. J. Heeger, M. A. Druy, and A. G. MacDiarmid, to be published.

Transport Quantities in One-Dimensional Disordered Systems

Barbara S. Andereck and Elihu Abrahams

Rutgers University, Serin Physics Laboratory
Piscataway, NJ 08854, USA

In discussing transport quantities in one-dimensional disordered systems, care must be taken to assure that averaged quantities are statistically meaningful. ANDERSON, THOULESS, ABRAHAMS, and FISHER [1] have recently pointed to α, the inverse localization length, as a proper quantity to average. The mathematical properties of α have been discussed in the past by FURSTENBURG [2] and O'CONNOR [3], for example. The connection of α to resistance and conductance was made by LANDAUER [4]. He considered a collection of random barriers of total length L, and described by transmission coefficient T and reflection coefficient R = 1-T. By using the Einstein relation between conductivity and the diffusion constant, he showed that the conductance, G, is given by

$$G = \frac{e^2}{2\pi h}\frac{T}{R} \quad , \qquad (1)$$

where e is the electronic charge and h is Planck's constant. LANDAUER [4] and ANDERSON et al. [1] agree that averaging the resistances of a collection of random systems gives a quantity proportional to $\exp(2\alpha L)-1$, where α is a parameter independent of length. While this accurately describes average resistance its usefulness is questionable, for if we instead average G, we will find that $1/\langle G\rangle \neq \langle 1/G\rangle$. In fact, $\langle G\rangle$ will not even decrease exponentially with L. Remarking that average resistance and average conductance are not physical quantities that are representative of the ensemble in one-dimensional disordered systems, ANDERSON et al. [1] used scaling arguments to find a quantity that is statistically meaningful. Such a quantity is

$$\alpha = \frac{1}{L}\,\ell n(1+\rho) - \frac{1}{L}\,\ell n\, T \quad , \qquad (2)$$

where $\rho = R/T$ is the dimensionless resistance [5]. They argued that α obeys the central limit theorem for large L.

To elucidate these ideas, we studied a system of uniformly spaced, delta-function potentials of random strengths described by the Schrödinger equation

$$-\frac{\partial^2\psi(x)}{\partial x^2} + \sum_{n=1}^{L} q_n\,\delta(x-an)\psi(x) = E\psi(x) \quad , \qquad (3)$$

where a is the lattice spacing, hereafter be taken to be unity. The q_n's are distributed according to a rectangular probability distribution

$$P(q_n) = \begin{cases} \frac{1}{2q} & \text{if } -q \leq q_n \leq q \\ 0 & \text{otherwise} \end{cases} \quad . \qquad (4)$$

Let an electron with energy $E = k^2$ impinge upon the L-site system from the left. The transfer matrix, M, relates the wavefunction on the right, $\psi_R(x) = Ce^{ikx} + De^{-ikx}$, to the wavefunction on the left, $\psi_L(x) = Ae^{ikx} + Be^{-ikx}$ by $\psi_R = M\psi_L$. Here

$$M = \begin{pmatrix} 1/t^* & r/t \\ r^*/t^* & 1/t \end{pmatrix} \tag{5}$$

if ψ_L and ψ_R are represented by the spinors $\binom{A}{B}$ and $\binom{C}{D}$ respectively. r and t are the complex reflection and transmission amplitudes whose absolute squares are R and T.

The matrix, M_i, which relates the wave functions on right and left of the ith delta function potential is easily found to be

$$M_i = \begin{pmatrix} 1 - iq_i/2k & -iq_i/2k \\ iq_i/2k & 1 + iq_i/2k \end{pmatrix} . \tag{6}$$

Propagation of an electron from one delta-function to the next can be described by a matrix X

$$X = \begin{pmatrix} e^{ik} & 0 \\ 0 & e^{-ik} \end{pmatrix} . \tag{7}$$

Applying M_i and X repeatedly allows us to construct the total transfer matrix for the system.

A computer was programmed to generate random chains described by (3) and (4) for arbitrary k and q. At regular intervals, as the length increased, α and G were calculated using the square of the off-diagonal element of M. A large number of chains were generated simultaneously in order to study the distributions of these quantities.

It was immediately clear that resistance and conductance were improper quantities to average as the standard deviations of their ensemble distributions were consistently greater than their means. (For a mathematical discussion of the moments of the distribution of resistances, see ABRAHAMS and STEPHEN [6].) α, on the other hand, proved to be an ideal statistical quantity. The standard deviation, σ_α, of the distribution of α's, was always less than the mean, $\langle\alpha\rangle$. $\langle\alpha\rangle$ for a given k and q was independent of length, indicating that there is no L dependence in the coefficient of the exponential describing the behavior of the resistance of a particular system. χ^2 fits on distributions such as shown in Fig.1 showed them to be Gaussian with confidence levels greater than 90% [7]. σ_α was proportional to $L^{-1/2}$, as is expected for a Gaussian distribution. σ_α was independent of the number of systems comprising the ensemble, however. More systems give better values of σ_α, but not different values.

Large fluctuations (independent of the number of systems being averaged) were seen for $\langle\rho\rangle$ and $\langle G\rangle$. For finite numbers of systems, the average resistance is not weighted sufficiently by the anomalously large values, and thus the data for $\ell n\langle\rho\rangle$ vs. L did not fall around a line with slope 2α as predicted by ANDERSON et al. [1]. In order to check their prediction we calculated $\langle\rho\rangle = R/T$ analytically by averaging

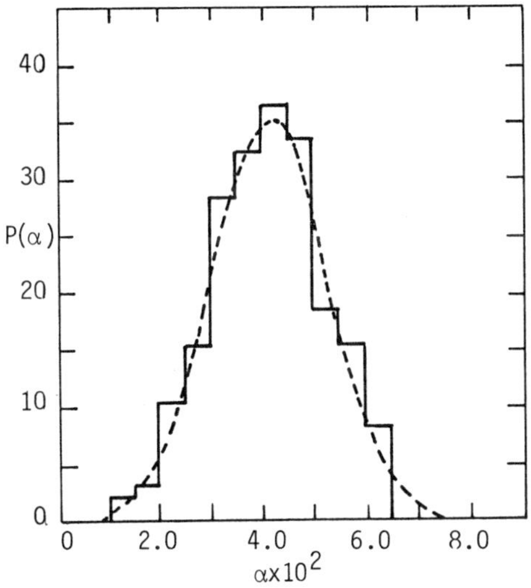

Fig.1 A histogram (solid line) of α for 200 samples with k=1.4, q=1.0, and L=700. The dashed line shows the best Gaussian fit to this data, which has a χ^2 confidence level of 98%

$$M^\dagger M = \begin{pmatrix} 1 + 2R/T & 2r/t^2 \\ 2r^*/t^{*2} & 1 + 2R/T \end{pmatrix} \quad . \tag{8}$$

Since we have assumed the probability distributions for different delta functions to be equivalent and independent, we can average the q_i's on each site consecutively.

$$< M^\dagger \; M> \; = \; <M_1^\dagger X^\dagger \; \ldots \; <M_{L-1}^\dagger X^\dagger <M_L^\dagger X^\dagger X M_L > X M_{L-1}> \; \ldots \; X M_1> \quad . \tag{9}$$

After n such averages

$$< M^\dagger \; M>_n \; = A_n \quad + B_n \sigma_x + C_n \sigma_y \quad , \tag{10}$$

where   is the identity matrix and σ_x and σ_y are Pauli matrices. Multiplying (10) on the left and right by $M_i^\dagger X^\dagger$ and XM_i respectively and averaging allows us to determine the recursion matrix, S, for A_n, B_n, and C_n. The eigenvalues, λ, of S describe the behavior of their corresponding eigenvectors, V

$$V_{n+1}(A,B,C) = \lambda^n V_1 = V_1 \exp(n \,\ell n\, \lambda) \quad . \tag{11}$$

Since A = 1 + 2R/T can be written as a linear combination of the three eigenvectors of the recursion matrix, we will find A for large n to be proportional to $\exp(n \,\ell n\, \lambda_1)$, where λ_1 is the largest eigenvalue. The eigenvalues are given by

$$\lambda^3 - (1+w+\cos 2k(2-w))\lambda^2 + (1-w+\cos 2k(2+w))\lambda - 1 = 0 \quad , \tag{12}$$

where $w = <q_i^2>/2k^2 = q^2/6k^2$. If we define b by $<\rho> \sim \exp(bL)$, then at the center of the band (i.e. at $k = \pi/2$)

$$b = n[2q^2/3\pi^2 + [(2q^2/3\pi^2)^2 + 1]^{1/2}] \quad , \tag{13}$$

or for small q, $b = 2q^2/3\pi^2$. Comparison of b and the $\langle\alpha\rangle$ obtained numerically shows that b is indeed 2α. This is seen in Fig.2, where both quantities are plotted for energies in the first band.

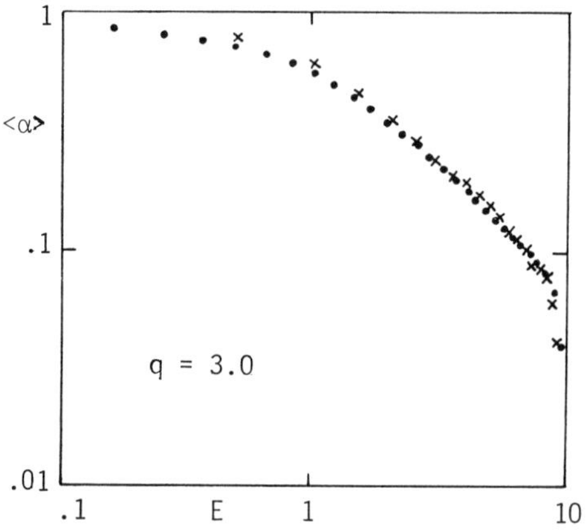

Fig.2 Comparison of the numerical results for $\frac{1}{L}\langle \ell n(1+\rho)\rangle$ (indicated by x), and the analytical results for $\frac{1}{2}\frac{1}{L}\ \ell n\langle\rho\rangle$ (indicated by •) for energies in the first band

We thank P. W. Anderson and M. J. Stephen for discussions. This work was supported in part by the National Science Foundation.

References

1. P.W.Anderson, D.J.Thouless, E.Abrahams, D.S.Fisher: to be published
2. H.Furstenberg: Trans. Amer. Math. Soc. 108, 377 (1963)
3. A.J.O'Connor: Commun. Math. Phys. 45, 63 (1975)
4. R.Landauer: Phil. Mag. 21, 863 (1970)
5. α, the transmission coefficient, has been shown to be equivalent to the inverse localization length by K.Ishii: Sup. Prog. Theor. Phys. 53, 77 (1973)
6. E.Abrahams, M.J.Stephen: J. Phys. C 13, L377 (1980)
7. For $L \gtrsim 1/\langle\alpha\rangle$, α should be described by a Γ-distribution (T.Lukes, preprint) which is indistinguishable from a Gaussian in the large L limit.

Part VII

Superionic Conductors, Coulomb Systems, Molecular Systems and Fractals

One-Dimensional Superionic Conductors

H.U. Beyeler
Brown Boveri Research Center
CH-5405 Baden-Dättwil, Switzerland

1. Introduction

Superionic conductors are solids with exceptionally high ionic conductivities. Whereas "usual" ionic compounds show ionic conductivities of the order of 10^{-10} to 10^{-15} $(\Omega cm)^{-1}$ an increasing number of compounds is now known with conductivities of the order of 10^{-2} to 1 $(\Omega cm)^{-1}$. These values differ from electronic conductivities in metals essentially by a factor given by the mass ratio between ions and electrons.

One-dimensional superionic conductors represent a rather small and special group of little technical interest. Their study has on the other hand yielded a very valuable insight into some of the central mechanisms of superionic conduction. This is mainly due to the inherent complexity of the many particle processes involved. The recurrence to one-dimensional systems - and the thereby reduced number of parameters - is for many aspects a necessary prerequisite for obtaining analytically solvable models. Furthermore the partial information available about important structural features such as the short range order among the mobile species permits only in one-dimensional systems an unequivocal interpretation in terms of microscopic parameters.

In addition the transport properties of such systems exhibit "genuine" dimensionality effects which stimulated the development of a novel microscopic model of "anomalous" diffusion, a model which may be of very general relevance for problems of mass and charge transport in disordered systems.

The aim of this paper is to give a survey on the many interesting facets of the physics of one-dimensional superionic conductors. The discussion of the microscopic models used and of their inherent assumptions must therefore be rather brief, for further details the interested reader is referred to the original papers.

2. Essential features of superionic conductors

In the following we summarize the features of superionic conductors which are essential for the further discussions. Comprehensive reviews on the subject are given in Refs. [1-4].

It is convenient to consider a superionic conductor to be composed of an immobile, periodic framework structure and the subsystem of the mobile ions. These occupy substoichiometrically sites in the conduction paths surrounded by the framework. The conduction paths may have the topology of linear channels, planar or complex 3-dimensional networks. The high ionic mobility is due both to the availability of energetically equivalent empty sites in the paths and to the weakness of the diffusion barriers between sites (∿ 0.1 - 0.2 eV). The density of mobile charge carriers is generally very high (> 10^{21} cm^{-3}). The subsystem of the mobile ions often shows a considerable occupational and displacive disorder; the jump probability of a mobile ion is significantly influenced by the nature of the short range order exhibited by the neighboring mobile particles.

3. One-dimensional superionic conductors

In some superionic conductors the ionic motion is confined to linear, noncommunicating channels and the ions are left with no chance to pass each other. Examples of such linear ionic conductors are β-eucryptite ($LiAlSiO_4$) [5], hollandite [6-11] (a large family of compounds, e.g. $K_{2x}Mg_xTi_{8-x}O_{16}$, $0.75 \leq x \leq 1$), $Ag_2Tl_6I_{10}$ [12], $Li_2Ti_3O_7$ [13] and $Na_{0.7}Ga_{4.72}Ti_{0.29}O_8$ [14], whereby most compounds show some residual lateral conductivity and should more correctly be regarded as highly anisotropic three-dimensional conductors.

In addition, if we want to consider a single channel of such a compound as an independent one-dimensional subsystem we must require that not only as far as diffusion is concerned, but for all relevant properties such as the state of order there is no crosstalk, i.e. no noticeable interaction between ions in different channels. Little is known in this respect about most channel conductors with the exception of hollandite (see Fig. 1) which shows no sign of inter-channel diffusion and exhibits a planar diffuse x-ray scattering [8], characteristic for a one-dimensional state of order with no lateral correlation. This compound is somewhat unique as it is composed of an inpenetrable framework of edge sharing

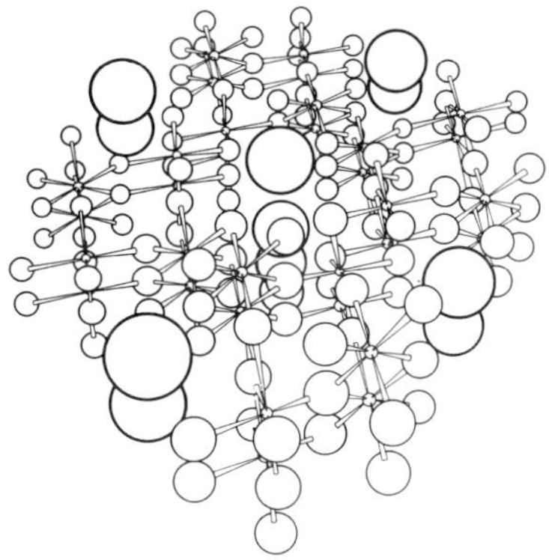

Fig. 1 Perspective view of the hollandite structure. Potassium ions (large spheres) reside in separate, linear channels formed by TiO_6 and MgO_6 octahedra. Small spheres represent Ti, Mg ions, intermediate ones oxygen ions.

TiO_2 octahedra whose very high polarizability ($\varepsilon \sim 100$) provides a very effective screening of the electrostatic interaction between channels. It is the only system for which the static and dynamic properties of interest have been extensively studied. In the following discussion we shall mostly discuss model results taylored to describe experimental data on hollandite of the composition $K_{1.54}Mg_{0.77}Ti_{7.23}O_{16}$; the models themselves are of course more general and could in principle be applied to any type of one-dimensional superionic conductor.

4. Microscopic model of a one-dimensional superionic conductor

We describe a one-dimensional ionic conductor by considering a set of particles to be free to move within a linear channel, the particles are assumed to feel a periodic potential due to their interaction with the framework (for simplicity we assume in the following this potential to be piecewise parabolic with a total amplitude given by C_p); the interaction between the particles we describe with effective, harmonic nearest neighbor forces given by the parameter C_I (see Fig.2, as shown in Ref. [11] analogous models with different types of periodic potentials and interactions lead essentially to the same conclusions). We further restrict ourselves to a finite system and imply periodic boundary conditions, i.e. our system is equivalent to a ring. The model is completely specified by the two parameters C_p and C_I. In a broader context this simple model is a specific example of a generalized lattice gas model for a superionic conductor as dicussed in Ref. [15].

If the number of potential wells is larger than the number of particles -as is the case in superionic conductors - such a system exhibits many

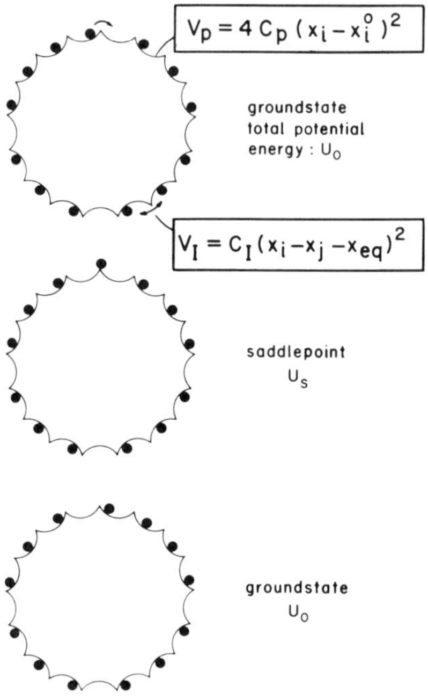

Fig. 2 Top: example of a ring model configuration and the definition of the model parameters C_p and C_I; C_p is equal to the amplitude of the parabolic potential V_p, C_I is a measure of the nearest neighbor interaction V_I; x_i, x_j are particle coordinates in lattice units of the framework, x_i^o is the position of the minimum of the well occupied by particle i, x_{eq} is the average separation between particles. Central and bottom rings: saddlepoint and final configuration of a diffusion process starting from the top configuration. The effective barrier height for this process is equal to U_s - U_o.

equilibrium configurations whereby the particles are not exactly located at the minima of the parabolic wells.

It could be shown that by using an analogy to an Ising model with exponential interaction the relevant properties of this model such as the equilibrium coordinates of the particles for all configurations and their energies can be computed analytically. In particular, using a model with the average of 0.77 particles per well - corresponding to the stoichiometry of the investigated hollandite - the diffuse x-ray scattering due to the configurational disorder was determined. The comparison with experimental data shows, that the model very well reproduces the rather complex diffuse scattering and permits an unambiguous determination of the parameters C_p and C_I.

4.1 Configurational specific heat

At low temperatures only very few configurations, those with the most regular distribution of the vacancies achievable with a stoichiometry of 0.77 are populated. At higher temperatures disorder increases as more and more configurations become populated. The model makes the predic-

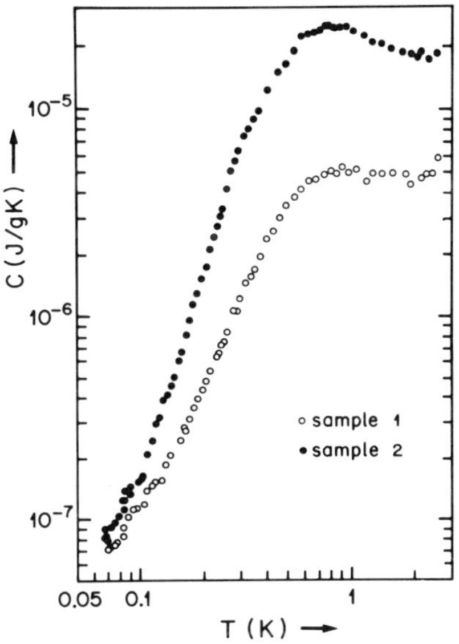

Fig. 3 Specific heat of two samples of hollandite with slightly different particle densities (from Ref. [16]). The observed anomalies differ in position and magnitude in agreement with the configurational model.

tion that this results in an anomaly in the specific heat between 50 K and 100 K [16]. Unfortunately in this range the specific heat of hollandite is dominated by structures due to phonons which impede the experimental investigation of the relatively weak configurational contribution. At still lower temperatures (below 1 K) a further ordering process sets in whose contribution can also be evaluated quantitatively. Recent experiments [17] have in fact evidenced the existence of a specific heat anomaly at exactly the predicted temperature (see Fig. 3). It is certainly somewhat astonishing to see such a simple model fitted to room temperature structural data making a correct prediction about the specific heat below 1 K.

In particular the detailed study of our model clearly shows that the configurational contributions to the specific heat occur in the form of separate anomalies, each related to a specific ordering process involving successively smaller characteristic energies. It may be conjectured that this also holds for superionic conductors of higher dimensions. The effect of the periodic framework potential thus causes the disorder in the mobile subsystem of a superionic conductor to be qualitatively different from that of amorphous solids which exhibit at low temperature a structureless linear specific heat. We then conclude, that the linear specific heat observed in superionic conductors such as β-alumina (and also the linear background observed in hollandite) is not an intrinsic feature of the mobile subsystem but is due to its interaction with compensating charges in the framework and crystal defects.

4.2 Diffusion barriers

As the ion-ion interaction shifts the ions off the minima of the framework potential - in particular those adjacent to a vacancy which are in a position to perform a jump - the effective diffusion barrier differs from the total amplitude C_p of the periodic potential. The effective barrier height is the difference between the potential energy of the total system in the saddlepoint configuration and that of the initial state as depicted for a specific jump in Fig. 2. Due to the repulsive interaction the effective barrier is always lower than C_p, the "naked" barrier. The dependence of the effective barrier on the ratio C_I/C_p is shown in Fig.4.

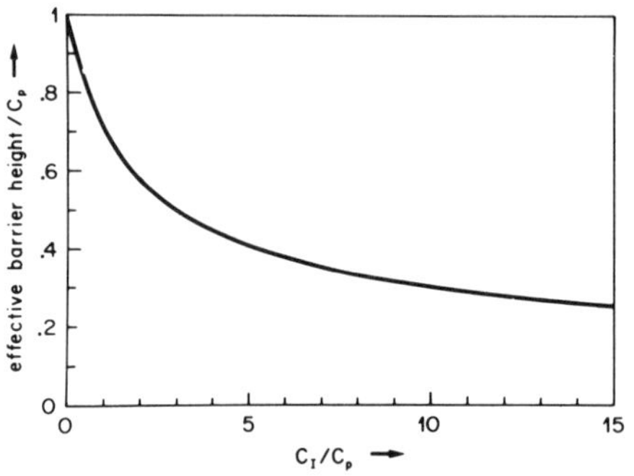

Fig. 4 The height of the effective diffusion barrier compared to the "naked" one for the diffusion process shown in Fig. 2. With increasing C_I/C_p, i.e. with increasing repulsive interaction between the particles the barrier is greatly reduced. For hollandite $C_I/C_p \sim 3.2$, the effective barrier is thus only 40% of the naked one.

For hollandite $C_I/C_p \sim 3.2$ which means that the diffusional barrier is only about 0.4 C_p. In absolute values we find for hollandite $C_p \sim 0.23$ eV and therefore an effective barrier height of ~ 0.09 eV. This is, of course, still only an upper limit because the real framework potential certainly does not exhibit the pointed barrier structure of our piecewise parabolic potential. Estimates with more realistic potentials (see Ref.[11]) yield effective barrier heights of the order of 0.05 eV.

In the framework of absolute rate theory applied to diffusion in our model we would expect this height to show up as activation energy of the ionic conductivity.

For reasons discussed further on, the intrinsic conductivity of a one-dimensional system is rather difficult to measure, it is necessary to recur to microwave frequencies. According to recent measurements by

KHANNA et al. [18] (Fig. 5) the conductivity is thermally activated, with an activation energy of ∼ 0.038 eV. In view of the rather crude simplifications this result is in very satisfactory agreement with our model prediction. For no other superionic conductor structural and transport data could so far be related by such a simple model. The model furthermore demonstrates the many body aspects of a diffusional jump and shows how these may drastically enhance diffusion rates by lowering the barriers.

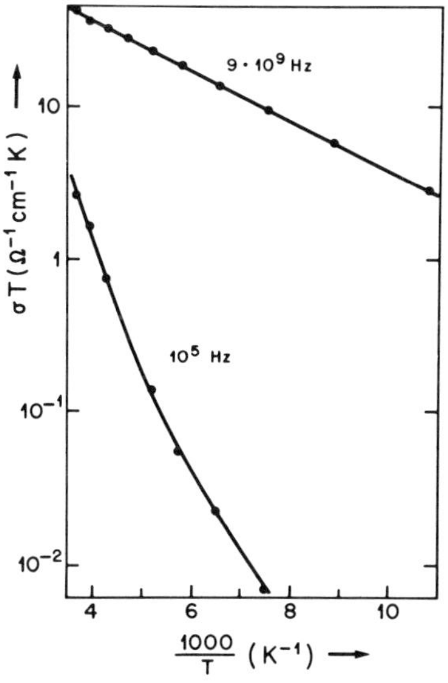

Fig. 5 Arrhenius plot of the conductivity of hollandite for two frequencies (microwave data from Ref. [17]). At 10^5 Hz σ is dominated by the extrinsic barriers and shows an anomalous, non Arrhenius behavior. At $9 \cdot 10^9$ Hz particle motion is short enough to be unperturbed by the extrinsic barriers, at this frequency the intrinsic σ of hollandite is observed, the activation energy corresponds to 400 K.

5. Dimensionality effects in the transport properties

In contrast to the high (frequency independent) ionic conductivity predicted by the model (room temperature conductivity is expected to be $0.2\ \Omega^{-1}\mathrm{cm}^{-1}$) one observes in hollandite a vanishingly small d.c. conductivity. The complex a.c. conductivity $\sigma(\omega)$, however, is finite and shows between 50 Hz and 10^8 Hz a power law of the form:

$$\sigma(\omega,T) = C(T)(-i\omega)^{\nu(T)} \quad . \tag{1}$$

This discrepancy is due to the fact that transport in such a one-dimensional system is ultrasensitive to any type of perturbation of the ideal channel geometry. The effect of a single impurity or defect in one channel is tremendously amplified as it essentially affects the motion of all the ions in that channel. Thus the "normal" level of defects and impurities present in a real specimen creates additional barriers for the motion of the ions which in higher dimensions are easily circumvented but in one dimension greatly affect the transport properties. We know neither the density nor the heights of these barriers; the latter may vary between that of the intrinsic barriers discussed in sect. 4 and several eV which would correspond to a completely blocking barrier.

Assuming only completely blocking barriers to be of importance leads to the interrupted strand model, the corresponding $\sigma(\omega)$ is, however, significantly different from that observed in hollandite. The dispersion observed in $\sigma(\omega)$ for many decades of frequencies calls for a model with a very broad spectrum of characteristic time constants.

Such a model has recently been developed by BERNASCONI et al. [19-22] by studying the effect of barriers with a broad distribution of heights or transition rates. The essentials of this model are as follows: the "extrinsic" barriers are assumed to occur at regular intervals (a generalization to random occurrence was shown to yield no new features); the heights Δ are statistically distributed according to the probability $\rho(\Delta)$:

$$\rho(\Delta) = W_o \exp(-\Delta/T_m) \qquad (2)$$

where W_o is a normalizing constant and T_m an adjustable model parameter. As the intrinsic conductivity is very high the perfect segments between extrinsic barriers are assumed to be instantaneously polarized. The dynamics is then determined by the relatively slow inter segment transfer. This is described by a system of rate equations for the number of particles in each segment. As the main result an analytical expression for $\sigma(\omega)$, valid in the range of experimental frequencies could be given. It turns out to be equal to the empirically found form of Eq. (1), and the exponent ν is very simply related to the model parameter T_m:

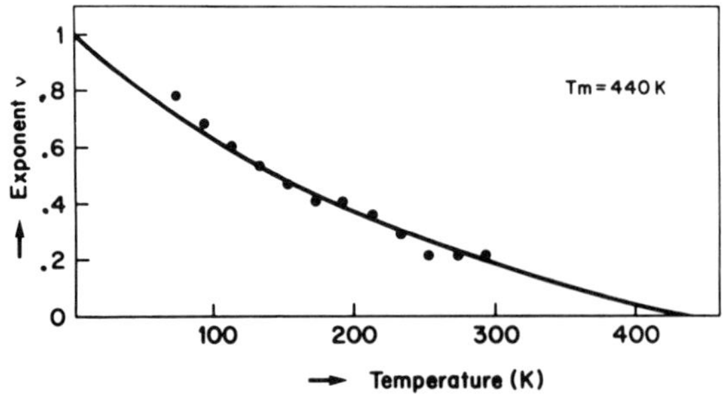

Fig. 6 Temperature dependence of ν (Eq. (1)) for the hollandite $K_{1.54}Mg_{0.77}Ti_{7.73}O_{16}$. The agreement between theory (Eq. (3) with T_m = 440 K) and experiment indicates that the exponential barrier spectrum (2) is at least for this range of temperatures a valid assumption.

$$v = \frac{1 - T/T_m}{1 + T/T_m} \qquad (3)$$

As shown in Fig. 6 experiments are in full agreement with this temperature dependence of the exponent and T_m turns out to be about 440 K. The model relates the prefactor C(T) to such microscopic quantities as density of barriers and attempt frequency. The analysis of experimental data on C(T) yields very reasonable values for these parameters.

There is no physical basis for the very specific form of the barrier spectrum (Eq. (2)), it has been introduced solely for mathematical convenience. It is reasonable to assume a decreasing probability for higher barriers and it turns out that in a wide range of temperatures well below T_m the qualitative features of the model are not noticingly altered when some other form of wide distribution of barrier heights is assumed. As shown in the next section the exponential spectrum has, however, a very interesting consequence.

6. The mobility transition

The most intriguing question raised by the Bernasconi model is, what happens if T approaches and passes T_m? The model predicts at T_m a transition from the anomalous behavior with no d.c. conductivity to a normal regime with a finite σ_{dc}; T_m may thus be called a mobility transition temperature. The existence of this transition, however, hinges entirely on the specific form of Eq. (2): a cutoff of ρ at some upper barrier height leads to a finite σ_{dc} at all temperatures, a weaker than exponential decrease of $\rho(\Delta)$ results in a sufficient number of very high barriers to suppress σ_{do} for all temperatures.

Although this transition might at this point be considered to be an artifact of the model, let us consider its eventual experimental manifestation. According to the model the mean square displacement $\langle x^2 \rangle$ of a particle is proportional to

$$\langle x^2 \rangle \quad \alpha \quad t^{2T/(T+T_m)} = t^{1-v} \qquad (4)$$

Below T_m the particles diffuse thus slower than in a normal random walk regime. Approaching T_m their behavior reaches gradually the normal diffusion behavior. At the transition no changes on a local scale occurs and no local experimental probe is expected to see the transition.

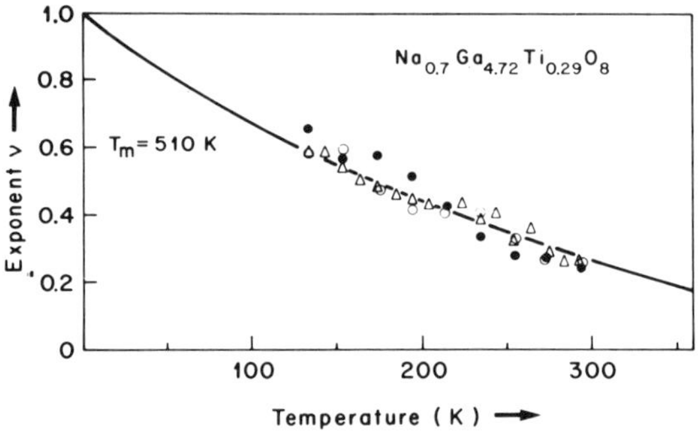

Fig.7 Temperature dependence of ν (Eq. (1)) for $Na_{0.7}Ga_{4.72}Ti_{0.29}O_8$ for three different samples. The data scatter around the theoretical curve (Eq. (3)) for $T_m \sim 510$ K.

Recent theoretical investigations [22,23] have further shown that for usual experimental frequencies Eqs. (1) and (3) are no more valid for $T \geqq 0.9T$ so that one can not follow experimentally the approach to T_m by determining ν from audio frequency conductivity measurements. An undisputable proof of the existence of the transition is only obtainable from d.c. conductivity experiments. Ionic d.c. conductivity measurements, however, require carefully prepared "reversible" electrodes containing the mobile species. Hollandite itself is, unfortunately, unstable in contact with metallic potassium, ruling out such measurements on this material.

First experimental results on $Na_{0.7}Ga_{4.72}Ti_{0.29}O_8$ indicate that this compound might be a better candidate to investigate the mobility transition. High temperature measurements have already established the existence of a finite d.c. conductivity above 518 K [14]. We have recently investigated the a.c. conductivity at lower temperatures and found a behavior in agreement with Eq. (1). The temperature dependence of the exponent ν is shown in Fig. 7. Due to the small size of the available samples the data scatter appreciably, but cluster around the theoretical curve for T_m = 510 K. All presently available experimental information is thus consistent with the existence of a mobility transition in that compound around 500 K. Yet one has to be aware that this compound is less perfectly one-dimensional than hollandite. A residual "lateral" conductivity may effectively help the ions to circumvent the barriers and give rise to a d.c. component of the conductivity parallel to the channels persisting at all temperatures. Further investigation of this system, in particular of σ_{dc} at lower temperatures is thus of great interest.

7. Conclusions

The simple two-parameter model consistently describing many aspects of superionic conductors demonstrates clearly the power of one-dimensional models. It is conjectured that its principle features are qualitatively valid also for systems of higher dimensions.

The very peculiar transport properties of one-dimensional ionic conductors are akin to those observed in many disordered and amorphous solids [22] where conductivities are often found to follow Eq. (1), too. The microscopic model for anomalous diffusion may thus also be of relevance for a number of related phenomena so far only described by phenomenological concepts such as a "universal law" [24].

As a most intriguing consequence the model predicts the existence of a novel type of transition in the transport properties of a one-dimensional system. Experimental evidence for such a transition is, however, still incomplete.

Acknowledgements: The author is greatly indepted to G. Gruner and R. Orbach, J. Bernasconi, L. Pietronero and S. Strässler for many stimulating discussions and to Miss V. Greznaryk for very valuable technical assistance.

References

1. "Superionic Conductors". ed. by G.D. Mahan and W.L. Roth (Plenum Press; New York, 1976)
2. "Solid Electrolytes", ed. by S. Geller, Topics in Applied Physics, Vol.21 (Springer, Berlin, Heidelberg, New York 1977)
3. "Physics of Superionic Conductors", ed. by M.B. Salamon, Topics in Current Physics, Vol.15 (Springer, Berlin, Heidelberg, New York 1979)
4. "Fast Transport in Solids", ed. by P. Vashishta, J.N. Mundy and G.K. Shenoy (North Holland New York, 1979)
5. U. v. Alpen, E. Schönherr, H. Schulz and G.H. Talat: Electrochim Acta 22, 805 (1977)
6. A. Byström and A.M. Byström: Acta Crystallogr. 3, 146 (1950)
7. J. Singer, H.E. Kautz, W.L. Fielder and J.S. Fordyce in: "Fast Ion Transport in Solids", ed. by W. van Gool (North Holland Amsterdam, 1973), p. 653
8. H.U. Beyeler: Phys. Rev. Lett. 37, 1557 (1976)

9. H.U. Beyeler, L. Pietronero, S. Strässler and H.J. Wiesmann: Phys. Rev. Lett. 38, 1532 (1977)
10. H.U. Beyeler and C. Schüler: Solid State Ionics 1, 77 (1980)
11. H.U. Beyeler, L. Pietronero and S. Strässler: to appear in Phys. Rev.
12. Chen Li-chuan, A. Rabenau and W. Weppner: Appl. Phys. 17, 233 (1978)
13. J.B. Boyce and J.C. Mikkelsen, Jr.: Solid State Commun. 31, 741 (1979)
14. G.V. Chandrasekhar, A. Bednowitz and S.J. La Placa in Ref. 4, p. 447
15. H.U. Beyeler, P.Brüesch, L. Pietronero, W.R. Schneider, S. Strässler and H.R. Zeller: "Status and Dynamics of Lattice Gas Models", in Ref.3, p.77
16. L. Pietronero, W. R. Schneider and S. Strässler: to be published
17. H. v. Löhneysen, Mui-Xieng Ding, W. Arnold, H.U. Beyeler, L. Pietronero and S. Strässler: to be published
18. S.K. Khanna, G. Grüner, R. Orbach and H.U. Beyeler: to be published
19. J. Bernasconi, S. Alexander and R. Orbach: Phys. Rev. Lett. 41, 185 (1978)
20. J. Bernasconi, H.U. Beyeler, S. Strässler and S. Alexander: Phys. Rev. Lett. 42, 819 (1979)
21. J. Bernasconi, W.R. Schneider and W. Wyss: Z. Physik B37, 175 (1980)
22. S. Alexander, J. Bernasconi, W.R. Schneider and R. Orbach: "Excitation Dynamics in Random One-Dimensional Systems", in this volume, p.277
23. P.N. Richards and R.L. Renken: Phys. Rev. B21, 3740 (1980)
J. Bernasconi and H.U. Beyeler: Phys. Rev. B21, 3745 (1980)
24. A.K. Jonscher: Nature 267, 673 (1977)

One-Dimensional Coulomb Systems

Ph. Choquard, H. Kunz, Ph.A. Martin, and M. Navet*
Laboratoire de Physique Théorique, Ecole Polytechnique Fédérale de Lausanne
CH-1000 Lausanne, Switzerland

Introduction

The basic role played by Coulomb interactions in physics makes them of interest even in one dimension. This explains the effort devoted over the last years to the elucidation of physical properties of some one-dimensional Coulomb systems. Two possible classes of models could be considered. In the first one, the system consists of charged particles moving on a line and interacting with the potential $1/|x|$. In the second one, the charged "particles" interact with a potential whose Fourier transform is $2/|k|^2$, i.e. $-|x|$ in real space in one dimension. Surprisingly no work has been done (to our knowledge) on the first class of models. We will summarize here the extensive work done on models of the second kind. A physical realization of such a system could be a large assembly of charged sheets. We will furthermore distinguish between two kinds of such Coulomb models: the first ones, usually called two- (or more-) component systems, consist of a neutral assembly of two (or more) species of charged "particles", whereas the second one, usually called one-component plasma (jellium), consists of one species of charged particles immersed in a homogeneous neutralizing background. Their properties appear to be rather different, especially in one dimension. Attention has concentrated on the equilibrium properties of such systems, and most of the time at the classical level. These studies have shown that despite their somewhat artificial character, such systems display unusual and interesting properties: two-component systems are *dielectric*, in fact an imperfect *dipole gas*, whereas one-component systems crystallize at all temperatures, forming a *Wigner lattice*.

* Permanent address: CRPE/CNET-CNRS - F-45045 Orléans Cédex, France

1. The Two-Component Plasma

1.1 Basic Formulas and Tools

The system is made of $N_\pm$ particles of charges $e\sigma_j$ with $\sigma_j = \pm 1$. It is globally neutral, i.e. $\Sigma\sigma_j = 0$, but contains boundary charges $(q, -q)$ in $\Lambda = [0,L]$. Equivalently these boundary charges can be seen as creating an homogeneous external field $\theta = 2q$. The energy of the system is given by

$$U_\theta = -e^2 \sum_{i<j}^{N} \sigma_i\sigma_j |x_i - x_j| - 2qe \sum_{i=1}^{N} \sigma_i x_i + q^2 L$$

or

$$U_\theta = e^2 \sum_{i,j=1}^{N} \sigma_i\sigma_j \min(x_i, x_j) - \theta e \sum_{i=1}^{N} \sigma_i x_i + \frac{\theta^2}{4} L$$

where we used the identity $\min(x,y) = -|x-y|/2 + (x+y)/2$ and the charge neutrality $\Sigma\sigma_j = 0$.

The grand-canonical partition function corresponding to this system is

$$Z_\theta^L = \sum_{N_\pm=0}^{\infty} \frac{z^{N_+ + N_-}}{N_+!\, N_-!} \sum_\sigma \int_{\Lambda^{N_+ + N_-}} d^{N_+ + N_-}x \exp[-\beta U_\theta(x,\sigma)] \quad , \qquad (\Sigma\sigma_j = 0) \quad ,$$

and the grand-canonical pressure

$$p(\beta,\theta,z) = \lim_{L\to\infty} \frac{1}{L} \left(\frac{1}{\beta} \log Z_\theta^L\right) \quad .$$

Correlation functions are defined in the usual way and will be denoted by $\rho_\theta^\Lambda(x_1\sigma_1, \ldots, x_k\sigma_k)$, which represents the probability to have a particle of charge σ_1 at x_1, another of charge σ_2 at x_2 and so on.

The basic tool discovered by LENARD and EDWARDS [1] to discuss the properties of this model is functional integration. They only considered however the case $\theta = 0$, and another boundary condition. Consider the gaussian process defined by $\langle\varphi(x)\rangle = 0$, $\langle\varphi(x)\varphi(y)\rangle = D \min(x,y)$ with $D = 2\beta e^2$. This is nothing else but the Brownian motion, in one dimension. The conditional probability that a trajectory starting from $\varphi(0) = \varphi_0$ goes through $\varphi(x_1) = \varphi_1, \ldots, \varphi(x_n) = \varphi_n$ is

$$P(\varphi_1 x_1, \ldots, \varphi_n x_n | \varphi_0, 0) = \prod_j P(\varphi_i - \varphi_{j-1}, x_j - x_{j-1}) \quad ,$$

where $P(\varphi,x)$ is the fundamental solution of

$$\left(\frac{d}{dx} + \frac{D}{2}\frac{d^2}{d\varphi^2}\right) P(\varphi,x) = 0 \quad , \quad P(\varphi,0) = \delta(\varphi) \quad .$$

Denoting by $\langle\cdot|\varphi_0\rangle$ an average with respect to this conditional probability distribution P, we get

$$\langle\exp[i \sum_j \sigma_j\varphi(x_j)]|\varphi_0\rangle = \exp[-\beta e^2 \sum_{i,j} \sigma_i\sigma_j \min(x_i,x_j) + i\varphi_0 \sum_j \sigma_j] \quad .$$

Using the identity

$$\delta_{\Sigma\sigma_j,0} = \frac{1}{2\pi}\int_{-\pi}^{\pi} d\varphi_0 \exp(i\varphi_0 \sum_j \sigma_j)$$

we can rewrite the partition function as

$$Z_\theta^L = \exp\left(-\beta\frac{\theta^2}{4} L\right) \sum_{N_\pm=0}^{\infty} \frac{z^{N_+ + N_-}}{N_+!\,N_-!}\frac{1}{2\pi}\int_{-\pi}^{\pi} d\varphi_0 \sum_\sigma \int_{\Lambda^{N_+ + N_-}} d^{N_+ + N_-}x$$

$$\times \langle\exp[i \sum_j \sigma_j\varphi(x_j)]|\varphi_0\rangle \exp(\beta e\theta \sum_j \sigma_j x_j) \quad .$$

We can now perform the summation over N_+, N_- and σ to get

$$Z_\theta^L = \exp\left(-\beta\frac{\theta^2}{4} L\right)\frac{1}{2\pi}\int_{-\pi}^{\pi} d\varphi_0 \langle\exp\{2z \int_0^L dx \cos[\varphi(x) - i\beta e\theta x]\}|\varphi_0\rangle \quad .$$

Feynman-Kac formula then tells us that

$$Z_\theta^L = \exp\left(-\beta\frac{\theta^2}{4} L\right)\frac{1}{2\pi}\int_{-\pi}^{\pi} d\varphi_0 \int_{-\infty}^{\infty} d\varphi Q(L\varphi|0\varphi_0) \quad ,$$

where $Q(L\varphi|0\varphi_0)$ is the fundamental solution of the equation $(\partial/\partial L)Q_L = -H(L)Q_L$ with

$$H(L) = -\frac{D}{2}\frac{d^2}{d\varphi^2} - 2z\cos(\varphi - i\beta e\theta L) \quad .$$

In order to solve this equation, let us introduce the operator $V(L) = \exp(\beta eLp)$ where $p = -i(\partial/\partial\varphi)$. This operator translates functions of φ by $-i\beta eL$, hence

$$H(L) = V(L)\,(\frac{D}{2}p^2 - 2z\cos\varphi)V^{-1}(L) \quad .$$

We can now easily solve the equation for Q

$$Q(L\varphi|0\varphi_0) = \exp\left(\beta\frac{\theta^2}{4} L\right)\left(\varphi|V(L)\exp\left\{-L\left[\frac{D}{2}\left(p + \frac{\theta}{2e}\right)^2 - 2z\cos\varphi\right]\right\}|\varphi_0\right)$$

The problem has now been reduced to the study of the "hamiltonian"

$$h_\theta = \frac{D}{2}\left(P + \frac{\theta}{2e}\right)^2 - 2z\cos\varphi$$

which represents a generalization of the Matthieu hamiltonian. Since h_θ is periodic in φ, we can introduce its Bloch states $|k,n)$, with $(\varphi|k,n) = \exp(-ik\varphi)\bar{V}_{k,n}(\varphi)$ as usual. $V_{k,n}(\varphi)$, the Bloch amplitude of h_θ, is however related to the Bloch amplitude $U_{k,n}(\varphi)$ of h_0, the usual Matthieu hamiltonian, by $V_{kn}(\varphi) = U_{k+(\theta/2e),n}(\varphi)$. Finally:

$$\exp\left(-\beta\,\frac{\theta^2}{4}\,L\right)Q(L\varphi|0\varphi_0) = \int_{-\frac{1}{2}}^{\frac{1}{2}} dk \sum_n \exp[-ik(\varphi - \varphi_0 - i\beta eL)]\,\bar{U}_{k+(\theta/2e),n}(\varphi - $$

$$- i\beta e\theta L)\,U_{k+(\theta/2e),n}(\varphi_0)\,\exp[-L\varepsilon_n(k+\theta/2e)] \quad ,$$

where $\varepsilon_n(k)$ denotes the energy of the band n of the Matthieu hamiltonian. The grand partition function is then given by

$$Z_\theta^L = \frac{1}{2\pi}\int_{\pi}^{\pi} d\varphi \int_{-\pi}^{\pi} d\varphi_0 \sum_n \bar{U}_{\theta/2e,n}(\varphi)U_{\theta/2e,n}(\varphi_0)\,\exp[(-L\varepsilon_n(\theta/2e)] \quad .$$

The thermodynamic limit is easily taken and we get for the grand canonical pressure $\beta p(\beta,z,\theta) = -\varepsilon_0(\theta/2e)$, $\varepsilon_0(k)$ denoting the energy of the first band of the Matthieu hamiltonian

$$h_0 = \frac{D}{2}\,p^2 - 2z\cos\varphi \quad .$$

Other thermodynamic quantities of interest are the *average internal* field E (θ) and the *polarization* $P(\theta)$. These are defined, in a finite box, by:

$$E^L(\theta) = \frac{1}{L}\int_0^L E^L(\theta,x)dx$$

where

$$E^L(\theta,x) = \theta + e\int_0^L sg(x - y)[\rho_\theta^L(x,+) - \rho_\theta^L(x,-)]$$

the first contribution corresponding to the external field, the second one to the field created by the charges in the system.

The polarization in a finite box is given by

$$P^L(\theta) = \frac{1}{L}\int_0^L \exp[(\rho_\theta^L(x,+) - \rho_\theta^L(x,-)]dx \quad .$$

In the thermodynamic limit, we simply have:

$$E(\theta) = \lim_{L\to\infty} E^L(\theta) \quad , \qquad P(\theta) = \lim_{L\to\infty} P^L(\theta) \quad .$$

These are easily computed by noting that $P^L(\theta) = \frac{1}{2}\,[(\theta - E^L(\theta)]$ and

$$\frac{d}{d\theta}\left(\frac{1}{L}\log Z_\theta^L\right) = \beta P^L(\theta) - \frac{\beta}{2}\theta = -\frac{\beta}{2}E_\theta^L \quad , \qquad \text{therefore}$$

$$E(\theta) = \frac{2}{\beta}\frac{d}{d\theta}\varepsilon_0(\theta/2e) \quad \text{and}$$

$$P(\theta) = \frac{1}{2}\,[(\theta - E(\theta)] \quad .$$

Using the same technique, we can also study the correlation functions and get explicit expressions for them in terms of the "eigenvectors" and "eigenvalues" of the Matthieu hamiltonian, in the thermodynamic limit [1,4]. These limiting correlation functions $\rho_\theta(x_1\sigma_1, \ldots, x_k\sigma_k)$ define the states of the system in the presence of an external electric field θ. We can call them θ-states.

1.2 Physical Properties

Let us now analyze the main qualitative properties of the system and elucidate their physical meaning.

Since $\varepsilon_0(k)$ is nondegenerate, one can show that the grand-canonical pressure is a real analytic function of z, β and θ. Therefore the system presents no phase transition, like all one-dimensional systems with short range interactions. In the weak coupling regime ($\beta e^2 \ll z$), the pressure is that of a dilute plasma and is given by Debye-Huckel theory $\beta p \simeq 2z - \sqrt{\beta e^2 z}$. In the strong coupling regime ($\beta e^2 \gg z$), the pressure becomes that of a perfect gas of density $\rho/2$, which can be interpreted by saying that the pressure is that of a perfect gas of dipoles of charges (+e,-e).

Each θ-state is translation invariant and neutral [4]. For example

$$\rho_\theta(x,+) = \rho_\theta(x,-) = \rho/2 \quad .$$

The correlation functions show exponential clustering [1,4]. For example:

$$|\rho_\theta(x_1\sigma_1, x_2\sigma_2) - \rho_\theta(x_1\sigma_1)\rho_\theta(x_2\sigma_2)| \leqq \exp(-\gamma|x_2 - x_1|) \quad , \qquad \gamma > 0 \quad ,$$

when x_2 is sufficiently far apart from x_1. This is a consequence of the existence of a gap between the first and the second band of the Matthieu hamiltonian. This exponential clustering could at first sight be interpreted as a manifestation of the usual screening in a plasma. We will see however that this interpretation is not correct.

Finally, since our Matthieu hamiltonian has a periodic potential of period 2π, $\varepsilon_0(k)$ is periodic in k of period 1 and therefore $\beta p(\beta,z,\theta) = -\varepsilon_0(\theta/2e)$ is periodic of period 2e. Consequently, the pressure, the average internal field, and the correlation functions are *periodic in the external field of period 2e* [4]. It is easily seen that $E(\theta) \neq 0$ for $\theta \neq 0,e$ and $E(\theta) = 0$ if $\theta = 0$ and e. Except for $\theta = 0,e$, the θ-states $\rho_\theta(.)$ are not invariant under charge conjugation and space inversion.

We will now see that all these properties can be understood as a manifestation of the fact that our system is not a plasma, but a dielectric gas, made of dipoles of charge (+e, -e). For this purpose, we will now discuss four criteria enabling us to discriminate between a plasma and dielectric or conducting and insulating phase, and show that they give in all cases the same characterization for our system.

The most natural one concerns the dielectric constant [5,6]. The field θ corresponds to the displacement field D and therefore the dielectric constant is given by

$$\mathcal{E}^{-1} = \frac{dE}{d\theta}(\theta)\Big|_{\theta=0} \quad .$$

A system is a plasma if $\mathcal{E} = \infty$ and a dielectric if $1 < \mathcal{E} < \infty$. In our case, since

$$E(\theta) = -2\frac{d}{d\theta}P(\theta) \quad , \quad \text{and}$$

$$\varepsilon_0(k) \simeq \frac{1}{2m^*}k^2$$

for small k, m* being the effective mass of the first band, we see that the dielectric constant can be written as

$$\mathcal{E} = \frac{m^*}{m} \quad , \quad m = (2\beta e^2)^{-1}$$

being the bare mass. We see that the dielectric sonstant is indeed finite and tends to 1 at very strong coupling ($\beta e^2 \gg z$). Conversely in the very weak coupling regime ($\beta e^2 \ll z$), the dielectric constant is very large.

Another characterization of a conductor (or plasma) is that the electric field E inside it vanishes. We expect therefore that for a plasma $E(\theta) = 0$, whereas for a dielectric $E(\theta) \neq 0$. We have seen that this condition is satisfied in our case. A consequence of this last condition is that since in 1 dimension $\theta = E + 2P$, the susceptibility

$$\eta = \frac{dP}{d\theta}(\theta)\Big|_{\theta=0}$$

is such that $\eta = 1/2$ for a plasma, but $\eta \neq 1/2$ for a dielectric. On the other hand, we have from its definition

$$\eta = \lim_{\theta\to 0}\lim_{|\Lambda|\to\infty} \frac{\beta}{|\Lambda|} \langle(\sum_j \sigma_j x_j)^2\rangle_\Lambda(\theta)$$

and therefore in a translation invariant neutral one-dimensional state

$$\eta = -\frac{\beta}{2}\int x^2 S(x)dx \quad \text{where}$$

$$S(x) = \lim_{|\Lambda|\to\infty} \langle[\sum_i \sigma_i \delta(xi)]\,[\sum_j \sigma_j \delta(x - x_j)]\rangle_\Lambda(\theta)\Big|_{\theta=0}$$

is the charge density correlation function. Hence $-\beta/2 \int x^2 S(x)dx = 1/2$ in a plasma. This condition is known as the Stillinger-Lovett sum rule. It is violated in the one-dimensional case as we have seen, because the system is a dielectric.

A third way to discriminate between a dielectric and a plasma is to look at the effective potential between particles in the system $v(x)$ [7]. It should be screened in a plasma, but remains Coulomb-like in a dielectric. This effective potential is defined in terms of the dielectric function $E(x)$ in the usual way $v(x) = \int E^{-1}(x-y)V(y)dy$.

On the other hand, the dielectric function is given from linear response theory by $\hat{E}(k) = 1 - \beta\hat{V}(k)\hat{S}(k)$, $\hat{S}(k)$ being the Fourier transform of $S(x)$. Therefore, since $\hat{S}(k) \simeq \eta k^2/\beta$ for k small, we see that $\hat{E}^{-1}(k) \simeq 1 - 2\eta$, $k \simeq 0$, and consequently, according to the previous result, $\hat{v}(k) \simeq \hat{\varepsilon}^{-1}(0)/k^2$ in a dielectric, where $1 - 2\eta \neq 0$ but $\hat{v}(0)$ is finite, in a plasma where $1 - 2\eta = 0$. This shows that in our system, the effective potential between particles remains Coulomb-like at large separation.

Finally, it is interesting to characterize the phase of a system by the presence of a symmetry and a related order parameter. In our case, the symmetry physically corresponds to the fact that all the charges of the particles are multiple of each other. This is reflected mathematically in the periodicity of the potential appearing in Matthieu equation: $2z\cos\varphi$. The symmetry group is [Z].

Were we considering a system with particles of charge $(+e, -\sqrt{2}e)$, the corresponding potential would be quasi-periodic, and new phenomena might appear. An order parameter corresponding to this symmetry [3] is the following. Define $m(q,x)$ to be the statistical average of $\exp[-\beta qU(x)]$ where $qU(x) = -qe\sum_j \sigma_j |x - x_j|$ is the interaction energy of an additional charge q in x. In the functional integral representation we have $m(q,x) = \langle \exp[iq\varphi(x)]\rangle$.We now say that $m(q,x) = 0$, $q \neq 0$, $\pm ne$, $n \in [Z]$ in a dielectric, since $m(q,x) = \exp(2i\pi q)m(q,x)$ in a symmetric state, but $m(q,x) \neq 0, \forall q$ in a plasma. Therefore, in the dielectric phase, the symmetry group [Z] is not broken and $m(q,x) = 0, q \neq ne$. This is indeed the situation in one dimension at all temperatures. But in a plasma phase, the symmetry is broken and $m(q,x) \neq 0$. This is the situation in three dimensions. In two dimensions, Kosterlitz-Thouless predicted a transition from a dielectric to a plasma phase.

In conclusion, all these criteria indicate that our system is in dielectric phase for all temperatures. We would like now to show that this can be explained by looking at the system as an imperfect dipole gas. We saw already that at strong coupling (or low temperatures), thermodynamic quantities and correlation functions were those of such a gas. One can also understand this as follows:

The potential energy of a pair of particles $x_1\sigma_1$ and $x_2\sigma_2$ in the field θ is

$$V(x_1, x_2) = -\sigma_1\sigma_2|x_1 - x_2| + \theta(\sigma_1 x_1 + \sigma_2 x_2) \quad .$$

Therefore, for a pair $(+e, -e)$ in relative coordinates $y = x_2 - x_1$, we have

$$V(y) = e(e|y| + \theta y) \qquad \text{for} \qquad 0 \leq \theta \leq e \quad .$$

We see that this potential is confining for $0 \leq \theta < e$. Therefore, there is a gas of polarized dipoles $(+e, -e)$ and $E(\theta) > 0$. When $\theta = e$, $V(y)$ is no more confining and the molecules dissociate, so that the internal field $E(e) = 0$. For $e < \theta \leq 2e$, one $(+e, -e)$ pair is broken. The corresponding charges go to the boundary and overscreen the applied field, the remaining "molecules" are again bound with opposite polarization (hence the change of sign of $E(\theta)$ at $\theta = e$). For $\theta = 2e$ the surface charge cancels the applied field exactly and the $\theta = 2e$ state is identical to the $\theta = 0$ state. As θ increases the same mechanism repeats. Whenever θ passes an odd multiple of e, another "molecule" dissociate and the surface charge increases by one unit. It remains to understand however why the charge correlation functions decay exponentially like in a plasma. This is because Newton's theorem implies that in one dimension two "molecules" $(+e, -e)$ and $(-e, +e)$ have only a contact interaction. For such gas of dipoles with a short range interaction, the correlation functions decay exponentially.

All these features are very nicely seen in a molecular dynamics study of the system [16].

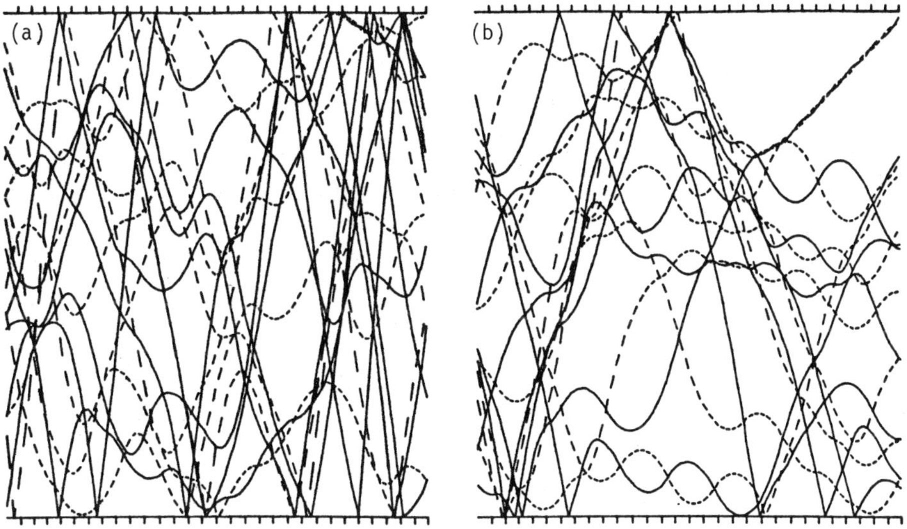

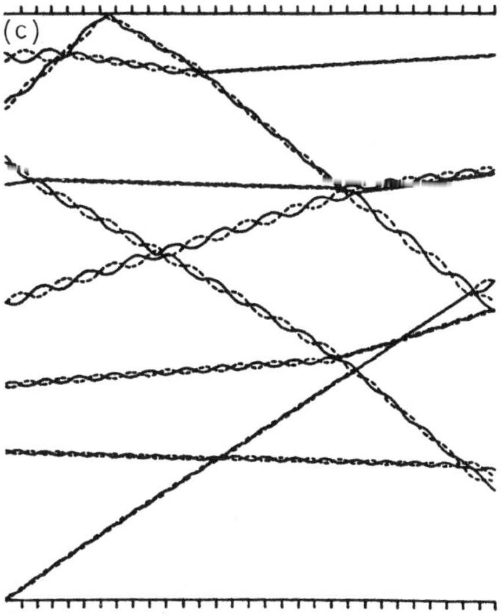

Fig.1a-c. Typical trajectories for a system of 16 particles. The x_i are plotted as functions of the time (in units of ω_p^{-1}. $\omega_p^2 = 2\rho e^2/m$). Dotted and continuous lines differentiate the 2 species of particles. The value of the coupling parameter $g = (2\beta e^2/\rho)^{\frac{1}{2}}$ is imposed by the thermostatic walls; i.e. when a particle hits the wall it is reflected not with opposite velocity, but with a velocity choosen at random in the suitable gaussian distribution. (a) g = 0.5 almost uncorrelated trajectories: imperfect gas of particles; (b) g = 1 intermediate regime: tendency to form neutral molecules; (c) g = 4 almost perfect gas of molecules

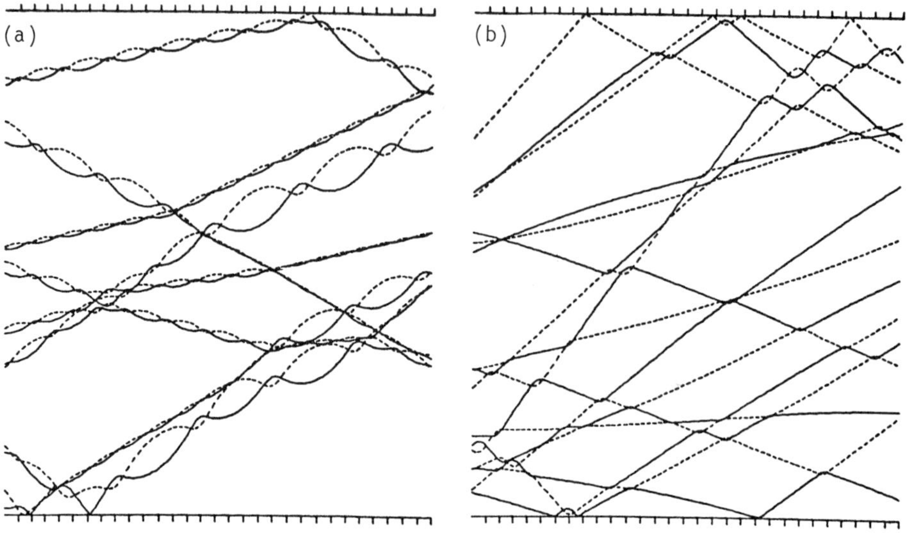

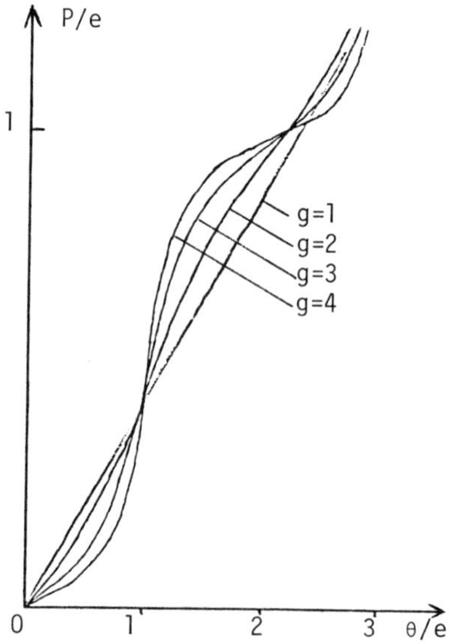

Fig.2a-c. The same system as in Fig.1, with an exteranl applied electric field, for g = 3; (a) θ = 0.6e polarization of the molecules appears clearly. (b) θ = 1.01e just after the first dissociation of a molecule: trajectories are nearly straight lines because of almost vanishing internal field; (c) θ = 2e the external field is exactly canceled by one dissociated molecule

Fig.3. Polarization of the two-component plasma as a function of applied external field. The imperfect periodicity of P- θ/2 is due to the finite size of the system ($N_+ + N_- = 16$)

2. The One-Component Plasma

In this model, we have N particles of charge -e in a box

$$\Lambda = \left[-\frac{L}{2}, \frac{L}{2}\right]$$

immersed in an homogeneous neutralizing background of density $\rho = N/L$. Taking the usual periodic boundary conditions, the hamiltonian takes the familiar form

$$U = \frac{1}{L} \sum_{k \neq 0} \frac{2e^2}{k^2} \left| \sum_j \exp(ikx_j) \right|^2$$

whereas for rigid walls boundary conditions, we have

$$U = -\frac{e^2}{2} \sum_{i,j} |x_i - x_j| + e^2\rho \sum_j \int_\Lambda |x - x_j| dx - \frac{e^2\rho^2}{2} \int_\Lambda \int_\Lambda dxdy|x - y| \quad .$$

2.1 The Classical Case

Although it is possible to treat this model in the grand canonical ensemble, by means of functional integration techniques [9,10,11], the analysis is subtle and less transparent than the original one [12], which uses a transfer operator technique. Let us illustrate this by looking at the canonical partition function, for rigid walls boundary conditions

$$Q_\Lambda^N = \frac{1}{N!} \int_\Lambda dx, \dots \int_\Lambda dx_N \exp[-\beta U(x_1 \dots x_N)] \quad .$$

Noting [8] that in the ordered sector $-\frac{L}{2} \leqq x_1 \leqq x_2 \dots \leqq x_N \leqq L/2$ the potential energy can be written as

$$U = e^2\rho \sum_j \left[x_j - \rho^{-1}\left(j - \frac{N+1}{2}\right)\right]^2 + \frac{e^2 L}{12} \quad ,$$

i.e. a sum of *ordered* Einstein oscillators having their equilibrium positions at $x_j^0 = \rho^{-1}[j - (N+1)/2]$ and possessing the same plasma frequency $\omega_P = \sqrt{2e^2\rho/m}$, m being the mass of the particles. The second term represents the Madelung energy of this lattice. Making the change of variables $y_j = \rho x_j - [j - (N+1)/2]$ we can rewrite Q_Λ^N as:

$$Q_\Lambda^N = \rho^{-N} \exp\left(-\frac{\lambda^2 N}{24}\right) \int_{-\infty}^{\infty} \dots \int_{-\infty}^{\infty} dy_1 \dots dy_N K\left(\frac{1}{2} \middle| y_1\right) \prod_{j=1}^{N-1} K(y_j|y_{j+1}) g(y_N) \quad ,$$

$$\lambda^2 = \frac{2e^2}{\rho} \beta \quad ,$$

where $K(x|y)$ is the kernel of an operator K (transfer operator given by

$$K(x|y) = \exp\left(-\frac{\lambda^2}{2} y^2\right) \theta(y - x + 1) \qquad \text{with } \theta(x) = \begin{cases} 1, & x \geqq 0 \\ 0, & x < 0 \end{cases}$$

$$\text{and} \quad g(y) = \theta(\tfrac{1}{2} - y) \quad .$$

This suggests putting the partition function in the form:

$$Q_\Lambda^N = \rho^{-N} \exp\left(\frac{\lambda^2 N}{24}\right) \left(K^N g\right) \left(\frac{1}{2}\right) \quad .$$

The usefulness of such a representation comes from the fact that it is possible to prove that for a suitable class of functions f

$$\lim_{N\to\infty} Z_0^{-N} (K^N f)(x) = \nu(f)\psi_0(x) \quad .$$

Here, Z_0 denotes the largest eigenvalue (which can be proved to be simple) and $\psi_0(x)$ the corresponding eigenvector of the integral equation

$$K\psi_0 = Z_0\psi_0$$

and

$$\nu(f) = \frac{\int_{-\infty}^{\infty} dx \exp\left(-\frac{\lambda^2}{2} x\right)\psi_0(-x)f(x)}{\int_{-\infty}^{\infty} dx \exp\left(-\frac{\lambda^2}{2} x\right)\psi_0(-x)\psi_0(x)} \quad .$$

Therefore, the free energy density Φ is given by

$$\Phi(\rho,\beta) = \Phi_0(\rho,\beta) + k_B T \frac{\lambda^2}{24} - k_B T \log\left(\frac{Z_0}{e}\right) \quad ,$$

$\Phi_0(\rho,\beta)$ being the free energy of the perfect gas. Such a result was first obtained, using a different technique, by BAXTER [8] some years ago. It is possible to prove [12] that the free energy is analytic in ρ and β, and therefore, no phase transition occurs. In the small coupling limit ($\lambda \ll 1$), Debye-Hückel theory is recovered [8], whereas in the strong coupling one, we get the free energy of independent Einstein oscillators with the frequency ω_f [8]. Finally, it is possible to treat the case of periodic boundary conditions in the same way and to show that the free energy density is the same as for rigid walls boundary conditions [12].

A real understanding of the nature of the equilibrium state, however, is only reached by looking at the correlation functions. Looking at the simplest one, the one-point correlation function $\rho(x)$, one finds in the thermodynamic

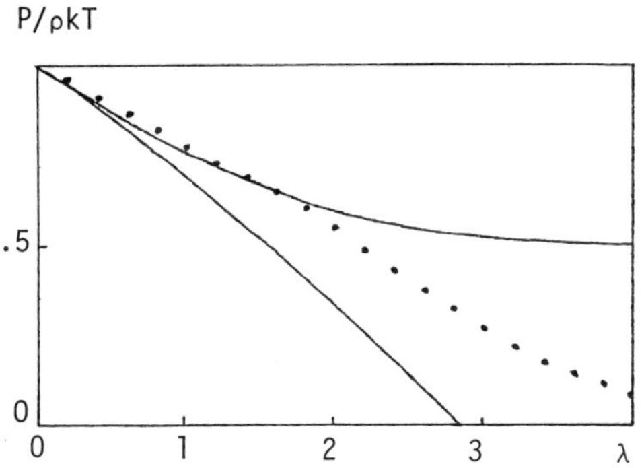

Fig.4. The equation of state of the one-component plasma. Lower line shows the thermal pressure as defined in [12], upper line shows the thermal pressure as defined in [8], circles show the non-negative kinetic pressure as measured in Monte-Carlo or molecular dynamics experiments (see [16])

limit, again starting from rigid walls boundary conditions [12]:

$$\rho(x) = \rho \sum_{n=-\infty}^{\infty} \exp\left[-\frac{\lambda^2}{2}(\rho x - n)\right] \frac{\psi_0(\rho x + n)\,\psi_0(-\rho x - n)}{\int_{-\infty}^{\infty} dx\,\exp\left(-\frac{\lambda^2}{2}x^2\right)\psi_0(x)\psi_0(-x)} \quad .$$

Such a formula shows immediately that $\rho(x)$ is periodic of period ρ^{-1}. We need to know however whether it is constant ($= \rho$) or not. In order to see that, one proceeds as follows: when $\lambda >> 1$ (low temperatures), $\psi_0(x)\psi_0(-x)$ is shown to be approximately equal to 1 when $-1 < x < 1$ and 0 otherwise and since $\lambda \exp[-(\lambda^2/2)x^2]$ is strongly peaked at the origin $\rho(0) - \rho(1/2\rho) > 0$ and $\rho(x) \neq \rho$. But since it is possible to prove that $\rho(x)$ is analytic in λ, it follows that $\rho(x) \neq \rho$ for almost all λ. In physical terms, we have proven that the system is in a crystalline state for all temperatures and densities. This crystal is nothing but the classical analogue of the Wigner lattice. Notice in particular that its entropy varies smoothly from that of an Einstein crystal to that of a perfect gas. In the case of periodic boundary conditions, however one finds a constant density.

If we turn to the two-point correlation function $\rho(x_1, x_2)$, we find that in the case of rigid walls boundary conditions $|\rho(x_1, x_2) - \rho(x_1)\rho(x_2) \leqq \exp(-\gamma|x_2 - x_1|)$ for sufficiently large values of $|x_2 - x_1|$. This exponential decay is due to the presence of a gap (ω_p) in the excitation spectrum of this crystal. Conversely, for periodic boundary conditions, we find that if $a = \rho^{-1}$

$$\rho^{p.b.c.}(x_1, x_2) = \frac{1}{a}\int_0^a dy\, \rho^{f.b.c.}(x_1 + y, x_2 + y)$$

which reflects the fact that the center of gravity of the crystal has extensive fluctuations in the case of periodic boundary conditions. We thus see that in this case, correlation functions are translation invariant, but the crystalline nature of the system reflects itself in a breakdown of clustering for the two-point correlation function, i.e. $\rho^{p.b.b.}(x_1, x_2) \not\to \rho^2$ when $|x_2 - x_1| \to \infty$. See [12] for a proof of these statements.

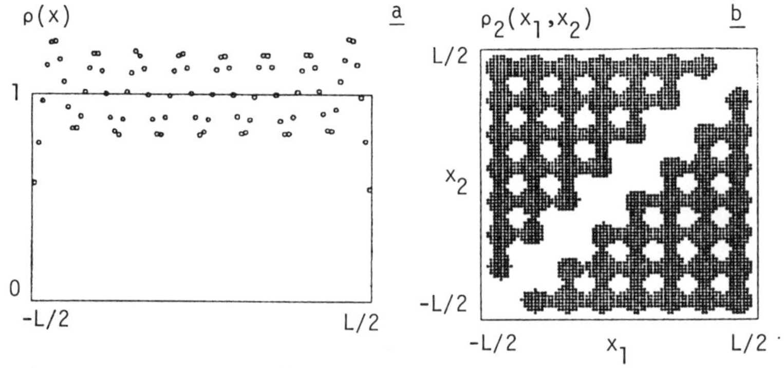

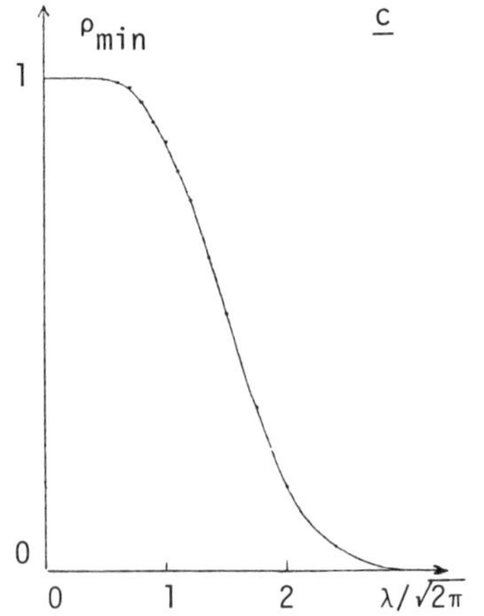

Fig.5a-c. The one-component plasma with rigid walls boundary conditions. Monte-Carlo experiment (average over 4 10^6 states): N = 8 particles, $\lambda = 2\sqrt{2}$. (a) The one-point correlation function (note the mechanical effect of the boundary). (b) The two-point correlation function: shaded area indicates the points where $\rho_2(x_1, x_2)$ is larger than its mean value. (c) An index of crystallinity: the amplitude of the minimum of $\rho(x)$ as a function of the coupling parameter. The Fig. shows that the intermediate zone between the homogeneous plasma and the crystal of ordered Einstein oscillators is actually very small, although no phase transition occurs

If we apply an external electric field θ, to the system, then we find that the free energy is independent of θ, which implies that the internal field $E(\theta)$ vanishes, and the dielectric constant is infinite. The effect of the field θ on the correlation functions is simply to translate their arguments by $\theta\rho^{-1}/2$. We conclude therefore that the Wigner lattice is not dielectric, but a plasma.

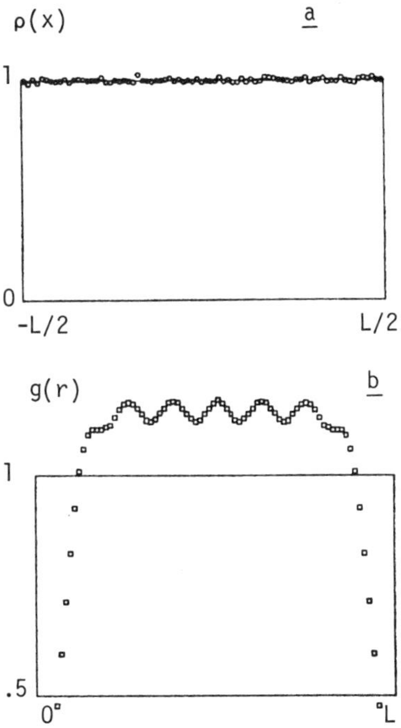

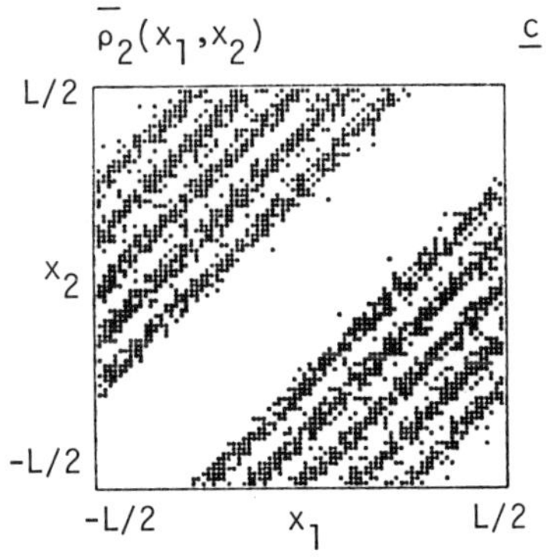

Fig.6a-c. The one-component plasma with periodic boundary conditions. Monte-Carlo experiment (average over 4 10^6 states): N = 8 particles, $\lambda = 2\sqrt{2}$. (a) The one-point correlation function is a constant. (b) The pair distribution function g(r) reveals the crystalline character of the system. (c) The underlying two-point correlation function is only a function of $|x_1 - x_2|$.

2.2 The Quantum Case

In the case of the jellium model, some of the conclusions drawn in the classical case can be extended to the quantum one. If the particles are treated as *spinless fermions*, then BRASCAMP-LIEB [13] proved that the system crystallizes and forms a Wigner lattice at low enough temperatures. More precisely, they show by means of a functional integral representation that the one point-correlation function can be represented like in the classical case as a Poisson sum

$$\rho(x) = \rho \sum_{n=-\infty}^{\infty} \exp[-\pi\bar{\lambda}^2(\rho x - n)^2 - v_{\bar{\lambda}}(\rho x - n)] \quad \text{where}$$

$$\pi\bar{\lambda}^2 = \frac{m\omega_p}{\hbar\rho^2} \, \text{th}\left(\frac{\beta\omega_p}{2}\right) \quad .$$

The important point is that they can prove that the unknown function $v_{\bar{\lambda}}(x)$ is convex. This is enough to guarantee that when λ is large enough, $\rho(x)$ is a non trivial periodic function of period ρ^{-1}, i.e. $\rho(x) \neq \rho$. This shows the crystalline nature of the system, for large enough λ, but cannot rule out the possibility that $\rho(x) = \rho$ for small λ. The question of a Wigner transition remains therefore unsettled. It would be interesting in any case to study the magnetic and transport properties of the Wigner lattice.

It is also possible to study a modified jellium model, where the particles are treated now as *bosons*. The modified model is defined by adding to the potential energy U of the jellium a special contact interaction

$$\tfrac{1}{2}\,\hbar\omega_p \sum_{ij} \delta(x_i - x_j)$$

which should not affect too much the properties of the system in the low density limit [14]. One then checks that the ground state wave function Ψ of the system can be taken as

$$\Psi = \exp\left(-\frac{1}{\hbar\omega_p} U\right) \quad .$$

It appears therefore that the one-point correlation function of this system is given by that of the corresponding classical system, where the inverse temperature β is given by $2/\hbar\omega_p$. We can then conclude that this model will be in a crystalline state for all densities.

In the same spirit, SUTHERLAND [15] has studied another modified jellium model, where the particles are treated as spinless fermions, and reaches a similar conclusion.

References

1 S. Edwards, A. Lenard: J. Math. Phys. *3*, 778 (1962)
2 A. Lenard: J. Math. Phys. *2*, 682 (1961)
3 J. Fröhlich, T. Spencer: "On the Statistical Mechanics of Classical Coulomb and Dipole Gases". IHES preprint
4 M. Aizenman, Ph.A. Martin: "Structure of Gibbs States of One-Dimensional Coulomb Systems", to appear in Comm. Math. Phys. (1980)
5 J. Zittartz: Z. Phys. B*31*, 63 and 79 (1978)
6 W. Apel, H.U. Everts, H. Schulz: Z. Phys. B*34*, 183 (1979)
7 K.D. Shotte, T.T. Truong: Z. Phys. B*37*, 193 (1980)
8 R. Baxter: Proc. Cambridge Philos. Soc. *59*, 779 (1963)
9 R. Baxter: Phys. Fluids *7*, 38 (1964)
10 Ph. Choquard: Helv. Phys. Acta *48*, 585 (1975)
11 Ch. Gruber, Ph.A. Martin, Ch. Lugrin: Helv. Phys. Acta *51*, 829 (1978)
12 H. Kunz: Ann. Phys. *85*, 303 (1974)
13 H.J. Brascamp, E. Lieb: In *Functional Integration and its Applications*, ed. by A.M. Arthurs. Clarendon Press (1975)
14 Ph. Choquard, H. Kunz: In *One-Dimensional Conductors*, ed. by H.G. Schuster, Lecture Notes in Physics, Vol.34 (Springer, Berlin, Heidelberg, New York 1975)
15 B. Sutherland: Phys. Rev. Lett. *35*, 185 (1975)
16 M. Navet: Doctoral Thesis, Orléans, France (1980)

Exciton Dynamics in Quasi-One-Dimensional Molecular Systems

Ahmed H. Zewail[1,2] and Duane D. Smith[3]

Arthur Amos Noyes Laboratory of Chemical Physics, California Institute of Technology, Pasadena, CA 91125, USA

Introduction

Quasi-one-dimensional systems offer great challenges to theoreticians as well as experimentalists. Because of the low dimensionality one expects the dynamics of excitation (electron) transfer in these systems to be relatively simple when compared with the dynamics of higher-dimensionality solids. In fact, if one is really optimistic, we may anticipate the use of 1-D physics formulaes, found in most text books, to explain experimental findings. However, the problem is not that simple and indeed it requires much more understanding than that presented in standard textbooks. Two interesting questions are relevant here: First, do 1-D systems exist? Second, if they exist, what determines the dynamics in such prototypical systems?

For many years molecular crystals have been exploited to answer these questions. This class of solids is prototypical for at least two reasons. First, the molecules (e. g., napthalene) stack in the crystal in unique ways making the intermolecular interactions highly anisotropic. As a result, an effective dimensionality (1-D, 2-D, or 3-D) for the excitation transport may exist. Secondly, from the pioneering work of FRENKEL and DAVYDOV [1, 2] we expect, at least theoretically, the spectroscopy of these excitons to be rather simple and to reveal the combined properties of the molecules and the crystal structure. Armed by the Frenkel-Davydov theory, experimentalists studied molecular excitons focusing on two major problems--the band structure and dynamics of the exciton.

To probe the band structure, primarily optical and magnetic resonance spectroscopy have been used. Basically, two approaches were adopted. Either one measures the exciton density of states ρ (E) or the resonance interaction matrix elements, β, between molecules. Both ρ and β give information on the band structure. To illustrate the point, let us consider the simplest case of a linear chain crystal. The exciton is made of

---

[1] Alfred P. Sloan Fellow and Camille and Henry Dreyfus Teacher Scholar.

[2] To whom correspondence should be addressed.

[3] Present address: The James Franck Institute, University of Chicago, Chicago, Illinois 60637 USA.

different $\underset{\sim}{k}$ states, where $\underset{\sim}{k}$ is the wavevector of the quasi-particle and the dispersion, energy vs $\underset{\sim}{k}$, is simply

$$E(k) = 2\beta \cos \underset{\sim}{k}\,\underset{\sim}{c} \quad (1)$$

In obtaining this expression, we assume that the molecules stack along the crystallographic $\underset{\sim}{c}$-axis and only nearest-neighbor resonance interactions between the molecules are allowed. From the equation, we see that an experimental measurement of β or $\rho(E) = \partial k/\partial E$, will in principal, provide E(k) and the exciton bandwidth, 4β. However, precisely what determines β is unclear.

The coupling β is known to depend on the vibronic state and is generally an exchange or multipolar interaction for molecular triplet and singlet states, respectively. The unique way in which molecules stack in the chain provides a comfortable way for the π and/or σ electrons on neighboring molecules to overlap, hence a nonzero β value. Believing the geometry and overlap integrals to be of primary importance, several laboratories have studied exciton band structure in a variety of crystals, hoping to establish relationships among band structure, stacking and σ-π electron density in the excited states.

Out of the "structural studies" it may be concluded that the β's can be measured but cannot be accurately calculated. The difficulty lies, in part, with the complexity of the σ-π interactions between molecules and that, for some crystals, in the triplet state the interaction is rather small, being in the micro to millivolt range.

More importantly, however, several solids with well-defined dimensionality have emerged, as we shall discuss later, from the band structural studies. But, <u>are the band structure determinations sufficient to establish the precise nature of the exciton dynamics</u>? The answer is no.

Dynamically speaking, one is seeking information on the transport after the exciton is created by light absorption. Unlike the exciton structure studies, the dynamics are more intricate, requiring a detailed understanding of a large number of communication channels. It is fair to say that despite an enormous amount of work, we are still learning about the dynamics of exciton transport, especially in disordered lattices.

In this article, a prototypical 1-D system called DBN (1,4-dibromonapththalene) will be discussed in some detail. I shall emphasize the theoretical and experimental aspects of the work.

DBN Crystal and Band Structure

DBN ($C_{10}Br_2H_6$) crystallizes at room temperature with a $P2_1/a$ space group and eight molecules per unit cell. The crystallographic axes are a = 27.45, b = 16.62, and c = 4.09Å with $\beta = 91° 51'$. The most important structural feature is that the molecules stack along the shortest crystallographic axis ($\underset{\sim}{c}$) with the planes of their rings parallel, thus favoring good overlap of the π-electron systems. The dimensionality of DBN was first established using optical spectroscopy [3,4]. These experiments indicated that the Davydov splitting was immeasurably small with respect to the intrinsic exciton line broadening (the Davydov splitting is a measure of the interaction between translationally inequivalent molecules in the lattice) and that the in-chain translationally equivalent interaction is 6.2 cm^{-1}. Magnetic resonance and Stark effect studies [5,6] have confirmed DBN to be quasi-1-D with the cross-chain coupling matrix element to be at least 600 times less than the in-chain matrix element. It is known that

$$\beta_c = -6.2\,cm^{-1}$$

$$|\beta_{cross\text{-}chain}| \leq 0.01\,cm^{-1} \quad .$$

It is also known from the optical spectra that (a) the dispersion follows (1) with a negative β (importantly, this places $\underset{\sim}{k} = 0$ at the bottom of the band) and (b) the density of states is essentially that of a 1-D system. Further, with a good oscillator strength for the triplet transition, a lifetime in the millisecond range and large excitonic bandwidth, DBN is a paradigm experimental system.

The Two-Molecule Limit

When, for example, 5% DBN-H_6 is doped into DBN-D_6, impurity n-mers are formed in a statistical manner. The protonated DBN molecules form traps 65 cm^{-1} below the deutero host band. Due to the nearest-neighbor coupling of 6.2 cm^{-1}, the proto dimer has two states, split by 12.4 cm^{-1}. The lowest dimer (+) state, forming a 6 cm^{-1} trap for the proto monomer exciton, is the emitting state.

There is little doubt that the (±) dimeric states of DBN are at least partially coherent. This is because the separation between the states is 12.4 cm^{-1} while the optical linewidth is ~ 2 cm^{-1}. If one defines the degree of coherence as

$$\eta_c = \frac{\text{splitting}}{\text{linewidth}}$$

then a value of approximately 6 for DBN will be obtained. If η_c is infinite, one has the perfectly coherent limit. The key quantity, therefore, is the optical homogeneous linewidth of the dimer. It may turn out that η_c is much larger than 6 because these optical lines (~ 2 cm^{-1} width) are inhomogeneously broadened [7]. The value of $\eta_c \cong 6$ is at low temperatures ($\leq$ 4.2 K) and decreases at higher temperatures. The bottom line is that, at low temperature, the small exciton (or dimer) optical transition exhibits some coherence and the exciton is not completely localized on one molecule.

Magnetic resonance spectroscopy also yields information on dimeric coherence but on a different time scale than the optical experiments. Being a large subject in itself, we shall not discuss the details, deferring to the review of BURLAND and ZEWAIL [8].

The Pure Crystal Limit and Band-to-Band Scatterings

In DBN, there are eight molecules per unit cell that are grouped into two subunits. Two triplet exciton bands, separated by 50 cm^{-1} have been identified by CASTRO and HOCHSTRASSER [9] and we shall refer to them as E1 and E2 (see Fig. 1).

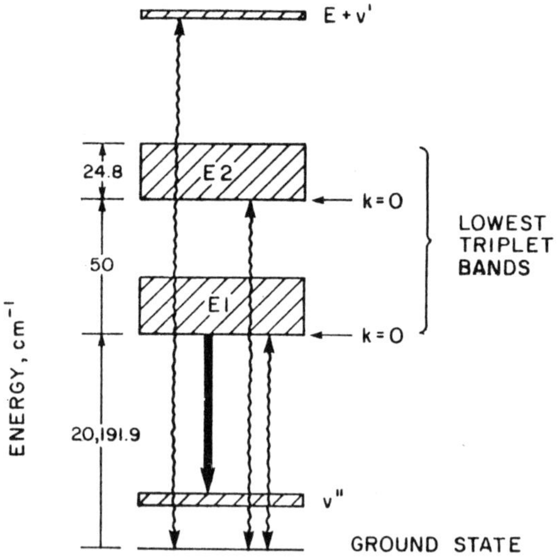

Fig. 1 Energy band diagram illustrating the position of some important states in the DBN H_6 crystal. The first triplet exciton (E1) is 50 cm^{-1} from the second triplet exciton (E2). The quantum number v is for molecular vibrational excitons and the $\underset{\sim}{k} = 0$ states lie at the bottom of the E1 and E2 bands. The bold arrow pointing down from E1 indicates the band-to-band transition observed in the DBN experiments. The lines with two arrowheads indicate the states pumped by the laser in our experiments.

In principle, there are 24 triplet bands (8 molecules × 3 spin sublevels per molecule) but only some branches are optically allowed because of symmetry. Both bands (separated by 50 cm^{-1}) have the same width, i. e., $4\beta = 24.8$ cm^{-1}.

When the neat DBN crystal is optically pumped, only the near $\underset{\sim}{k} = 0$ region (hereon we shall simply call it $\underset{\sim}{k} = 0$) is excited, due to the vanishingly small photon momentum. Similarly, emission to the zero vibrational level of the ground state occurs primarily from the $\underset{\sim}{k} = 0$ state (neglecting phonon sidebands) resulting in one line. The absorption and emission 0,0 spectra are displayed in Fig. 2.

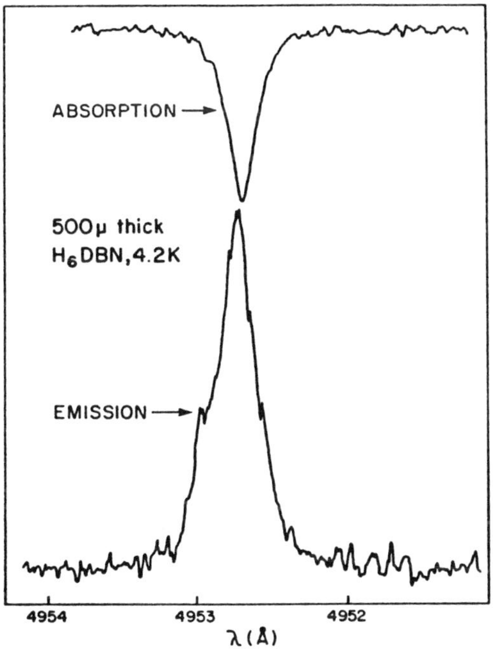

Fig. 2 The absorption and emission spectra of a thin crystal of nominally pure H_6 DBN at 4.2K. There is a noticeable red shift and broadening of the emission with respect to the absorption as has been reported by other workers in the field. This is probably due to thermalization among the states. The absolute position is uncertain to 1 cm^{-1}. The absorption lineshape is usually asymmetric with a tail to higher energy, but is sensitive to the manner of crystal preparation.

On the other hand, transitions to higher energy vibrational states in the ground state, which form narrow ($\leq 2\ cm^{-1}$) vibrational excitons, result in an emission lineshape displaying the joint occupation and density of initial and final band states. Further, labeling the vibrational exciton wavevector as q, these optical band-to-band transitions (BTBT) occur with the selection rule $k - q \cong 0$. Since the ground vibrational exciton has a width of, at most, $2\ cm^{-1}$ and the electronic triplet exciton band is $25\ cm^{-1}$, the joint density of states is essentially that of the triplet exciton. Additionally, at low temperature, the high energy molecular vibrations are unpopulated, thus, in steady state experiments with $kT \cong 4\beta$, the 0,0 emission will be $\sim 2\ cm^{-1}$ wide and emission to (say) the $527\ cm^{-1}$ mode is $25\ cm^{-1}$ wide with a lineshape, in principle, the same as the density of states of 1-D systems. However, due to homogeneous and inhomogeneous broadenings the k-states have finite widths and are distributed about a mean energy such that a "double hump" spectrum is observed for the band-to-band transition.

To optically measure the dynamics of scattering the $\underset{\sim}{k}$-states, we have recently used the technique of time-resolved laser line-narrowing spectroscopy on the BTBT [10]. The experiments separate the inelastic and elastic scatterings and determine their contributions to the homogeneous width. The experiments can be described as follows: using a narrow-band laser, one pumps the triplet $\underset{\sim}{k} = 0$ level of DBN to perform phosphorescence line narrowing (PLN) following the laser excitation. The emission from the entire band to a ground state vibrational exciton (see Fig. 3) is then detected as a function of delay time after the laser pulse (≤ 8 nsec duration).

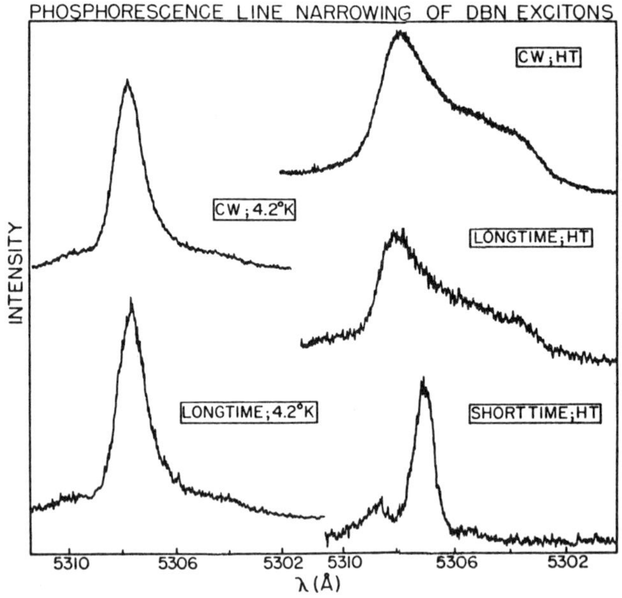

Fig. 3 Phosphorescence line narrowing and CW emission spectra (moderate resolution) of doped DBN at different temperatures and delay times. HT means high temperature (16.8 K). The most important thing to notice is that only the $\underset{\sim}{k} \simeq 0$ region is populated at short time and high temperature, demonstrating the unequilibrated population among the $\underset{\sim}{k}$-states.

In principle, if the exciton scattering time is, say, 100 nsec from $k=0$ to other $\underset{\sim}{k}$ states, then detecting the BTBT at $t = 10$ nsec after the pulse, one will only see the $\underset{\sim}{k} = 0$ emission. On the other hand, observing the emission at $t = 100$ nsec, or longer, one will see the entire band-to-band transition, as observed in steady state. Furthermore, if the pump laser bandwidth is narrow enough, the short time emission lineshape should give directly the total dephasing time. Indeed, the experiments of SMITH et al. have been successful in finding the scattering time this way.

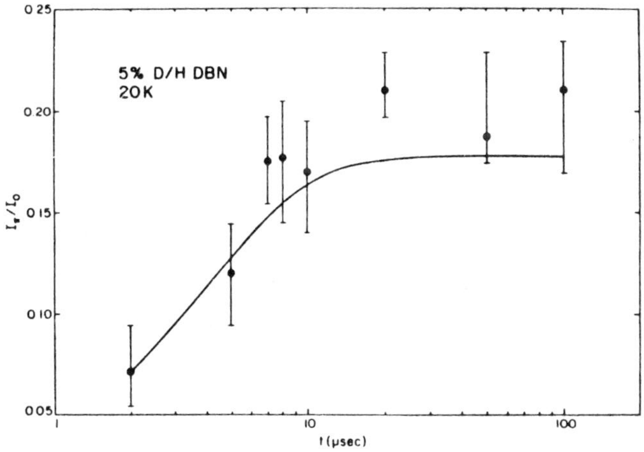

Fig. 4 The time-resolved emission intensity ratio as a function of time after excitation for the $\underset{\sim}{k} = 0$ and $\underset{\sim}{k} = \pi/c$ for 5% D/H DBN at 20 K. The solid line is a theoretical fit to a simple two level model with $T_{1+} = 4$ and $T_{1-} = 22$ microseconds.

Following the time dependence of the signal we obtain a $\underset{\sim}{k}$-to-$\underset{\sim}{k}$ scattering time of the order of microseconds for DBN at 20K. To obtain the inelastic dephasing time, we at the moment assume that the population is scattering from $\underset{\sim}{k} = 0$ to π/c by a simple T_{1-} process and scattered back to $\underset{\sim}{k} = 0$ by a T_{1+} process. We make a detailed balance by assuming $T_{1+}/T_{1-} = \exp -(\epsilon/KT)$ where ϵ is the energy difference, and

$$\frac{1}{T_{1-}} = \sum_{\underset{\sim}{k} \neq 0} \frac{1}{T_{1\underset{\sim}{k}}} , \tag{2}$$

in a simplified "two-level" description. In other words, we have taken advantage of the near degeneracy of $\underset{\sim}{k}$ states around $\underset{\sim}{k} = \pi/c$. From the date of Fig. 4, we obtained $T_{1\pm}$ (see caption, Fig. 4). Using the short time linewidth of the $\underset{\sim}{k} \cong 0$ states, one concludes the pure dephasing time T_2' is $\geqslant$ 5 picoseconds.

To better our understanding of the physics of interband scattering, we have carried out PLN experiments on different initially prepared band states. (The reader should note that only intraband scattering has been discussed so far). Two interesting cases studied were (a) pumping the $\underset{\sim}{k} = 0$ state of the triplet exciton in the higher energy sublattice (exciton E2 of Fig. 1) and (b) pumping a vibrational exciton in the excited triplet state. When pumping the higher energy sublattice, relaxation to the lower sublattice is rapid enough that on our time scale only emission from the lower sublattice is detectable. We find that the time dependence of the PLN spectra are essentially the same whether the first or second exciton band is pumped. We conclude (tentatively) that a resonant phonon of approximately 50 cm^{-1} is involved in the interband scattering. The exciton created in the higher energy sublattice with $\underset{\sim}{k} = 0$ is selectively transferred to $\underset{\sim}{k} = 0$ of the lowest band, without $\underset{\sim}{k}$-to-$\underset{\sim}{k}'$ scattering! If the relaxation was a multiphonon process, one might expect the $\underset{\sim}{k}$-to-$\underset{\sim}{k}'$ scattering not to be $\Delta\underset{\sim}{k}$ specific, due to a large number of possible intermediate states.

If one pumps a vibrational exciton in the triplet manifold, no large PLN is observed (although our S/N at the moment is not high). That is, after rapid relaxation of the vibrational exciton, observing emission from the zero vibrational level of the E1 exciton at 500 nsec after the laser pulse, all band states were occupied (within the signal to noise). This may indicate that multiphonon processes are involved which scramble the $\underset{\sim}{k}$-to-$\underset{\sim}{k}'$ scattering or that the T_{1k} of the vibronic exciton is very short such that vibrational relaxation is from equilibrated vibrational band states. The latter is less probable because vibrational relaxation generally occurs on the picosecond time scale.

The above results raise the following interesting question: Why is the T_{1k} slow? In an intuitive Fermi Golden rule sense, one can see that for elastic and inelastic scattering, the number of available final states and suitable phonons is dramatically increased for 2-D and 3-D versus 1-D. With this sort of reasoning, one might expect experiments on 1-D materials to be the most tractable, and anticipate the relatively large value of T_{1k}.

In summary, these new experiments provide a number of interesting findings. First, it is clear that even at 20-30K, the homogeneous broadening due to inelastic scattering of the exciton is only of the order of Mhz. At the same temperatures, the (0,0) exciton line shape (steady state) is ~2 cm^{-1} wide. So, if it is homogeneous, this implies that the prepared $\underset{\sim}{k} = 0$ exciton packet dephases (T_2') in a pure sense in approximately 5 psec. Consequently, the theoretical treatments of exciton scattering [2, 11] should now separate the T_1 and T_2' processes. We are currently examining thermal and structural effects on the total dephasing, hoping to be able to relate the origin of T_1 and T_2' to specific Hamiltonians.

Impurity Band Transport: Heavy Disorder and "Mobility" Edges

In the previous section, we discussed exciton motion in the nominally pure and dilute barrier limit. In what follows, we discuss the dual: transport in the relatively dilute trap limit. At low temperature, all host excitation is trapped and transport properties are determined by energy transfer amongst the impurity clusters. The isotopically mixed DBN crystals are excellent for studying such energy transfer processes since the H_6 dimer forms a 6 cm^{-1} trap for monomer excitation and is considered to be present in statistical quantities. That is, since the number of n-mers in a 1-D chain N long is $NC^n(1-C)^2$, C being the concentration, then the ratio of the number of dimers to monomers is always C. There is little tendency for the "supertraps" to cluster, making the experiments much easier to interpret.

We have probed impurity band transport by directly pumping the triplet state of proto monomer (or dimer) with a pulsed laser, recording the monomer and dimer emission spectrum as a function of time after the pulse [12]. By performing such experiments as a function of temperature, concentration and position of the laser in the inhomogeneous envelope, one can gain valuable information on the impurity-impurity coupling and the direct role of inhomogeneous broadening on spatial and spectral transport. Steady state experiments, on the other hand, gave us an overall view of the transport at very long times [13].

For the case where the monomer is pumped, a typical set of time-resolved spectra are shown in Fig. 5. As time increases, the dimer

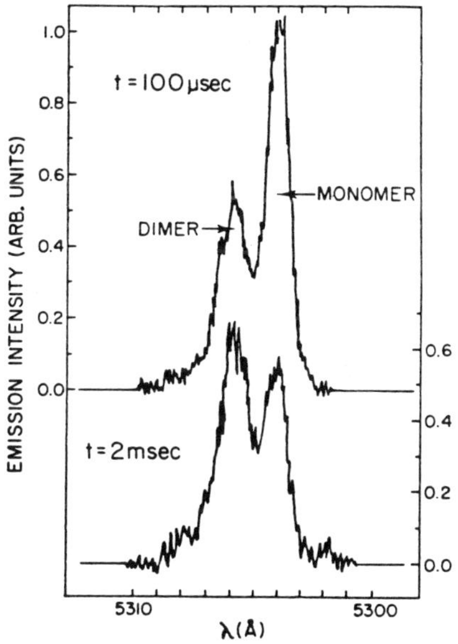

Fig. 5 Moderate resolutions emission spectrum of 11.9% H/D DBN at 1.8 K for two delay times. One can clearly see that as time increases, population is transferred from the monomer to the dimer.

emission increases relative to the monomer. We have studied the time-dependent dimer-to-monomer phosphorescence intensity ratio (or monomer-to-dimer for pumping the dimer) for several samples at temperatures of 1.3, 1.8, and 4.2 K. At all three temperatures the 11.9 and 24.1% H_6 in D_6 samples exhibit time-dependent ratios, while in the lightly doped samples, time-dependence can be detected only at the highest temperature. In the lightly doped samples, the monomer and dimer lifetimes were found to be approximately 4 milliseconds and independent of temperture from 1.3 to 4.2 K.

The model used to interpret the data is as follows. Neglecting correlated motion, excitation migration among the monomers, with the eventuality of being trapped at a dimer may be describable as a random walk. In such a case, the energy transfer rate at which distinct sites are sampled in the random walk is: $\omega_{md} = C_d \dot{S}(t)$, where C_d is the fraction of dimers (and larger clusters which can also trap the monomer excitation) and S(t) is the so-called sampling function. In the limit of many steps and uniform step time, $\dot{S}(t)$ has the asymptotic form $\dot{S}_{1-D} \sim (2/\pi\tau t)^{1/2} + \ldots$ where τ is the mean hop time. The leading term in the transfer rate is $\omega_{md} = C_d (2/\pi\tau t)^{1/2}$.

Thus, using the sampling function to relate the microscopic transfer to the macroscopic observables (e. g. , the dimer/monomer population ratio) one can write down a system of differential equations for the population (or probabilities), find the transient solution and fit it to the data. SMITH et al [12] have done this, fitting the data to the solution by nonlinear regression for time-dependent and time-independent transfer rates. From the fits it was discovered that the time-dependent forward transfer rate (m to d) predicted by the sampling function described the data more accurately than a constant rate. Further, the concentration dependence of the energy transfer rate was found to be consistent with a Monte Carlo ensemble-averaged 1-D superexchange coupling.

In a simplified form, the superexchange coupling matrix element coupling two host-isolated impurities in a perfectly 1-D crystal is

$$J = \beta\left(\frac{\beta}{\Delta}\right)^n \chi \; ; \quad \langle J \rangle = \beta\chi \left\langle \left(\frac{\beta}{\Delta}\right)^n \right\rangle \qquad (3)$$

where β is the nearest-neighbor exchange coupling (neighboring guest and host), Δ is the trap depth, n is the number of host molecules intervening between the two impurities, and χ is a guest-host Franck-Condon overlap constant.

From the ensemble-averaged J [14], denoted by $\langle J \rangle$, it is clear that one can test the transport on 1-D chains by this mechanism in two ways. Either we vary n and measure the rate of energy transfer, or for fixed n, vary Δ. These two experiments have been reported in the paper by SMITH et al. In varying the concentration, n and the distribution of clusters changes. The second parameter, Δ, was varied by tuning the laser within the inhomogeneously broadened line of the monomers.

To calculate the rates of energy transfer, we computed the distributions of clusters and $\langle J \rangle$ using Monte Carlo methods. The predicted energy transfer rate as a function of concentration agrees with the data to within experimental error. One finds that an inverse power law dependence on the trap depth is predicted for the coupling $\langle J \rangle$. Thus, tuning the pump laser into different portions of the monomer inhomogeneous profile should select different effective trap depths and transfer rates. Further, for certain Hamiltonians, it has been proposed that there should be so-called mobility edges in the inhomogeneous line that separate the local states in the wings of the line from the extended states in the central portion of the line. The position of the mobility edges (see Fig. 6) is a function of the impurity coupling

(dopant concentration). As the impurity concentration increases, the band of extended states becomes broader and the mobility edges are pushed further in to the wings. There are numerous theoretical calculations of the size of the states near the mobility edges for 2-and 3-D systems and calculations of the localization length in 1-D. In perfectly 1-D systems all states are localized, though with non-nearest-neighbor interactions, the possibility of extended states exists.

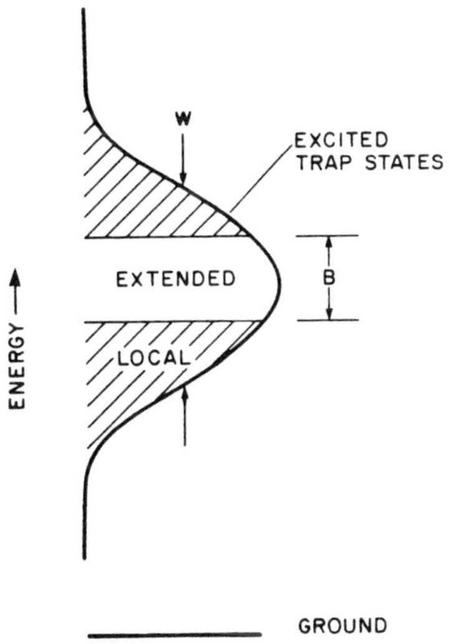

Fig. 6 Representation of the important variables for energy transfer in disordered systems. W is the inhomogeneous linewidth, B is average impurity band width. The mobility edges separate the extended and local states.

For the 11.9% H_6 in D_6 DBN sample, the experimental transfer rate coefficient, ω, has an approximate Δ^{-2} trap depth dependence. Three things are significant about the Δ dependence. First, the experiments are consistent with and quantify the superexchange effect. Secondly, the transfer rate is a decreasing function of Δ and no "mobility edges" were observed. Possibly a different concentration would show the edges or homogeneous broadening obscures it. Of course, on our experimental time scale DBN may be close to being perfectly 1-D and we, therefore, do not expect such abrupt changes in the transfer rate. Finally, our results clearly indicate the inhomogenity of the monomer band.

Acknowledgments

This work was supported in part by the National Science Foundation under grant DMR77-19578, and in part by the Department of Energy.

References

1. J. Frenkel, Phys. Rev. 37, 17 (1931); ibid 37, 276 (1931).
2. A. S. Davydov, Theory of Molecular Excitons, (Plenum Press, New York, 1971).
3. R. M. Hochstrasser and J. D. Whiteman, J. Chem. Phys. 56, 5945 (1972).
4. R. M. Hochstrasser and A. H. Zewail, Chem. Phys. 4, 142 (1974).
5. R. Schmidberger and H. C. Wolf, Chem. Phys. Lett. 16, 402 (1972); Chem. Phys. Lett. 25, 185 (1974); Chem. Phys. Lett. 32, 21 (1975).
6. R. M. Hochstrasser, L. W. Johnson, and C. M. Klimcak, J. Chem. Phys. 73, 156 (1980).
7. T. E. Orlowski and A. H. Zewail, J. Chem. Phys. 70, 1390 (1979).
8. D. Burland and A. H. Zewail in, Advances in Chemical Physics, eds. I. Prigogine and S. A. Rice (John Wiley & Sons, 1979) Vol. 40, p. 369.

9. G. Castro and R. M. Hochstrasser, J. Chem. Phys. 48, 637 (1968).
10. D. D. Smith and A. H. Zewail, J. Chem. Phys. 71, 3533 (1979).
11. R. Silbey, Ann. Rev. Phys. Chem. 27, 203 (1976) and references therein.
12. D. D. Smith, R. C. Powell, and A. H. Zewail, Chem. Phys. Lett. 68, 309 (1979).
13. D. D. Smith, D. P. Millar, and A. H. Zewail, J. Chem. Phys. 72, 1187 (1980).
14. J. Bernasconi, S. Alexander, and R. Orbach, Phys. Rev. Lett. 41, 185 (1978); J. Physique Colloque C6, Supp. an n° 8, p. C6-706; D. D. Smith and A. H. Zewail (to be published).

Critical Phenomena and Fractals with Dimensionality Near 1*

Y. Gefen and A. Aharony

Department of Physics and Astronomy, Tel Aviv University, Ramat Aviv, Israel

B.B. Mandelbrot

IBM Thomas J. Watson Research Center, Yorktown Heights, NY 10598, USA

1. Introduction

This work is concerned with the critical properties of Ising models on self similar fractal lattices [1]. Fractal lattices are *not* translationally invariant, in contrast to the usual integer dimensional lattices. A fractal's simplest characteristic is its *fractal dimensionality*, D, which is ordinarily *not* an integer. (The ε-expansion arguments are predicated upon the existence of lattices that combine non integer dimensionality with translational invariance, but no such lattice has been actually exhibited). Numerous real physical systems, e.g. polymers and critical percolation clusters, are usefully viewed as self similar fractals [1,2].

Using techniques of real space renormalization group, we find that the critical properties of the Ising model vary with *both* the fractal dimensionality D of the lattice, and diverse topological characteristics [3] such as the *ramification order* R or the *connectivity* Q (both measure the 'connectivity' in the intuitive sense, with Q applying when $R=\infty$), and the *lacunarity* L (which, in a way, is a measure of the failure of the lattice to be translationally invariant). These factors are largely independent of each other, a fact that should be incorporated in an extended concept of *universality*.

2. Systems with $R<\infty$

By repeating iteratively the construction scheme whose first 2 stages are sketched on Fig.1, we obtain a typical *branching Koch curve* [3]. For the corresponding Ising model, the smallest microscopic scale interactions are of the form $-Js_is_j$ between spins placed at *a* and *b*, *b* and *c*, *c* and *d*, *d* and *e*, and *b* and *d*. We solved the Ising model exactly for this and other branching Koch curves, as well as for diverse homogeneous Koch curves and for the Sierpiński gasket. We find that in all cases, $T_C=0$. This result is consistent with entropy arguments analogous to those which explain why $T_C=0$ for phase transitions in 1D systems with short range interactions [4]. We obtained the correlation length exponent ν defined through $\xi\sim\exp[(-2J/k_BT)^{-\nu}]$,and found that ν decreases with D ($1<D<2$) and increases with R, approaching infinity for some $R_C<\infty$. For $R=R_C$, we find $\xi\sim\exp[A\exp(4J/k_BT)]$.

---

* Supported by the U.S.-Israel Binational Foundation

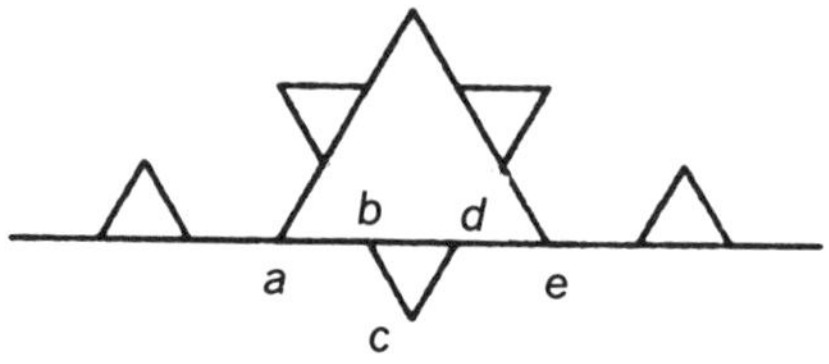

Fig.1 An example of branching Koch curve with $D=ln5/ln3$, with interactions between nearest-neighbor spins, and $R<\infty$.

3. Systems with $R=\infty$

The *Sierpiński carpets* are lattices constructed recursively. Fig.2 sketches the first 2 stages of an illustrative example: here, each square is subdivided into b^2 subsquares, of which ℓ^2 are cut out, yielding $D=ln(b^2-\ell^2)/ln\ b$, hence $1<D<2$.

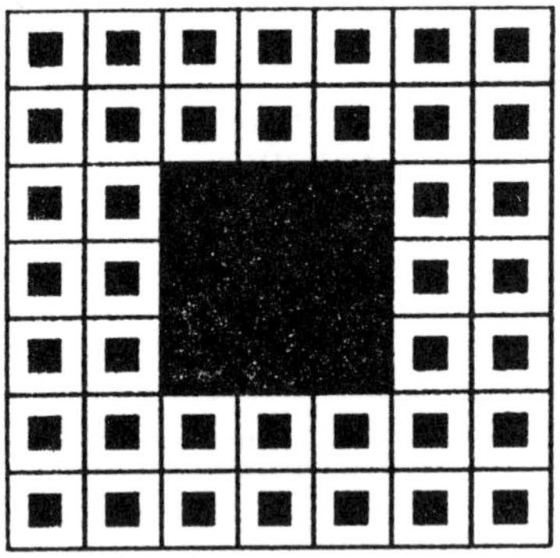

Fig.2 A Sierpiński carpet with $b=7$ and $\ell=3$. The fractal dimensionality is $D=ln40/ln7$.

First, we show *exactly* that $T_C>0$ for these systems. We apply a majority rule position space renormalization group, and compare a low-temperature expansion of the partition function in terms of the new interaction couplings, K', to an expansion in terms of the old couplings, K[3,5], where $K=J/k_BT$. The exact asymptotic result near $T=0$ is $K'>K$, and the fixed point $T=0$ is stable. Since the paramagnetic fixed point at $T=\infty$ is also stable, there must exist at least one unstable fixed point at a finite value of T, hence $T_C>0$.

Additional quantitative results at $T>0$ are obtained by a modified Migdal-Kadanoff bond-moving method [3,6]. It is necessary to distinguish between nearest-neighbor bonds on the boundary of a cutout, K_w, and internal bonds, K. Moving all the bonds within a dedecorated square to its perimeter, and then decimating, yields recursion relations for K and K_w. We thus obtain approximate renormalization group schemes in a two-dimensional parameter space, with flow lines, fixed points, critical exponents and values for T_C. They all vary with D as well as with Q and L.

4. Remark

The fact that for any $1<D<2$ there exist systems with $T_C=0$ as well as systems with $T_C>0$ means that no lower critical dimension (in the conventional sense) can be defined.

References

[1] B. B. Mandelbrot, Fractals: *Form, Chance and Dimension* (W. H. Freeman, San Francisco 1977) and forthcoming sequel.

[2] H. E. Stanley, R. J. Birgenau, P. J. Reynolds, and J. F. Nicoll, J. Phys. *C9*, L553 (1976); B. B. Mandelbrot, Ann. Israel Phys. Soc. *2*, 226 (1978); D. Stauffer, Phys. Reports *54*, 1 (1979); S. Kirkpatrick, *Les Houches Summer School on Ill-Condensed Matter 1978*, ed. by R. Balian *et al* (North Holland, 1979), p. 323.

[3] Y. Gefen, B. B. Mandelbrot and A. Aharony, to be published.

[4] See e.g. the papers by R. B. Griffiths and by C. J. Thompson in *Phase Transitions and Critical Phenomena*, edited by C. Domb and M. S. Green (Academic, New York, 1976) vol. 6, p. 357.

[5] See also P. Suranyi, Phys. Rev. Lett. *38*, 1436 (1977), who used a similar approach to obtain approximate results near T_C for the triangular lattice.

[6] L. P. Kadanoff, Ann. Phys. (N.Y.) *100*, 359 (1976).

Index of Contributors